Flavor Development for Functional Foods and Nutraceuticals

Nutraceuticals: Basic Research and Clinical Applications

Series Editor: Yashwant V. Pathak, PhD

Herbal Bioactives and Food Fortification: Extraction and Formulation, *D. Suresh Kumar*

Nanotechnology in Nutraceuticals: Production to Consumption, *edited by Shampa Sen and Yashwant V. Pathak*

Handbook of Metallonutraceuticals, *edited by Yashwant V. Pathak and Jayant N. Lokhande*

Nutraceuticals and Health: Review of Human Evidence, *edited by Somdat Mahabir and Yashwant V. Pathak*

Marine Nutraceuticals: Prospects and Perspectives, *Se-Kwon Kim*

Nutrigenomics and Nutraceuticals: Clinical Relevance and Disease Prevention, *edited by Yashwant V. Pathak and Ali M. Ardekani*

Food By-Product Based Functional Food Powders, *edited by Özlem Tokuşoğlu*

Flavors for Nutraceuticals and Functional Foods, *M. Selvamuthukumaran and Yashwant V. Pathak*

Antioxidant Nutraceuticals: Preventive and Healthcare Applications, *Chuanhai Cao, Sarvadaman Pathak, and Kiran Patil*

Advances in Nutraceutical Applications in Cancer: Recent Research Trends and Clinical Applications, *edited by Sheeba Varghese Gupta and Yashwant V. Pathak*

Flavor Development for Functional Foods and Nutraceuticals, *edited by M. Selvamuthukumaran and Yashwant V. Pathak*

For more information about this series, please visit: https://www.crcpress.com/Nutraceuticals/book-series/CRCNUTBASRES

Flavor Development for Functional Foods and Nutraceuticals

Edited by
M. Selvamuthukumaran
Yashwant V. Pathak

CRC Press
Taylor & Francis Group
Boca Raton London New York

CRC Press is an imprint of the
Taylor & Francis Group, an **informa** business

CRC Press
Taylor & Francis Group
6000 Broken Sound Parkway NW, Suite 300
Boca Raton, FL 33487-2742

Printed on acid-free paper

International Standard Book Number-13: 978-1-138-59954-3 (Hardback)

Visit the Taylor & Francis Web site at
http://www.taylorandfrancis.com

and the CRC Press Web site at
http://www.crcpress.com

I thank
Almighty
My Family
My Friends
and
Everybody

Who have encouraged, motivated and whole-heartedly supported me to finish this book.

M. Selvamuthukumaran

Dedicated to all the Rishis, sages, Shamans, medicine men and women, and people of ancient traditions and cultures who contributed to the development of drugs and nutraceuticals worldwide and kept the science of health alive for the past several millennia.

Yashwant V. Pathak

Contents

Series Preface

A growing preference for personalized nutrition and lower success rate of pharmaceuticals in dealing with chronic diseases has driven the growth of the nutraceuticals market significantly. Nowadays the consumers are very much interested in dietary supplements and nutraceuticals, which can combat lifestyle diseases as like drugs, but the advantage is being without causing any side effects.

This emerging important trend is positively impacting the nutraceuticals market. Many people with medical conditions of the gut are increasingly opting for a personalized nutrition approach as the generic treatment is not effective in solving their medical problems. Many people are making dietary changes to improve their health as well as to lower the risk of conditions such as obesity, type 2 diabetes, and heart diseases.

As the market size of nutraceuticals is growing by leaps and bounds, the consumer appeal needs to be addressed with appropriate taste and smell of the products to increase product acceptance, so flavoring of the nutraceuticals and functional foods is becoming more and more important. This needs specialized attention from the scientist and formulators. There are very few titles which are addressing this need of the nutraceuticals market.

The scope of the CRC series on *Nutraceuticals: Basic Research/Clinical Applications* aims at bringing out a range of books edited by distinguished scientists and researchers who have significant experience in scientific pursuit and critical analysis. This series will address various aspects of nutraceuticals products including the historical perspective, traditional knowledge base, analytical evaluations, green food processing and applications. The series will be useful to not only the researchers and academicians but will also be valuable reference books for personnel in the nutraceuticals and food industries.

After the first book published in this area titled *Flavors for Nutraceuticals and Functional Foods* dated July 2018. This book titled *Flavor Development for Functional Foods and Nutraceuticals*, which is focused mainly on developing flavors especially for functional and nutraceutical based food industries. Flavor, which is being an integral part of nutraceutical formulations and it offer significant advantage to nutraceuticals when it comes to palatability and getting an edge over other products in an extremely competitive nutraceutical market. This book addresses different natural ingredients/botanicals used in various functional foods and nutraceutical products. The techniques of incorporating flavors in nutraceutical products can be classified as conventional and using recently developed modern techniques are covered in different chapters. These techniques are mainly used for masking the taste of nutraceutical and functional food products. The book discusses the basics of flavors and the significance of the flavor industry in relation to nutraceuticals. This book covers various processes involved in incorporating flavor and improving product acceptability. It provides an overview on the potential applications of the main terpene-based flavors as part of nutraceuticals formulations. This book will serve as a reference to academicians and industry people who are involved in nutraceutical formulations and marketing.

Our second book in this series addressing the flavors is titled *Flavor Development for Functional Foods and Nutraceuticals*, which is being published dated November 1, 2019. We are very proud to present this second book on flavors as part of the nutraceutical series.

I am very positive that readers will enjoy this book and will gain a lot of interesting information for their applications.

I also take this opportunity to invite the scientists from industry and academia to contribute new books in this series with special focus on nutraceuticals basic research and clinical evidence.

I will also welcome constructive comments from the readers to make this series more useful to the scientific community.

Yashwant V. Pathak
Series Editor

Preface

Nutraceuticals, with its growing market size, is a multibillion-dollar industry, and the consumption of nutraceuticals is ever increasing. Nutraceuticals are catering to the needs of people representing all ages from children to elderly populations. Flavoring of nutraceuticals is becoming of prime importance nowadays due to its wide-spread acceptance.

It is said that the smell preferences for flavors are developed at an early age for human beings. Normally they say between age 2 and 4 the flavor preferences are formed. A person selects a flavored product with a particular flavor depending on his or her upbringing. It is the type of flavor they were served at an early age that decides what kind of flavor the person will like. If your household is buying strawberry-flavor ice cream or cakes, then in old age the people will prefer the strawberry flavor in their product of choice even though they are in their latter part of the life. Adding flavors and providing a range of flavors not only increases product acceptability but also in due course of time helps in building dedicated consumers due to their flavor choices.

Flavors play a major role in food products, dietary supplements and also pharmaceuticals. A similar pattern is seen in nutraceutical products as well. Bringing new flavors can pump up the sales of the products provided we keep in mind that consumers do have a built in choice and new flavors should match closely with their first choice which they have built since childhood. The aroma, texture, attractive appearance, and customer-friendly packaging are some of the criteria leading to the success of the product. This book is the second part of our efforts to discuss the role of flavors in nutraceuticals.

This book is about the crucial role added flavors play in any nutraceutical product. The book has its first chapter, History of Flavors Associated with Functional Foods and Nutraceuticals. The second chapter discusses the basics of flavors and the significance of the flavor industry in relation to nutraceuticals and functional foods. **Chapters 3** through **6** discuss flavor manufacturing, flavor challenges in legume-based products, flavor challenges in cereal-based products and encapsulation of flavors for food and dairy nutraceuticals. **Chapters 7** through **9** talk about natural flavors and various aspects related to their applications in functional foods and nutraceuticals. **Chapters 10** and **11** discuss the recent trends in the flavor industry and how to mask bad taste products using appropriate flavor ingredients. **Chapters 12** through **14** provide insight on taste assessment trials for nutraceutical products, quality control of flavors, and legislation for natural and synthetic flavors in functional food products and nutraceuticals.

We believe that this will be a very good reference book for scientists, students and academicians who are working in this area and will be very useful information helping them to select appropriate flavors for their products.

We would like to express our sincere thanks to Mr. Steve Zollo and the team of CRC Press/Taylor & Francis Group who have helped to bring the book into market. We also wish to express our sincere thanks to all the chapter authors, as without their support this would not have been possible. Last but not least, we thank our institutions, our administrators and colleagues where we are working for their support and our families for their unconditional support to get this book see the sunlight.

M. Selvamuthukumaran and Yashwant V. Pathak

Editors

Dr. **M. Selvamuthukumaran** is presently associate professor and department head of the Department of Food Technology, Hindustan Institute of Technology & Science, Chennai. He was a visiting professor at Haramaya University, School of Food Science & Postharvest Technology, Institute of Technology, Dire Dawa, Ethiopia. He received his PhD in Food Science from Defence Food Research Laboratory affiliated to University of Mysore, India. His core area of research is processing of underutilized fruits for development of antioxidant rich functional food products. He has transferred several technologies to Indian firms as an outcome of his research work. He received several awards and citations for his research work. He has published several international papers and book chapters in the area of antioxidants and functional foods. He has guided several national and international postgraduate students in the area of food science and technology.

Dr. **Yashwant V. Pathak** completed his PhD in Pharmaceutical Technology from Nagpur University, India and EMBA & MS in Conflict Management from Sullivan University. He is professor and associate dean for Faculty Affairs at the College of Pharmacy, University of South Florida. Tampa, Florida. With extensive experience in academia as well as industry, he has more than 200 research publications, including research papers, chapters and reviews; 2 patents and 2 patent applications; and 26 edited books published, including 12 books in Nanotechnology and 10 in nutraceuticals and drug delivery systems. He has several books in cultural studies and conflict management. He has received several national and international awards. Dr. Yashwant V. Pathak is also an adjunct professor at Faculty of Pharmacy, Airlangga University, Surabaya, Indonesia.

Contributors

W. Abhay
National Institute of
Pharmaceutical Education and
Research (NIPER)-Raebereli
Lucknow, India

Umi Athiyah
Faculty of Pharmacy
Universitas Airlangga
Surabaya, Indonesia

Priyanka Bhatt
Department of Pharmaceutical
Sciences
College of Pharmacy
University of South Florida
Tampa, Florida

Nikhil Chauhan
National Institute of
Pharmaceutical Education and
Research (NIPER)-Raebereli
Lucknow, India

S. J. S. Flora
National Institute of
Pharmaceutical Education and
Research (NIPER)-Raebereli
Lucknow, India

Gulimirerouzi Fnu
Department of Pharmaceutical
Sciences, College of Pharmacy
University of South Florida
Tampa, Florida

Madan Mohan Gupta
School of Pharmacy
Faculty of Medical Sciences
The University of the West Indies
St. Augustine, Trinidad & Tobago

Mayank Handa
National Institute of
Pharmaceutical Education and
Research (NIPER)-Raebereli
Lucknow, India

Dewi Melani Hariyadi
Faculty of Pharmacy
Universitas Airlangga
Surabaya, Indonesia

Kanchan Kashyap
Department of Pharmaceutics
National Institute of
Pharmaceutical Education and
Research (NIPER)-Raebereli
Lucknow, India

Dheeraj Kumar
National Institute of Pharmaceutical Education and Research (NIPER)-Raebereli
Lucknow, India

J. Lalrengpuii
National Institute of Pharmaceutical Education and Research (NIPER)-Raebereli
Lucknow, India

B. Siva Lokesh
National Institute of Pharmaceutical Education and Research (NIPER)-Raebereli
Lucknow, India

Ankita Mehta
Department of Pharmacology
Institute of Pharmacy
Nirma University
Ahmedabad, India

Priya Narvekar
Department of Pharmaceutical Sciences
College of Pharmacy
University of South Florida
Tampa, Florida

Komal Parmar
Rofel Shri G M Bilakhia College of Pharmacy
Vapi, India

Anita Patel
Nootan Pharmacy College
Faculty of Pharmacy
Sankalchand Patel University
Visnagar, India

Bhoomika Patel
Department of Pharmacology
Institute of Pharmacy, Nirma University
Ahmedabad, India

Jayvadan Patel
Nootan Pharmacy College
Faculty of Pharmacy
Sankalchand Patel University
Visnagar, India

Mayur M. Patel
Department of Pharmaceutics, Institute of Pharmacy
Nirma University
Ahmedabad, India

A. Prabakaran
National Institute of Pharmaceutical Education and Research (NIPER)-Raebereli
Lucknow, India

Chetan Rajak
National Institute of Pharmaceutical Education and Research (NIPER)
ITI Compound
Rae Bareli, India

Rahul Shukla
National Institute of Pharmaceutical Education and Research (NIPER)-Raebereli
Lucknow, India

Raghuraj Singh
Department of Pharmaceutics
National Institute of Pharmaceutical Education and Research (NIPER)-Raebereli
Lucknow, India

K. Vignesh
National Institute of Pharmaceutical Education and Research (NIPER)-Raebereli
Lucknow, India

Deepak Yadav
Department of Pharmaceutics
National Institute of Pharmaceutical Education and Research (NIPER)-Raebereli
Lucknow, India

and

School of Pharmacy
University of Nottingham
Nottingham, United Kingdom

Kiran Yadav
Department of Pharmaceutical Sciences
Indira Gandhi University
Rewari, India

1

History of Flavors Associated with Functional Foods and Nutraceuticals

B. Siva Lokesh, Dheeraj Kumar, Mayank Handa, and Rahul Shukla

Contents

1.1 Introduction

The Society of Flavour Chemists (1969) formulated the following definition of flavor: "A flavor is a substance which may be a single chemical entity, or a blend of chemicals of natural or synthetic origin, whose primary purpose is to provide all or part of the particular effect to any food or other product taken in the mouth."

The International Organization of the Flavor Industry (IOFI) defined flavors from the industry's point of view: "Concentrated preparation, with or without solvents or carriers, used to impart flavor, with the exception of only salty, sweet, or acid tastes. It is not intended to be consumed as such."

According to the Council of Europe: "Flavoring is a substance which has predominantly odor-producing properties and which possibly affects the taste." A flavor is something that affects the sense of taste (Berger, 2007; Krings and Berger, 1998).

Nutraceuticals are products that are intended to supplement the diet because of enrichment in vitamins, minerals, amino acids, herbs or other botanicals; it is a concentrate, metabolite, constituent, extract, or combination of these ingredients (Ravichand, 2015).

Furthermore, it must also confirm to the following criteria:

- Be intended for ingestion in pill, capsule, tablet, powder or liquid form
- Not to be represented for use as a conventional food or as sole item of a meal/diet
- Be labeled as a "dietary supplement" (Augustin and Sanguansri, 2013)
- **Functional foods** are defined as products resembling traditional foods but possess demonstrated physiological benefits (Bech-Larsen and Grunert, 2003).

1.2 History of flavors

Flavors were added into the food and lives well before recorded history. Adding flavors into food and beverages to make them taste better and to add zest to the eatables is not dated yesterday or decades, it is a centuries-old concept. The concept of flavors started with the invention of essential oils in eleventh century. Before the synthesis of artificial flavors, ancient Egyptians extracted essential oils from the plants, which were used as flavoring agent.

1.2.1 Old world era

Persian philosopher and physician Avicenna, in eleventh century, figured out that essential oils can be extracted from plant parts in the same way as alcohol. This lead to the discovery of many more essential oils and gave birth to the essential oil industry. The discovery of America is also related to flavors. Christopher Columbus discovered America in 1492 when he was looking for a faster way to reach Far East spice sources. Centuries later, the advancement in the modern food industry leads to the discovery of different flavors to fulfill the needs of customer.

Germans and Swiss extended into the flavor market in 1800s, mainly by deriving the different flavoring substances that were naturally occurring in the food stuffs. Emphasis on dealing with the essential oils and their extracts paved the way for the opening of many flavor-based industries. In 1851, the first artificial flavor appeared in the market. Stepping of candies in the market with different flavors like grape, pineapple, pears, apple, and so on opened the gates for flavors. By the end of nineteenth century more than 20 artificial flavors appeared in the market, which can be mixed to make more blends of flavors. In 1858, Nicholas-Theodore Gobley isolated vanillin from vanilla beans, which is responsible for vanilla flavor. In 1874, two German scientists—Ferdinand Tiemann and Wilhelm Haarmann—synthesized vanillin from the bark of pine tree. Later on, they opened world's first vanillin factory in 1875. Implementation of the Pure Foods and Drug Act in 1906 places restriction on production, transport and sale of drug or any substance having pharmacological activity. After the imposition of restrictions, it was made compulsory that all the flavors in the product must be labeled properly and limits on the use of flavors, which was initiated by government authorities.

1.2.2 World War era

During the First World War (1914–1918), the US food industry started to grow at the significant rate. During this phase, the trade between Germany and the United States was banned, and most of the artificial flavor companies were in Germany. Due to the restriction in trade, US flavor companies that were earlier importing the flavoring substance from Germany now started to synthesize their own flavors. The commonly used flavor in Chinese food MSG (mono sodium gluconate) was discovered during the Second World War (1939–1945) when the supply of ration increased, which is the main reason for development of food technologies.

1.2.3 Modern world era

The first flavored potato chips were manufactured by Irish company Tayto in 1950. Earlier chips were available in two flavors: cheese and onion, and salt and vinegar. Soon after, other flavors also appeared in market.

In 2009, the concept of natural vanillin came into market. A new type of yeast was engineered by scientist that produced vanillin as by-product. It was called *natural* because no chemical was involved in it.

The current flavor industry acts as a stone for entrepreneurship and new ways of employment as there is always a demand for flavors in the food and beverage industry on a commercial level. Today, there are numerous flavors available in the market that can enhance the aesthetic and palatability of eatables. These flavors can be used in many creative ways to add different flavors to

Figure 1.1 *Schematic presentation of the history of flavors.*

juices, snacks, foods, nutritional products and even water. The various stages in the history of flavors are given in **Figure 1.1**.

1.3 Flavoring agents

The perfume of a rose, the tang of an ocean breeze, the aroma of sizzling steak—tastes and smells are two of our senses that characterize the world around us and yet we cannot adequately express, define or explain our taste and smell sensations. We can record the sounds we hear, we can photograph the sights, but we cannot store and retrieve the flavor of a food or the scent of a flower except in and from our mind (Zviely, 1997).

Flavor is one of the most important components responsible for overall sensory properties of taste and smell/aroma in any food product. Among the many organoleptic quality components, such as color, rheological properties or packaging, flavor takes a particular place through stimulating the odor and taste receptors when eating (Wyrwa and Barska, 2017). Therefore, flavor plays an important role in consumer satisfaction (Attokaran, 2011).

Flavor is an important component for nutraceuticals and functional foods, which makes them taste better and which increases consumer acceptance. The sense of smell is most important to flavor followed by taste and touch.

Flavor refers to a mixed sensation of taste, smell, and sight, all of which involve a combination of physicochemical and physiological actions that influence the perception of substances (Kollmannsberger et al., 2007). Retronasally aroma compounds can reach olfactory receptors through the throat after the mastication, or they can be sensed orthonasally (Turin, 1996). Unwanted bitterness tends to be the most objectionable flavor characteristic of these foods (Taylor and Roberts, 2004).

When food is consumed it is also influenced by sensations of pain, heat and cold, and by tactile sensations. Flavors are principally perceived by aroma receptors in the nose and taste receptors in the mouth. Flavor compounds are generally divided into two classes: flavor compounds responsible for taste and those responsible for odor. Flavors may be volatile or non-volatile and represent many different chemical classes (Small and Prescott, 2005).

The flavor compounds that arouse the taste perceptions should dissolve in the saliva before they can be perceived. As a result, they interact with taste receptors located in taste buds on the tongue. Flavor compounds responsible for odor are volatile compounds that are perceived by the odor receptor sites of the smell organ such as the olfactory tissue of the nasal cavity. They reach the receptors when drawn to the nose (nasal detection) and via the throat after release by chewing.

The human nose is very sensitive to characteristic flavors that are produced by many aroma compounds. Due to small chemical differences, variations in flavor can occur. Even two different enantiomers (molecules that are chemically identical, but are non-superimposable mirror images of each other) may yield entirely different sensory qualities (Van Der Schaft, 2007). For example, Carvone-S-carvone is known as caraway, R-carvone as spearmint (Tournier et al., 2007).

Food matrix was added with flavorings to achieve the desired flavor profile. Flavoring substances are selected in such a way that they should be generally recognized as safe (GRAS) and regulations in their use depend on country, regulatory authority and application.

Flavorings are a heterogeneous set of substances. They impart good tastes such as sweet, sour, bitter, umami and salty tastes. Aroma compounds are generally volatile and are often hydrophobic. Sources of these flavoring substances include fermentation, natural and artificial chemicals, enzymology, botanical extracts and certain chemical reactions. There are many forms of flavors that include water-soluble liquids, oils, emulsions, spray-dried powders and plated powders. The best form of flavor to use depends on the food matrix and processing conditions (Spence, 2016).

Sensitivity of taste buds with a wide range of basic tastes includes sweet, sour, bitter, salt and possibly metallic and alkaline. But these responses may be modified by some additional factors such as temperature, physical nature and some special characteristics like astringency and pungency of the flavored material.

Age is also a factor that affects flavor acceptance. In general, children prefer fruit-flavored syrup, adults like more acidic taste and old people accept mostly mint or wine flavors.

In the formulation, flavors must be non-toxic, soluble (for a clear product like syrup, elixir), stable and compatible with preparation. The establishment of an acceptable flavor is more of an art than a science.

Sweetened base is mixed with the flavor in syrup form. It may also be in the form of mucilage prepared from a cellulose ester, an alginate containing a synthetic sweetener. These preparations not only act as a sweetener but also give better feel in the mouth due to its high viscosity. Sustained-release flavor substances such as chocolate or apricot, or by using adjuvant such as glycine and monosodium glutamate, are added to the bitter drugs in order to prevent persistent and unpleasant taste (Baines and Brown, 2015).

Aroma chemicals require novel strategies for increasing the market share of flavored and fragranced products (convenience food, beverages, cosmetics, detergents). Nearly 80% of the flavors and fragrances used worldwide are produced chemically. However, about 70% of all food flavors used in Germany till 1990 were from natural sources.

Flavoring these compounds can be especially challenging because, again, the flavors used must work synergistically with the functional components. The right flavor choice depends on the type of juice used, in addition to the levels of herbs, vitamins and minerals incorporated to it. Many times, the flavor profile is dependent on the functional pieces of the compound (Vilgis, 2013).

One of the most significant trends in flavoring functional foods, especially beverages and confectionery products, is the use of flavors that are associated with health and function. Exotic fruits with health benefits are becoming mainstream due to changing demographics, including the growing ethnic population in the United States, as well as the aging baby boomer population.

With the dramatic increase of the US ethnic population, food and beverage manufacturers are constantly trying to infuse popular tropical and exotic flavors into their product lines. This is also the case for functional food and beverage products. In fact, many of the fruits have functional properties, and the flavors of these fruits lend themselves well to incorporation with other nutraceutical components. Examples include acerola, mangosteen, goji, jabotacaba and passion fruit (Stevenson, 2012).

Currently, flavors used in the functional foods and nutraceuticals are of great interest. Typically, aroma is an equilibrium mixture of aroma compounds. Generally, most of the aroma compounds have the molecular weight relatively less than 400 daltons. Aroma compounds vary widely in their chemical structures; these include acids, neutral compounds, sulfur and nitrogen compounds, alcohols, aldehydes, ketones, hydrocarbons and esters. Partition coefficient between the air phase and food matrix helps in determining the release of aroma compounds from functional foods and as observed in the retronasal case, by the partition coefficient between the water phase (saliva) and the food matrix. When an aroma compound is added to the water matrix in a closed system and allowed to reach equilibrium, it will distribute between the air and water phases according to its air-to-water partition coefficient. When it is taken, volatile flavor molecules pass via nasopharynx into the nose, and flavor molecules are released from the functional foods and nutraceuticals into the mouth, which give pleasant odor and good taste for intake. The sufficient concentrations of flavor molecules are released from the functional foods in order to stimulate the olfactory system and elicit a response (Naknean and Meenune, 2010).

There are majorly three flavors used in a blending process –

- Main
- Enhancing
- Accent

Main is the dominating flavor in the formulation forming the core of the blend and is supported by enhancing flavors.

Accent flavor is added in a small quantity, which serves to round up the flavor of the other two types and is normally very strong and assertive in its sensorial impact.

Flavors are segmented on the basis of the industry they serve. These include savory flavors, fruity flavors, cheese and other dairy flavors, and meat flavors (Kaitano, 2013).

- **Savory flavors**: These are used in snacks, cooking sauces and ready-to-eat meals. Demand for savory flavors is driven by increased consumer demand for convenience in food and drinks.
- **Fruity flavors**: These are used in fruit-flavored drinks such as alcohol and fruit juices. The demand is driven by increasing innovation in the food and beverage (F&B) industry.
- **Cheese and other dairy flavors**: These are used by flavored milk, yoghurt, cheese and other dairy product manufacturers. The demand is driven by increasing health consciousness of customers.

Flavors and fragrances (F&F) are very critical components of food and beverage and home and personal care products as they create a strong sensorial impact on the customer, contributing to a positive or negative brand recall. F&F blends use a large number of ingredients which can either be natural or synthetic or a combination. Blending involves mixing of different flavors or fragrances to obtain the desired flavor or fragrance. The major end-users of food and fragrances blends are personal care (soaps and shampoos, skin care products), household care (detergents), food and beverages (dairy, confectionery and bakery) and perfume manufacturers (Linforth, 2010).

The fragrance of a perfume is described by the notes that it is made up of. Notes are the descriptors of scents that can be sensed upon application of a perfume. These are of three classes: top/head notes, middle/heart notes and base notes. These notes differ with respect to the time and can be sensed after application of the perfume.

- **Top/head note**: This is generally the lightest note of a fragrance which is recognized immediately upon application. They have a light molecular structure and are the first to fade away. Common top notes include citrus (lemon, orange zest), light fruits (grapefruit, berries) and herbs (clary sage, lavender).
- **Middle/heart note**: This note is sensed once the top note evaporates. It lasts longer than the top note and has a strong influence on the base note. Typical middle notes include geranium, rose, lemongrass, coriander, nutmeg and jasmine.
- **Base note**: This is the final fragrance note that appears and leaves a lasting impression. It mingles with the middle note to create the full body of the fragrance. Common base notes include cedar wood, sandalwood, vanilla, amber, patchouli and musk.

1.4 Flavoring functional beverages

Fruit flavors are the most flavored for functional beverages. In these drinks, a "true to fruit" profile is essential (Krings and Berger, 1998). Consumers expect healthy products to taste natural and fresh. The flavors should truly complement the functional ingredients used in the product.

Citrus flavors and astringent berry flavors work especially well with antioxidant fortifications. These flavors blend nicely with the acidity that is inherent in vitamin C and other antioxidant ingredients. Another alternative is to pair a citrus flavor with a less acidic, milder fruit to create a tropical-type flavor-blend.

1.5 Functional confectionery flavors

Currently, a strong trend in the confectionery segment toward the development of products with a healthy halo (Kennedy and Cosnett, 2001). Numerous combinations with exotic fruits, vitamins, herbs and plant extracts from all over the world are now offering the added health benefits demanded by consumers. For the manufacture of functional confectionery, the use of compounds is particularly recommended. In addition to fruit or vegetable juice, they may also contain the required food coloring raw materials, flavors, vitamins, minerals, herbs and other functional ingredients. Flavors and sensations in body are given in **Figure 1.2** and their differentiation in **Figure 1.3**.

1.6 Chemistry of flavors

The receptors in the mouth perceive the flavor and olfactory regions of nose observe the chemicals in a variety of foods (Kennedy and Cosnett, 2001). Signals produced initiate the receptors to respond and activate, which are then translated by the brain as sensation of aroma and taste. Certain taste and

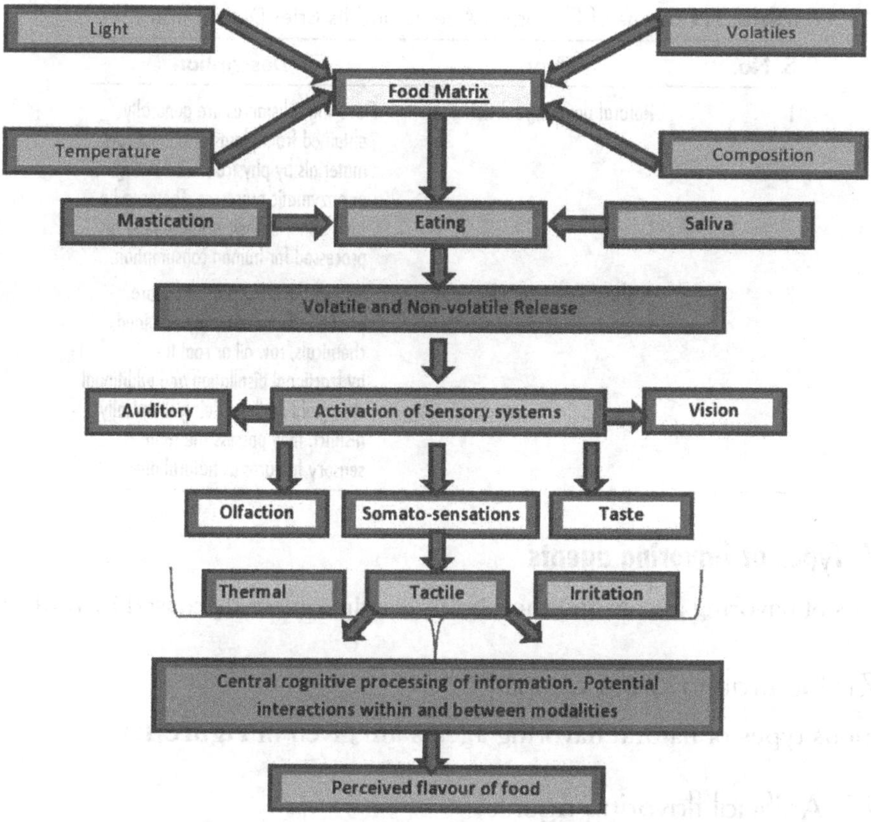

Figure 1.2 *Diagrammatic representation of flavors and their sensation in the body.*

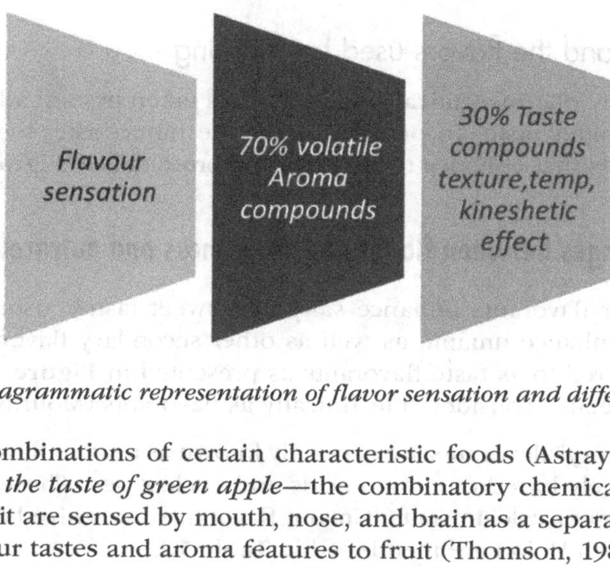

Figure 1.3 *Diagrammatic representation of flavor sensation and differentiation.*

aroma are combinations of certain characteristic foods (Astray et al., 2007). For example, *the taste of green apple*—the combinatory chemicals found normally within it are sensed by mouth, nose, and brain as a separate mixture of sweet and sour tastes and aroma features to fruit (Thomson, 1984).

Table 1.1 Types of Flavoring Agents and Its Brief Description

S. No.	Type	Description
1	Natural flavoring substances	Flavoring substances are generally obtained from plants or animal raw materials by physical, microbiological or enzymatic processes. They can be used either in their original state or processed for human consumption.
2	Artificial flavoring substances	Artificial flavoring substances are produced from naturally obtained chemicals, raw oil or coal tar by fractional distillation and additional chemical handling. Being chemically distinct, they possess the same sensory features as natural ones.

1.7 Types of flavoring agents

Types of flavoring agents and their brief description are discussed in **Table 1.1**.

1.7.1 Natural flavoring agents

Various types of natural flavoring agents are given in **Figure 1.4**.

1.7.2 Artificial flavoring agents

Artificial flavoring agents are prepared by various methods by using chemical agents, which are given in **Table 1.2**.

1.7.3 Tastes and the flavors used for masking

There are many pharmaceutical agents that are taken by oral administration, which has a bitter taste. In order to mask the bitter taste, there are many flavoring agents that enhance the taste of the product (see **Figure 1.5**).

1.8 Interlinkages between flavor and fragrances and nutraceuticals

Salt and sugar flavorants enhance salty and sweet tastes, usually the compounds that enhance umami, as well as other secondary flavors are considered and referred to as taste flavorants as presented in **Figure 1.6**. Artificial sweeteners are also considered technically as flavorants (Mouritsen, 2015).

Umami or "savory" flavorants are generally known as taste or flavor enhancers, which are largely based on amino acids and nucleotides. These are typically used as sodium or calcium salts. Umami flavorants recognized and approved by the European Union are mentioned in **Table 1.3**.

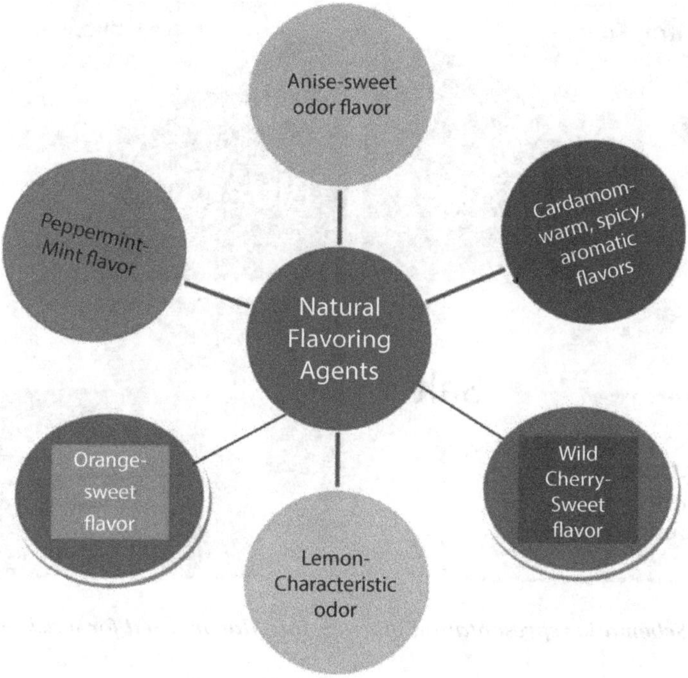

Figure 1.4 *Types of natural flavoring agents.*

Table 1.2 Examples of Chemically Synthesized Chemicals with Odors

Chemical	Odor
Diacetyl, acetylpropionyl, acetoin	Buttery
Isoamyl acetate	Banana
Benzaldehyde	Bitter almond, cherry
Cinnamaldehyde	Cinnamon
Ethyl propionate	Fruity
Methyl anthranilate	Grape
Limonene	Orange
Ethyl decadienoate	Pear
Allylhexanoate	Pineapple
Ethyl maltol	Sugar, cotton candy
Ethylvanillin	Vanilla
Methyl salicylate	Wintergreen
Manzanate	Apple

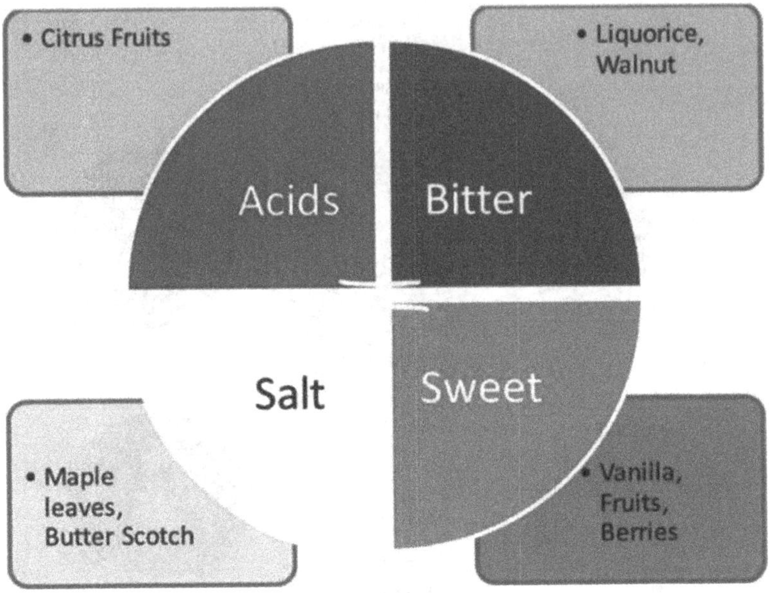

Figure 1.5 *Schematic representation of tastes and flavors used for masking.*

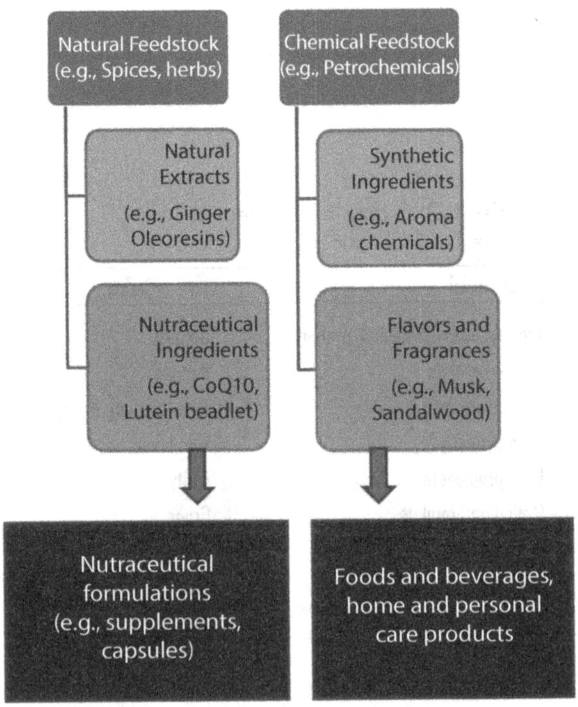

Figure 1.6 *Schematic diagram of interlinkages between flavors and fragrances and nutraceuticals values chain.*

Table 1.3 Umami Flavorants Approved by European Union

S. No.	Acid	Description
1.	Glutamic Acid Salts	Glutamic Acid sodium salt known as monosodium glutamate (MSG) is the most commonly used flavor enhancers in functional food processing. Example: Mono- and di-glutamate salts are used.
2.	Glycine Salts	These are the simple amino acid salts that are combined with glutamic acids as flavor enhancers.
3.	Guanylic Acid Salts	Nucleotide salts in combination with glutamic acids used as flavor enhancers.
4.	Inosinic Acid Salts	Inosinic salts are produced by breakdown of Adenosine Monophosphate (AMP); these are used in combination with glutamic acids as flavor enhancers.

Certain inorganic and organic acids can be used to enhance sour tastes, but like salt and sugar, these are usually not considered and regulated as flavorants under law. Each acid imparts a slightly different sour or tart taste that alters the flavor of a food (Bautista et al., 2013). Various types of acids and their flavors are described in **Table 1.4**.

Table 1.4 Some Generally Used Acids and Their Description

S. No.	Acid	Description
1.	Acetic acid	Vinegar gives sour taste and distinctive smell
2.	Ascorbic acid	Found in oranges and green peppers and gives a crisp, slightly sour taste and better known as vitamin C
3.	Citric acid	Found in citrus fruits and gives them their sour taste
4.	Fumaric acid	Not found in fruits, used as a substitute for citric and tartaric acid
5.	Lactic acid	Found in various milk or fermented products and give them a rich tartness
6.	Malic acid	Found in apples and gives them their sour/tart taste
7.	Phosphoric acid	Used in some aerated drinks to give an acidic taste
8.	Tartaric acid	Found in grapes and wines and gives them a tart taste

1.9 Taste *vs* flavors

The five primary tastes are:

- Sweet
- Umami
- Sour
- Salt
- Bitter

➢ Sweet is chiefly detected at the tip of the tongue, salt on the dorsum anteriorly, sour at the sides, and bitter at the back.

➢ Sweet and umami are detected by one receptor, whereas bitter is detected by about 25 receptors—called *T2Rs*.

Taste receptors are present in gustatory (oral) and non-gustatory tissues, which include the gut, brain, human airway smooth muscles and reproductive tissues.

Most of the flavors and odors are perceived retronasally. Odors (chemicals) reach the olfactory epithelium through the nose (orthonasal route) or mouth (retronasal route) and the information is sent to the olfactory bulb and then to mitral cells traveling to higher centers in the brain (Richardson, 2013).

Taste response varies from person to person, like bitter chemicals may occur due to large changes in the genetic variation in a bitter receptor gene known as the *TAS2R38* gene. Various studies showed that the phenotypic-genotypic relationship was modified with age. Genetic variation in bitter sensitivity may account for differences in medication compliance among children.

In conclusion, "bad taste" is going to be an on-going pediatric nutraceutical formulation problem because of the diverse number of receptors, the multiple transduction pathways and age-related sensitivity based on genotype. Infants and children live in different sensory worlds, and there is a need for validation of taste assessment methods.

By addressing the taste factor early in the product development stage can make nutraceutical company to save much. In doing so, they can get their medications to market more quickly, ensure patient compliance, gain market leadership and reap generous economic rewards. They can also stay in compliance with FDA's final rule (Lad et al., 2012).

1.10 Characteristics of flavor and fragrance materials

There are various characteristics of the flavoring agents that are included in pharmaceutical products. Some of the flavors and flavoring agents are represented in **Figure 1.7**.

***Figure 1.7** Illustrative representation of flavors and flavoring agents.*

- The value of flavors and fragrances is mainly dependent on their organoleptic quality and purity, i.e., the sale price was very sensitive to quality (Berger, 2007). In most cases, organoleptic properties are determined by the precise chemical and stereo chemical structure of molecules.

  Therefore, biotechnology processes that are capable of highly selective region and stereo chemical selective reactions are advantageous because flavor and fragrance perception is based on the precise stereo chemical properties of the molecules.
- Flavor and fragrance molecules are mostly detected in the nose or mouth at very low concentrations, and hence they are added or used in products at very low levels, giving them high-value with low-bulk performance chemicals status. For instance, the hexenals and hexanols, which are characteristic of the green taste of fresh fruits and vegetables, can be perceived at approximate concentrations of 1 ppm.
- The molecules that are commonly used as flavors and fragrances are chemically diversified but are usually small, with molecular weights of less than 300, because they have to be volatile to exert their effects.
- There will be large changes in organoleptic properties even if there are any minor changes in chemical structure, i.e., in terms of intensity, flavor and fragrance quality, and the threshold of detection by the nose or mouth.
- Different organoleptic characteristics can be possessed by molecules with same basic chemical functionality, e.g., aldehyde or ketones can often have very different organoleptic characteristics.
- The composition of flavors and fragrances is a complex mixture of different molecules as diagrammatically presented in **Figure 1.7**. Usually, a good-quality flavor or fragrance is due to the combined effect of number of different molecules, each with different taste or smell characteristics but mixed together in a complementary way by the flavorant or perfumer.
- Properties of flavor and fragranting molecules may be modified due to their interaction with other components such as food or cosmetic,

and hence, their effect may also be modified. In addition, other components of the product may have some smell or taste that might need to be balanced, for instance ethanol has a flavor, but this is only detected when present at relatively high concentrations.

- There are thousands of molecules known to have interesting tastes or aromas, but only a few hundreds are used regularly in flavors and fragrances, and only about 400 aroma chemicals are manufactured in quantities greater than one ton per annum (Cheetham, 2010).

1.11 Stability of flavor

Several mechanisms were involved in interaction of flavor compounds with food components.

In lipid system, solubilization and rates of partitioning control the interactions and partition coefficients, which determine the rate of release.

In polysaccharide system, polysaccharides interact with flavor compounds by non-specific adsorption and formation of inclusion compounds.

In protein system, protein involves adsorption, specific binding, entrapment, and covalent binding, and these mechanisms may account for the retention of flavor compounds.

Moisture affects diffusion and partition coefficients and macromolecular structures in the case of protein and polysaccharides, thereby affecting the rate of release of flavor (Marsh and Bugusu, 2007; Buffo and Holley, 2005).

1.12 Evaluation of flavors

Importance of flavor evaluation:

- Identification of food product properties
- Scientific measurement of food product properties
- Analysis and interpretation of the identified and measured food product properties (Nakai et al., 2012)

These are perceived through the five senses:

- Sight (e.g., color of a food product)
- Smell (e.g., presence of rancidity in a product)
- Taste (e.g., intensity of sweetness)
- Touch (e.g., firmness of food)
- Hearing (e.g., crunchiness of a potato chip)

1.13 Regulations

Safety assurance of the consumer in case of flavors are enforced by food laws only. Regulatory agencies across the globe are mainly mainly emphasized on information regarding additives, food inside the container, storage should be labeled. It is very difficult for the legislators to make this task complete as different countries have their own legislations with difference in approaches. Regulations are established to prevent any alteration of food due to addition of flavors or mixture of flavors. The United States was the first to regulate the standards for food safety through the enforcement of the Pure Food and Drugs Act in 1906. After the Second World War most nations set up their regulatory agency/bodies for maintaining the food safety and standards.

1.14 Flavor safety

Anything that is added for human consumption comes under the part of food, and to ensure their safety for human use, certain regulations are established. The following are types of systems that regulate flavor addition to food components.

1. Positive List System: It is the list comprised of flavors of raw materials that are safe and jejune to humans.
2. Negative List System: This list contains flavors that should be added in some concentration, which is specified, and above that concentration is not preferable.
3. Mixed System: This type of system is presently prevailing in European nations that include the combination of negative and positive list systems.

1.15 Conclusion

Nutraceuticals are products intended to supplement the diet that are enriched with vitamins, minerals, amino acids, herbs or other botanicals. Flavors were added into the food well before the recorded history. Flavors are added into food and other beverages to make them taste better and to add zest to the eatables. With the advent of increasing population there is an increasing demand for flavors. But the cost of natural flavors bears a pressure on the pocket of the consumer and synthetic flavors are not effective; to that extent that they can provide the simulated taste of natural flavors. It is time-consuming to create cost-effective extraction techniques for flavors so that the cost of flavors comes down and each segment of the population can enjoy specific ethnic taste flavors. These can be used to add different flavors to juices, snacks, foods, nutritional products and even water. Flavor is one of the most important components responsible for the overall sensory properties of taste and smell/aroma in any food product. The selection of flavoring substances is done on the basis of which are included in the generally recognized as safe (GRAS) list.

References

Astray, G., García-Río, L., Mejuto, J.C., Pastranac, L., 2007. Chemistry in food: Flavors. *Electron. J. Environ. Agric. Food Chem.* 6(2), 1742–1763.

Attokaran, M., 2011. Flavors, in: *Natural Food Flavors and Colorants*. Oxford, UK: Wiley-Blackwell, pp. 9–10. doi:10.1002/9780470959152.ch2.

Augustin, M.A., Sanguansri, L., 2013. Challenges in developing delivery systems for food additives, nutraceuticals and dietary supplements, in: *Encapsulation Technologies and Delivery Systems for Food Ingredients and Nutraceuticals*, pp. 19–48. doi:10.1533/9780857095909.1.19.

Baines, D., Brown, M., 2015. Flavor enhancers: Characteristics and uses, in: *Encyclopedia of Food and Health*. Burlington, VT: Elsevier Science, pp. 716–723. doi: 10.1016/B978-0-12-384947-2.00297-X.

Bautista, E.N., Tanchoco, C.C., Tajan, M.G., Magtibay, E.V.J., 2013. Effect of flavor enhancers on the nutritional status of older persons. *J. Nutr. Heal. Aging* 17, 390–392. doi:10.1007/s12603-012-0438-9.

Bech-Larsen, T., Grunert, K.G., 2003. The perceived healthiness of functional foods: A conjoint study of Danish, Finnish and American consumers' perception of functional foods. *Appetite* 40, 9–14. doi:10.1016/S0195-6663(02)00171-X.

Berger, R.G., 2007. *Flavours and Fragrances: Chemistry, Bioprocessing and Sustainability*. Berlin, Germany: Springer. doi:10.1007/978-3-540-49339-6.

Buffo, R.A., Holley, R.A., 2005. *Innovations in Food Packaging*. Oxford, UK: Elsevier Science. doi:10.1016/B978-012311632-1/50046-2.

Cheetham, P.S.J., 2010. Natural sources of flavours, In: *Food Flavour Technology*: Second Edition. Chichester, UK: Wiley-Blackwell. pp. 127–177. doi:10.1002/9781444317770.ch5.

Kaitano, T.E., 2013. Food additives: Flavors and flavor enhancers, in: *Encyclopedia of Food Safety*. Amsterdam, the Netherlands: Elsevier. pp. 466–470. doi:10.1016/B978-0-12-378612-8.00230-4.

Kennedy, J.F., Cosnett, I.R., 2001. Food flavours biology and chemistry. *Carbohydrate Polymers*. doi:10.1016/S0144-8617(01)00236-3

Kollmannsberger, H., Nitz, S., Blank, I., 2007. Advanced instrumental analysis and electronic noses, in: *Flavours Fragrances: Chemistry, Bioprocessing and Sustainability*. 313–361. doi:10.1007/978-3-540-49339-6

Krings, U., Berger, R.G., 1998. Biotechnological production of flavours and fragrances. *Appl. Microbiol. Biotechnol.* doi:10.1007/s002530051129

Lad, M., Hewson, L., Wolf, B., 2012. Enhancing saltiness in emulsion based foods. *Flavour* 1, 13. doi:10.1186/2044-7248-1-13.

Linforth, R.S.T., 2010. Modelling flavour release, in: *Food Flavour Technology*: Second Edition. pp. 207–228. doi:10.1002/9781444317770.ch8.

Marsh, K., Bugusu, B., 2007. Food packaging - Roles, materials, and environmental issues: Scientific status summary. *J. Food Sci.* doi:10.1111/j.1750-3841.2007.00301.x.

Mouritsen, O.G., 2015. The science of taste. *Flavour* 4, 18. doi:10.1186/s13411-014-0028-3.

Nakai, J., Kagawa-Nagamura, Y., Ohkura, M., Seimiya, H., Fujita, T., Matsunaga, T., Hasegawa, T., Gengyo-Ando, K., 2012. Methods for quantitative evaluation of flavors: evaluation of flavors using bioassay. *Aroma Res.* 13, 126–131.

Naknean, P., Meenune, M., 2010. Factors affecting retention and release of flavour compounds in food carbohydrates. *Int. Food Res. J.* 17, 23–34.

Ravichand, D.M., 2015. Neutraceuticals: Role of natural molecules in pharmacotherapy. *Int. J. Pharma Bio Sci.* 6, P444–P455.

Richardson, L., 2013. Flavour, taste and smell. *Mind Lang.* 28, 322–341. doi:10.1111/mila.12020.

Small, D.M., Prescott, J., 2005. Odor/taste integration and the perception of flavor, in: Experimental Brain Research. pp. 345–357. doi:10.1007/s00221-005-2376-9.

Spence, C., 2016. Multisensory flavour perception, in: *Flavour: From Food to Perception*. pp. 373–394. doi:10.1002/9781118929384.ch16.

Stevenson, R.J., 2012. The role of attention in flavour perception. *Flavour* 1, 2. doi:10.1186/2044-7248-1-2.

Taylor, A.J., Roberts, D.D., 2004. Flavor perception symposium, 2003. *Flavour Fragr. J.* 19, 469. doi:10.1002/ffj.1532.

Thomson, D.M.H., 1984. Flavour perception. *Nutr. Bull.* 9, 69–82. doi:10.1111/j.1467-3010.1984.tb01299.x.

Tournier, C., Sulmont-Rosse, C., Guichard, E., 2007. Flavour perception: Aroma, taste and texture interactions. *Food* 1, 246–257.

Turin, L., 1996. A spectroscopic mechanism for primary olfactory reception. *Chem. Senses* 21, 773–791. doi:10.1093/chemse/21.6.773.

Van Der Schaft, P.H., 2007. Chemical conversions of natural precursors, in: *Flavours and Fragrances: Chemistry, Bioprocessing and Sustainability*. pp. 285–301. doi:10.1007/978-3-540-49339-6_13.

Vilgis, T., 2013. Texture, taste and aroma: Multi-scale materials and the gastrophysics of food. *Flavour* 2, 1–5. doi:10.1186/2044-7248-2-12.

Wyrwa, J., Barska, A., 2017. Innovations in the food packaging market: Active packaging. *Eur. Food Res. Technol.* doi:10.1007/s00217-017-2878-2.

Zviely, M., 1997. Flavours and fragrances, in: *Trends in Food Science and Technology*. pp. 277–279. doi:10.1016/S0924-2244(97)01044-3.

2

Basics of Flavors and the Significance of the Flavor Industry in Relation to Nutraceuticals and Functional Foods

Raghuraj Singh, Kiran Yadav, and Deepak Yadav

Contents

2.1 Introduction

The flavor of the food could be a sensation with multidimensional elements involving subjective and objective perception of taste, feeling factors and odor. Humans are influenced by the senses of taste and odor. Human history is closely tied to the development and usage of flavors whereas in prehistoric times herbs and spices were used for flavoring purposes. Nowadays a broad spectrum of flavoring is available not only for use in the individual household, but for the production of food, nutraceuticals and functional food on an industrial scale. Flavor is a sensation created by a material taken in the mouth, perceived primarily by the senses of taste and smell, and also by overall pain, tactile, and temperature receptors in the mouth. So flavor is an integrated response to a complex mixture of stimuli not only based on the senses of smell and taste, but also on sight (color and appearance), tactile sensations (texture and mouthfeel) and pain (pungency) (Reineccius, 2013). Flavor conjointly denotes the total of all the characteristics of a material that produces that sensation in mouth. Flavoring agents have been used to flavor foods, nutraceuticals and functional foods to make them palatable. Flavors, like taste substances, are chemical messengers that vary from highly complex mixtures to a single chemical (Bauer et al., 2008). The flavor itself most likely does not deliver health benefits (unless the flavor has health implications in its own right, like certain spices).

Functional foods and nutraceuticals should not be considered as one homogeneous group, but rather as separate products among the various food categories. Increasing a functional food's functionality, like increasing concentration of bioactive compounds, may cause off-flavors that may cause a negative client reaction due to taste. In spite of the added value of heath functionality, consumers are unlikely to purchase functional foods and nutraceutical products over the long term if such products do not have good taste (McDonald, 2016; Yadav et al., 2018). A 2011 consumer trending survey on functional foods conducted by the International Food Information Council found that in the United States the most important factors affecting customers' decision to buy functional foods and nutraceutical products were their cost and taste (Freeland-Graves and Nitzke, 2013). Study on off-flavor in juice with specific health benefits found that even with information of health claims, consumers' degree of liking and possibilities of its consumption reduced with increasing severity of flavor defect. In medical foods also, where the consumer may be taking a product under a physician's recommendation, flavor is very important. Flavor has several benefits besides making functional food and nutraceutical products taste good, such as stimulation of digestion as well as the physiological well-being of the consumer (McDonald, 2016). Flavors added to functional foods and nutraceuticals serve a number of functions such as to impart flavor to an otherwise bland product, to modify or complement an existing flavor base, and to disguise or cover undesirable flavor attributes by blocking, masking, or otherwise tricking the taste buds into not recognizing the off-flavor (Keskar and Igou Jr, 2011).

2.2 Flavor fundamentals

Flavor is the combination of aroma and taste. A flavor may be a mixture combined from natural and/or artificial aromatic substances, designed to impart, modify, enhance or mask odor and/or taste in foodstuffs and excludes

Table 2.1 Flavor Selection for Different Taste Sensations

Taste Sensation	Recommended Flavor
Salt	Butterscotch, maple, apricot, peach, vanilla, wintergreen mint
Bitter	Wild cherry, walnut, chocolate, mint spice, mint combinations, passion fruit, anise
Sweet	Fruit and berry, vanilla
Sour	Citrus flavors, liquorice, root beer, raspberry

Source: Ebrahim, S., *The Pharmaceutical Development of a Fixed Combination Anti-tuberculosis Dosage Form*, University of the Witwatersrand, Johannesburg, South Africa, 1998; Janovsky, H.L., *Drug Cosmet. Ind.*, 86, 335, 1960; Wesley, F., *J. Am. Pharmaceutical Assoc.* (*Practical Pharmacy Edition*), 18, 674–677, 1957.

substances that have a completely sweet, bitter or salt taste. According to Hall, "flavor is the sum of the characteristics of any material taken in the mouth, perceived primarily by the senses of taste and smell, and additionally the overall pain and tactile receptors within the mouth, as received and interpreted by the brain" (Bruno and Pavani, 2018; Fisher and Scott, 1997; Jung et al., 2018). There are five well-known tastes, i.e., sweet, sour, salty, bitter and umami. There is a biological basis for taste discrimination. Salty taste detects salt that could be a necessary nutrient. Sweet indicates sugars that are sources of energy. Umami is an indicator of protein. On the other hand, sour may indicate spoilage, and bitter is a warning of toxic compounds. Some generalizations concerning the selection of flavors to mask specific types of taste have been suggested by Janovsky (1960) and Wesley (1957) as summarized in **Table 2.1**.

Food contains certain volatile chemicals. Once these chemicals escape the food and are inhaled through the nose, a number of these compounds stimulate the receptors related to first cranial nerve and are perceived as aroma. This route is called orthonasal. On the other hand when food is eaten, taste buds in the mouth notice its fundamental taste. At the same time, air containing volatile compounds of food is swept to the back of the mouth and into the nasal cavity. It is called retronasal route. Retronasal olfaction is the perception of odors originating from oral cavity during eating and drinking. On the contrary orthonasal olfaction occurs during sniffing. It is the retronasal pathway that contributes to the flavor of foods and drinks and is associated with the sense of taste (Frank et al., 2011).

2.3 Nutraceuticals and functional foods

The terms nutraceuticals and functional foods and many other similar terms are commonly used in marketing. Nutraceuticals, which have also been called medical foods, designer foods, phytochemicals, functional foods and nutritional supplements, include everyday products such as fortified breakfast cereals and "bio" yoghurts, as well as herbal remedies, vitamins and even genetically modified foods and supplements (**Figure 2.1**). The term "nutraceutical" was introduced as a hybrid of nutrition and pharmaceutical in 1989 by Dr. Stephen DeFelice,

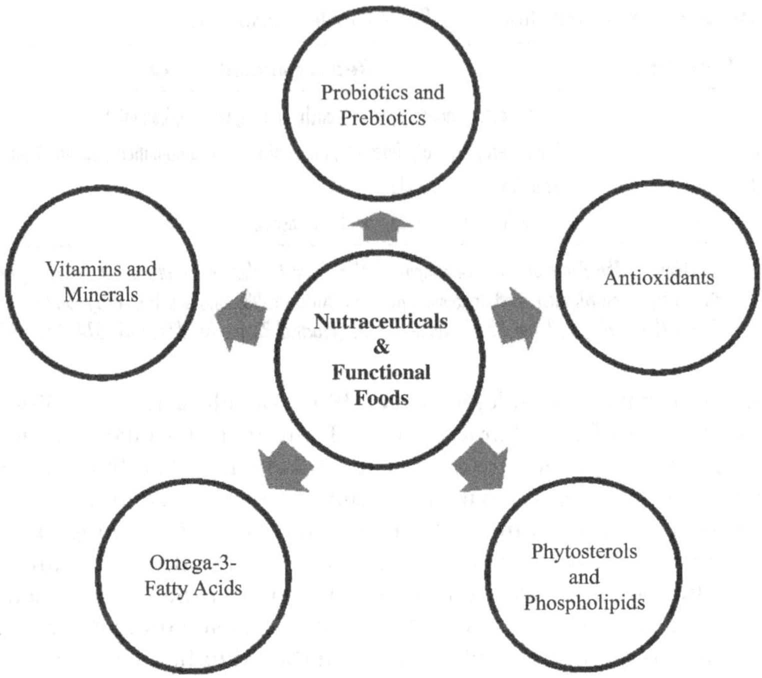

Figure 2.1 *Some common nutraceuticals and functional foods.*

the founder and chairman of FIM (Foundation for Innovation in Medicine), an American organization that encourages medical health research. He defined nutraceutical as a food or parts of a food that provide medical or health benefits including the treatment and prevention of disease. Functional foods are those foods that are intended to be consumed as part of the normal diet and that contain biologically active components which offer the potential of enhanced health or reduced risk of disease. Many different terms and definitions are used in different countries. Some of these terms are defined by regulatory agencies, while other terms are used commercially.

In Canada, a functional food has been defined as a food similar in appearance to conventional food consumed as a normal and regular part of the diet with demonstrable health benefits, whereas a nutraceutical is a product produced from foods but sold in pills, powders, or other medicinal forms not generally associated with food (Mazza et al., 2016). In Britain, the Ministry of Agriculture, Fisheries and Food has developed a definition of a functional food as "a food that has a component incorporated into it to give it a specific medical or physiological benefit, other than purely nutritional benefit" (Cockbill, 1994). Hence, both in Canada and in Britain, a functional food is essentially a food, but a nutraceutical is an isolated or concentrated form. In America, "medical foods" and "dietary supplements" are regulatory terms defined by United States Food and Drug Administration (US FDA). It defines medical food as "a food which is formulated to be consumed or administered enterally under the supervision of a physician and which is intended for the specific dietary management of a disease or condition for which distinctive nutritional requirements, based on recognized scientific principles, are established by medical evaluation." USFDA defines dietary supplements as products taken by mouth that contain a "dietary ingredient." Dietary ingredients include vitamins, minerals, amino acids, enzymes and herbs or botanicals, as well as other substances, that can be used to supplement the diet. Dietary supplements come in many forms, including tablets, capsules, powders, energy bars, and liquids. People use dietary supplements for a wide variety of reasons. Some seek to compensate for diets, medical conditions, or eating habits that limit the intake of essential vitamins and nutrients. Other people use them to boost energy or to get a good night's sleep. Postmenopausal women consider using them to counter a sudden drop in estrogen levels. The terms "nutraceuticals" and "functional foods" are not defined by USFDA and are used by consultants and marketers, based on consumer trends (Aarts, 1998).

Nutraceuticals are believed to provide physiological benefit or immunity against succeeding diseases/disorders such as cardiovascular disorders, obesity, diabetes, cancer, stress, Parkinson's and Alzheimer's diseases, neurologic diseases, respiratory and other organ diseases, and may also function as immunomodulators (Keservani et al., 2016). Several such nutraceuticals are phytonutrients, i.e., plant compounds possessing health protective and promoting properties. Nutraceuticals are usually used in nutrient premixes or formulated as nutrient systems within the foods and pharmaceuticals. Nutraceuticals that are present in foods are generally referred as "functional foods." However, there are probably a lot of accurately known components of foods and drinks that make them inherently functional; for example, red wine (containing resveratrol) is antioxidant (Bellaver et al., 2014) and anticholesterolemic, broccoli (containing sulforaphane) is cancer preventive (Dinkova-Kostova et al., 2007), and soy and clover (containing isoflavonoids) improve arterial health in women (Badger et al., 2002), inhibit undesirable blood vessel formation in tumorous tissues (antiangiogenic), and possess antitumor effects.

A functional food provides potential health benefits beyond the traditional nutrients it contains. Functional foods may include foods like cereals, breads, yogurts, snacks or beverages that are fortified with vitamins, herbs, or different specialty ingredients (**Table 2.2**). A specialty ingredient is also a naturally nutrient-rich whole food source such as spirulina, garlic, or soy, or a selected component of a food, like omega-3 fatty acids from salmon (Zhao, 2012).

At a basic level, there are two primary types of functional food/beverage products. "Better for you" foods that contain less or none of certain ingredients perceived by users to be less desirable. Examples include low-sodium soups, reduced fat, mayonnaise, cholesterol, and or cholesterol-free baked goods, no sugar added desserts. These products are not truly "functional" but do offer users a way to manage their nutrient intake and improve their health. Second types are "added functionality" foods in which nutrients are added in

Table 2.2 Different Types of Functional Foods and Nutraceuticals

Category	Example
Processed foods	Oat bran cereal
Processed foods with added ingredients	Calcium-enriched fruit juice
Food enhanced to have more of a functional component (via traditional breeding, special livestock feeding or genetic engineering)	Tomatoes with higher levels of lycopene (an antioxidant carotenoid) Oat bran with higher levels of beta glucan
Isolated, purified preparations of active food ingredients	Isoflavones from soy Beta glucan from oat bran

Source: Goldberg, I., *Functional Foods: Designer Foods, Pharmafoods, Nutraceuticals.* Springer Science & Business Media, Berlin, Germany, 2012; Gul, K. et al., *Crit. Rev. Food Sci. Nutr.*, 56, 2617–2627, 2016.

efficacious doses to food products for improved consumer health and wellness. Examples include high-fiber products, soy products, juices with added calcium or vitamins, fortified waters, and naturally antioxidant-rich juices. These types of products fit the classic profile of functional foods. Functional foods can be used according to specific health or disease conditions (**Table 2.3**).

Table 2.3 Examples of Some Functional Foods Serving Specific Physiological Conditions

Condition (Functional Effect)	Active Ingredients	Example
Energy	Green tea, caffeine, guarana phytochemicals	Perky jerky, Dynapep, Guru energy drink
Performance	Electrolytes, Amino acids, protein	O.N.E. coconut water, Nasoya tofu, Silk soy milk
Healthy aging	Omega-3s, antioxidants *trans*-resveratrol, sea buckthon	Vitamin water, Tofu plus, Welch's healthy star, Drench water, Sibu beauty
Heart health	Soy, fiber, cocona, phytosterols	Fiber One Soy joy, Barilla plus pasta
Joint and bone health (inflammation management)	Turmeric, soy, calcium, antioxidants	Yoplait, Joint Juice, POM Wonderful
Congnitive health	Citicoline, resveratrol, selenium, L-theanine (in tea leaves)	Nawgan, red wine extracts, green teas, Kaia foods, sprouted sunflower seeds
Digestion	Fiber, probiotics, whole grains, oats, enzymes	Barilla plus pasta, Mojo milk, GT's kombucha teas
Immunity	Camu camu, *Echinacea*	Wildly nutritious immunity blend, Praeventia, Yogi immunity tea, immune water
Weight management	Turmeric, soy, capsaicin, green tea, caffeine	Lychee wasabi capsaicin-spiced elixir, Naga chocolate bar
Diabetes	Stevia, yacon	Zevia natural sodas, Yerbamate royale tea, Terramazon yacon slices
Stress management and sleep support	Omega-3s, magnesium nuts, oats, melatonin	Tru2U Tart Cherry Juice

Source: Hilton, J., *Developing New Functional Food and Nutraceutical Products*, Elsevier, Amsterdam, the Netherlands, 1–28, 2017.

2.4 Flavor chemistry and its objective

Flavor chemicals are often either volatile or nonvolatile (or somewhere in between) compounds or mixtures. Chemicals that cause taste sensations tend to be nonvolatile and soluble. For instance, sucrose is dissolved in the saliva and is detected by certain taste buds, to be perceived as sweet. A particular stereo chemical arrangement of functional groups in a molecule can bind to specific receptors in the mouth to cause the sweet perception. Sucrose is standard for sweetness, alternative molecules are also detected as sweet, and some of them provide a much stronger sweetness response, like aspartame and saccharine (called high intensity sweeteners or HIS). Similarly, bitter receptors need molecules with a particular arrangement to elicit bitter response. Similarities between the sweetness receptors and bitterness receptors end in molecules, particularly several HIS, being both sweet and bitter (Kreutzmann et al., 2008). Umami is elicited most notably by the sodium salt of l-glutamic acid (monosodium glutamate or MSG) and 5′ nucleotides, such as 5′-inosine monophosphate and 5′-guanosine monophosphate. Bitterness, sweetness, and umami are all because of these molecules interacting with G-protein receptors (Rawson and Li, 2004). Saltiness and sour perception are related to ion channels. Salt taste is evoked by the movement of sodium ions, though other monovalent cations may also generate salty response, like lithium and potassium. Sour is related to the protons of acids.

Chemesthesis is often triggered by many nonvolatile and volatile compounds. Examples of nonvolatile pungent (hot) chemesthetic agents include capsaicinoids from chili peppers; gingerols, shogaols, paradols from ginger; piperine from black and white pepper. Sanshools are tingling compounds from szechuan peppers. Incidentally, szechuan peppers, chili peppers, and black and white peppers are not related. Volatile chemesthetic compounds include the pungent compounds cinnamaldehyde (from cinnamon and cassia), isothiocyanate (mustard and horseradish), and eugenol (numbing compound from cloves).

Most of what is perceived as flavor arises from volatile compounds that impact the sense of smell. This encompasses an enormous amount of compounds like phenolics, carbonyls, acids, esters, heterocyclic, and aromatic compounds, sulfur and nitrogen-containing chemicals, and many more (Bagchi and Nair, 2016). The sole characteristic they need in common is that they should be volatile to some extent and should elicit a response from olfactory receptors. There may be plenty of variation within the strength of flavor compounds. For example vanillin, the compound present in vanilla extract, has a recognition threshold of 4 ppm, whereas para-menthene-8-thiol, known as the grapefruit mercaptan, has a threshold of 10 ppm (Burdock, 2016; McDonald, 2016).

Flavor chemistry includes the identification of components liable for the odor and flavor of natural products of plant and animal origin, and to study their biogenesis, and deterioration as well as of artificial flavors, their synthesis, characterization and degradation (Bagchi and Nair, 2016). As a consequence, flavor chemistry in its broadest context is a specialized branch of chemistry that impinges on several alternative disciplines, including botany, zoology, anatomy, physiology, histology, cytology etc., for its complete understanding. Objectives of flavor chemistry include understanding the chemical composition of natural flavors and the mechanism of their formation to retard or prevent the development of the off-flavors in foods, nutraceuticals and functional foods, for example, reversion of rancid flavor in soybean oil; to generate new artificial flavorings and their use; restoration of fresh flavor to a processed food, improvement of food flavor by addition of natural, synthetic or semi-synthetic flavors; production of new foods with special flavor such as potato chip flavor; flavor improvement by acceleration of reactions that produce desirable flavor compounds (onion flavor: pH 5–7); to assist geneticist to breed new functional foods with improved flavor compounds or flavor precursors; to specify raw material and to control the quality of food nutraceuticals and functional food products (Anderson, 1965).

2.5 Flavor functions in nutraceuticals and functional foods

The role of a flavor is largely to impart sensory preference to a food, beverage, tobacco, pharmaceutical, nutraceutical or functional food. It is vital to know the function expected from the flavor so that the type best suited to this function can be delivered. Flavor serves mainly three functions, i.e., economic, physiological, and psychological (**Table 2.4**). Several flavor applications perform altogether these three classes of functions whereas others may offer primarily one or two. For example, some flavor applications are mainly for pleasure (psychological functions) such as a hard-boiled candy whereas in a flavored yogurt for breakfast flavor provides all the three functions—economic, physiological, and psychological applied to different degrees.

Table 2.4 Functions of Flavors

Economic	Physiological	Psychological
Simulate	Metabolic response	Nostalgia
Extend	Intestinal absorption	Association
Flavor the unflavored	Appetite and consumption	Intellect/belief, cognitive factors
Modify (cover taste)		Trend
Compensate for flavor losses		Flavor the flavored
Improve shelf-life		

Source: Sinki, G.S. and Gordon, R.J., *Flavoring Agents, Food Additives,* CRC Press, Boca Raton, FL, 354–413, 2001.

2.5.1 Economic function

Some nutraceuticals and functional food preparations might have undesirable taste (e.g., soya and vitamins). Flavors that can alter their taste and make them more palatable increase their consumer acceptance and offer economic benefits. "Flavor the unflavored" (Copyright Givaudan Roure) is a concept that can be applied to make bland nutraceuticals and functional foods more appealing to consumers. Also flavors can be used to compensate for flavor losses in food processing to preserve their freshness throughout storage.

2.5.2 Physiological function

Several studies on taste and fat digestion gave some indication that taste can alter the metabolic response to a fatty meal (Naim et al., 1985; Ramirez, 1985). In one study the administration of a small amount of oil or saccharin to the mouths of rats immediately before intragastric feeding (corn oil) prolonged the elevation in blood triglycerides 4 h later. This effect appeared only in rats having previous experience with oil paired with the taste (Ramirez, 1985). Studies on taste and intestinal absorption of glucose have also shown that oral stimulation affects the intestinal absorption of glucose (Threatte et al., 1986).

2.5.3 Psychological function

The main role of a flavor is to produce sensory pleasure, but psychological analysis of this pleasure can assist in proper selection of flavor. Examples of today's fashionable flavors are imaginary types that do not essentially represent a normally better-known food, however rather a unique combination, thus making a brand new identity. These are demonstrated by the various fruit punches, candies, toffees, etc. The idea of "flavor the flavored" is additionally helpful to add variety and raise sensory pleasure to foods that commonly do not need additional ingredient. The addition of flavors to tea and coffee is an example of this idea (de Graaf et al., 1996).

2.6 Types of flavors for nutraceuticals and functional foods

Early in the history of flavor creation, only natural extracts were on the market. Most sources were the extracts of spices and herbs. Several parts of plants were used, such as flowers, buds, seeds, bark, exudates, and rhizomes. A limited range of animal-sourced flavorings were conjointly available like musk from musk deer and castoreum from beavers. Plant extracts are separated from the flavorless portion of the plant material by either solvents or distillation. Ethyl alcohol is a commonly used solvent, and the resulting extract is used as is (without the removal of ethanol) to make a tincture. Alternative organic solvents may also be used, such as hexane or acetone. However, these solvents should be removed via vacuum to below a federally permitted level, and therefore the ensuing flavor extracts are referred to as oleoresins. If supercritical carbon dioxide (CO_2) is employed to extract the materials, the carbon dioxide is discharged and there is no residual solvent. The solvent extracts can have the entire flavor of the source material, as well as of volatile and nonvolatile components.

Flavorings are separated from the source material using distillation, making essential oils that have the aroma of the source material but lack the non-volatile portion of the flavors. For example, black pepper oleoresin (solvent extracted) has an aroma of black pepper, moreover as the pungent "bite" from piperine, and is pigmented and viscous. Black pepper essential oil, however, smells like black pepper but is clear, thin oil that does not contain piperine. Different ways of distillation extraction involve spinning cone technology that creates aqueous extracts. The development of organic chemistry led to creation of many artificial chemicals to be used for flavor industry such as compounds derived from petroleum or wood pulp. Also, chemicals not found in food or perhaps in nature were utilized for flavor. One such compound is ethyl methyl phenyl glycidate, or "aldehyde C_{16}," used for "red pop," a strawberry or raspberry-like flavor.

Before 1990s natural flavors were more common in use than artificial flavors. Another source of natural flavors is reaction flavors. This is often the formation of tasteful materials from typically low or non-flavorful precursors. For example, the amino acids phenylalanine and leucine, heated under specific conditions in the presence of a reducing sugar (fructose, dextrose, xylose, etc.), can form an affordable chocolate flavor.

This method is extremely common for making meat-type flavors without animal-derived ingredients, good for vegetarian foods. Another method is pyrolysis, good for making smoke and grill flavors. There are advances in fermentation, enzyme, and extraction technologies to supply natural flavorings (Bagchi and Nair, 2016).

2.6.1 Natural flavors

Natural flavoring substances are obtained from plant or animal raw materials, by physical, microbiological or enzymatic processes. They will be either employed in their natural state or processed for human consumption. The natural flavorants are initially extracted from the source substance by solvent extraction, distillation, or using force to squeeze it out. The extracts are sometimes pure and subsequently added as such to food products to flavor them. **Table 2.5** enlists natural food flavors used currently in nutraceuticals and functional foods.

Table 2.5 Different Natural Flavors Used for Nutraceuticals and Functional Foods

Flavor Types	Examples
1. Fruit flavors	
Citrus-type flavors	Grapefruit, orange
Berry-type flavors	Apple, raspberry, banana
2. Vegetable flavors	Lettuce, celery
3. Spice flavors	
Aromatic	Cinnamon, peppermint
Lachrymogenic	Onion, garlic
Hot	Pepper, ginger
4. Beverage flavors	
Unfermented flavors	Juices, milk
Fermented flavors	Wine, beer, tea
Compounded flavors	Soft drinks
5. Meat flavors	
Mammal flavors	Lean beef
Sea food flavors	Fish, clams
6. Fat flavors	Olive oil, coconut fat, pork fat, butter fat
7. Processed flavors	
Smoky flavors	Ham
Broiled fried flavors	Processed meat products
Roasted, toasted, baked flavors	Coffee, snack foods, processed, cereals

Source: Ohloff, G., *Flavour Ind.*, 3, 501–508, 1972; Priyadarshi, S. and Shukla, A., *Int. J. B. Pharm. Allied. Sci.*, 1, 1012–1019, 2012.

2.6.2 Artificial flavors or flavorings

An artificial flavor or flavorant is a substance that provides flavor of another substance making it to become sweet, sour or tangy etc. Completely different flavors have been imparted to foods, beverages, nutraceuticals or functional foods by utilization of various scents or fragrances in unnaturally flavored jellies and candies. Artificial flavorings are targeted on altering or enhancing the flavors of natural food product similar to meats and vegetables, or making flavor for functional food products or nutraceuticals or other food items that do not have the specified flavors such as candies and different snacks. Most forms of flavorings are targeted on odor and taste. Artificial flavors can be divided into the following two types.

2.6.2.1 Nature-identical artificial flavors

These are flavoring substances that are obtained by synthesis or isolated through chemical processes, which are chemically and organoleptically the same to the flavoring substances naturally present in products meant for human consumption.

2.6.2.2 Artificial flavors

These flavoring substances are not identified to any natural product meant for human consumption. These are produced by fractional distillation or any other physical or chemical process from naturally sourced chemicals or from crude oil or coal tar. They are chemically different, but in sensory character they are same as natural flavors (**Table 2.6**).

2.6.3 Flavor enhancers

A flavor enhancer is an ingredient that is added to a food, nutraceutical or functional food to improve its original taste or flavor. The most used ingredients in this category are monosodium L-glutamate (MSG), disodium 5′-inosinate (IMP) and disodium 5′-guanylate (GMP). Salt, although not classed as a food additive, is the most widely used flavor enhancer. E numbers are the codes given for substances that are permitted to be used as food additives for use within the European Union. The "E" stands for "Europe." The flavor enhancers are numbered from E620 to E640 in E numbering system.

Table 2.6 Different Artificial Flavors Used for Nutraceuticals and Functional Foods

Type of Flavor	Artificial Flavorant
Apple	Manzanate
Buttery	Diacetyl
Banana	Isoamyl acetate
Bitter almond	Benzaldehyde
Cinnamon	Cinnamic aldehyde
Fruity	Ethyl propionate
Grape	Methyl anthranilate
Orange	Limonene
Pear	Ethyl-2,4-decadienoate
Pineapple	Allyl hexanoate
Sugar, cotton candy	Ethyl maltol
Vanilla	Ethyl vanillin
Winter green	Methyl salicylate

Source: Wikipedia, T.F.E., *Flavor*, 2019.

Glutamic acid (E620) is a natural amino acid (building block of protein). It is commercially prepared from molasses by bacterial fermentation and also prepared from vegetable protein, such as gluten or soya protein. Glutamic acid and glutamates are used as flavor enhancers in sausages, and seasoning, savory snacks and many savory foods. Young children should avoid it because it can kill nerve cells, resulting in diseases such as Huntington's, Alzheimer's and Parkinson's. Its analogues include monosodium L-glutamate (MSG or E621) typically used in canned vegetables, canned tuna, dressings and many frozen foods. Pregnant women, children, hypoglycemic, elderly and those with heart disease are at risk from reactions. Monopotassium L-glutamate (E622) is less used and not as salty. It is a low sodium salt substitute flavor enhancer. It should not be used in products for babies under 12 months of age or those people with impaired kidneys. Other analogues include Calcium di-L-glutamate (E623), monoammonium L-glutamate (E624) and magnesium di-L-glutamate (E625). Calcium glutamate may have possible problems for asthmatics and aspirin-sensitive people.

Guanylic acid (E626) is a natural acid, which is part of RNA. Guanylic acid and guanylates do not have specific umami taste but strongly enhance many other flavors, thereby reducing the amounts of salt needed in such products. Guanlyic acid and guanylates are generally produced from yeasts, but partly also from fish, thus they may not be suitable for vegans and vegetarians. Examples include disodium guanylate (E627), dipotassium guanylate (E628), calcium guanylate (E629).

Inosinic acid (E630) is another natural acid that is mainly present in animals and used as flavor enhancer. It is commercially prepared from meat or fish (sardines). It may also be produced by bacterial fermentation of sugars. Its flavor enhancer analogues include disodium inosinate (E631), dipotassium inosinate (E632) and calcium inosinate (E633). These are not permitted in foods for infants and young children. Gout and asthma sufferers should also avoid such products. These are used in instant noodles, potato chips, snacks, savory rice, tinned vegetables, cured meats and packet soups. Mixture of different salts of guanylic and inosinic acid is also used as flavor enhancers in many products.

Maltol is derived from the bark of larch trees, pine needles, chicory wood, oils and roasted malt; it may be produced synthetically. It is an artificial sweetener and flavor enhancer used in baked goods to give a 'fresh baked' taste and smell in bread and cakes, chocolate substitute, soft and fizzy drinks, ice cream and jam. In large quantities it can cause Alzheimer's disease. Ethyl maltol is derived from maltol chemically. It is used as base for essences, synthetic artificial flavor and flavor enhancer. Their acceptable daily intake (ADI) is up to 2 mg/kg bodyweight. Some countries have banned these for babies and young children.

Glycine and its sodium salt are used as flavor modifier. Glycine is a natural amino acid, a building block of proteins. It can be mildly toxic if ingested. Glycine is produced mainly from gelatin, which is derived from animal bones. It is therefore not suitable for vegans and vegetarians.

2.7 Flavor dosage forms

Flavor potency and strength vary considerably. Some flavors are diluted with solvents and carriers, whereas others are aromatic compounds used without solvents or carriers. Flavors are available in a number of dosage forms such as liquid, powder, pastes and emulsions (o/w type). Liquids can be water, alcohol or oil soluble, and the solvents commonly used for them include alcohol, propylene glycol, triacetin, benzyl alcohol, glycerin, syrup, water and vegetable oils. Powders can be either spray dried or absorbates. Their mixtures are also used. The carriers used for these flavor powders include gum acacia, starch hydrolysates, simple carbohydrates and other hydrocolloids. All of the discussed carriers and solvents are also used for pastes and emulsions (Sinki and Gordon, 2001).

2.8 Product and processing considerations while flavoring nutraceuticals and functional foods

The flavors or flavor enhancers intended to be used should be compatible with the components of nutraceutical and functional food and the processing conditions involved in their preparation. For selection of appropriate flavor, details are required on the category and type of functional food or nutraceutical to be flavored. For example, if the product is a frozen dessert, it should be specified, whether it is pudding, sherbet, or ice cream. Information on packaging (glass, carton, plastic or metal) and storage conditions with its expected shelf should be there. Will the product have specific health claims? For example sugar-free, sodium-free, low cholesterol, fat-free products. There should be information on targeted population age group if any. What kind of matrix the product contains such as starches, gums, colors, preservatives, oils, type of sugar (sucrose, fructose, glucose), etc. If product matrix contains ingredients that might be reactive with flavors (proteins, vitamins, acidulants, minerals, stabilizers, oxidizing/reducing agents, etc.), active components, if any, as well as pH, viscosity and other related information on medium or base should be at hand.

Information on the type of equipment to be used for processing such as mixer, homogenizer, extruder etc. should be collected. Conditions of temperature and pressure, their holding times and ramps involved in pasteurization, sterilization, freeze drying or any other unit process must be considered. Type of pumps to be used and filling conditions should be considered while selecting flavors. Thermal or other hazardous processing conditions for such products may degrade the base of the flavor or the flavor itself. For example, Maillard reaction may add sweet or brothy flavors. Heat may also degrade unsaturated fatty acids. Since most of the flavors are volatile in nature, so in an open system the addition of flavor may cause flavor loss. pH, either inherent in the base (citrus juices) or from the addition of acids (for flavor, microbial stability, etc.), will affect the flavor. The best processing conditions for citrus and chocolate flavor are acidic and neutral condition, respectively. Oxidation of flavor and functional ingredients is also potentially caused by air contacts (Bagchi and Nair, 2016). Maximum information about the product and processing conditions should be collected, and it is very useful in selecting proper flavor and its dosage form for any nutraceutical or functional food.

2.9 Conclusion

The flavor industry is a service-based industry composed of small- to medium-sized companies developed primarily by entrepreneurs blessed with vision and creativity. The role of flavor is simply to impart sensory pleasure to different food, beverage, nutraceutical and functional food products. This simple role is associated with its multidimensional economic, physiological and psychological functions. It is important to understand the functions expected of the flavor especially for nutraceuticals and functional foods so that the type best suited for this function can be delivered.

Currently, flavor industry is moving toward more of the "fantasy" type flavors where there is no target provided. Hence the flavor created will not be the exact or modified image of nature but a combination of pleasant sensations for example tropical punches and fruit mixtures. Parallel the demand for natural flavors is increasing, which is providing thrust to research in this area. It ranges from preparing single chemicals to producing complex mixtures of aroma products. Also in flavor industry there is an ever-increasing use of computer technology to assist the flavorist in their creative task. The flavorist now has access to database of thousands of different flavor preparations through which he can do selective searches and can retrieve a starting formulation that meets specific parameters for the desired nutraceutical or functional food. So the time required to develop a new flavor has been reduced significantly.

Flavoring functional foods and nutraceuticals is complicated and must be done on individual basis. The most effective approach is to partner with a flavor house throughout all phases of product development. However, the previous trend was to "perfect" the base first, and then bring in the flavor house at later stages of development. This greatly decreases the chances of a successful product launch. The flavor houses have valuable skills in reducing off-flavors in the base, hence their input must be taken since the starting and throughout all phases of product development.

References

Aarts, T.D., 1998. How Long Will the "Medical Food" Window of Opportunity Remain Open? *Journal of Nutraceuticals, Functional & Medical Foods* 1, 45–57.

Anderson, D.F., 1965. Flavor chemistry of Blue cheese. PhD thesis, Oregon State University.

Badger, T.M., Ronis, M.J., Hakkak, R., Rowlands, J.C., Korourian, S., 2002. The health consequences of early soy consumption. *The Journal of Nutrition* 132, 559S–565S.

Bagchi, D., Nair, S., 2016. *Developing New Functional Food and Nutraceutical Products.* London, UK, Academic Press.

Bauer, K., Garbe, D., Surburg, H., 2008. *Common Fragrance and Flavor Materials: Preparation, Properties and Uses.* Hoboken, NJ, John Wiley & Sons.

Bellaver, B., Souza, D.G., Souza, D.O., Quincozes-Santos, A., 2014. Resveratrol increases antioxidant defenses and decreases proinflammatory cytokines in hippocampal astrocyte cultures from newborn, adult and aged Wistar rats. *Toxicology in Vitro* 28, 479–484.

Bruno, N., Pavani, F., 2018. *Perception: A Multisensory Perspective.* Oxford, UK, Oxford University Press.

Burdock, G.A., 2016. *Fenaroli's Handbook of Flavor Ingredients.* Boca Raton, FL, CRC Press.

Cockbill, C., 1994. Food law and functional foods. *British Food Journal* 96, 3–4.

de Graaf, C., van Staveren, W., Burema, J., 1996. Psychophysical and psychohedonic functions of four common food flavors in elderly subjects. *Chemical Senses* 21, 293–302.

Dinkova-Kostova, A.T., Fahey, J.W., Wade, K.L., Jenkins, S.N., Shapiro, T.A., Fuchs, E.J., Kerns, M.L., Talalay, P., 2007. Induction of the phase 2 response in mouse and human skin by sulforaphane-containing broccoli sprout extracts. *Cancer Epidemiology and Prevention Biomarkers* 16, 847–851.

Ebrahim, S., 1998. *The Pharmaceutical Development of a Fixed Combination Anti-tuberculosis Dosage Form,* Johannesburg, South Africa, University of the Witwatersrand.

Fisher, C., Scott, T.R., 1997. Flavour compounds. *Food Flavours.* Athenaeum Press Ltd, Gateshead, Tyne and Wear, UK, pp. 15–55.

Frank, D., Appelqvist, I., Piyasiri, U., Wooster, T.J., Delahunty, C., 2011. Proton transfer reaction mass spectrometry and time intensity perceptual measurement of flavor release from lipid emulsions using trained human subjects. *Journal of agricultural and food chemistry* 59, 4891–4903.

Freeland-Graves, J.H., Nitzke, S., 2013. Position of the academy of nutrition and dietetics: Total diet approach to healthy eating. *Journal of the Academy of Nutrition and Dietetics* 113, 307–317.

Goldberg, I., 2012. *Functional Foods: Designer Foods, Pharmafoods, Nutraceuticals.* Berlin, Germany, Springer Science & Business Media.

Gul, K., Singh, A.K., Jabeen, R., 2016. Nutraceuticals and Functional Foods: The Foods for the Future World. *Critical Reviews in Food Science and Nutrition* 56, 2617–2627.

Hilton, J., 2017. Growth patterns and emerging opportunities in nutraceutical and functional food categories: Market overview. *Developing New Functional Food and Nutraceutical Products.* Amsterdam, the Netherlands, Elsevier, pp. 1–28.

Janovsky, H.L., 1960. *Drug and Cosmetic Industry.* 86, 335.

Jung, Y.H., Park, B., Kim, J.U., Kim, T.I., 2018. Bioinspired electronics for artificial sensory systems. *Advanced Materials* 1803637.

Keservani, R.K., Kesharwani, R.K., Sharma, A.K., Gautam, S.P., Verma, S.K., 2016. Nutraceutical formulations and challenges, *Developing New Functional Food and Nutraceutical Products.* Elsevier, pp. 161–177.

Keskar, A., Igou Jr, W., 2011. Flavoring of nutraceuticals. *Handbook of Nutraceuticals,* Boca Raton, FL, CRC Press, 2, 485–512.

Kreutzmann, S., Christensen, L.P., Edelenbos, M., 2008. Investigation of bitterness in carrots (*Daucus carota* L.) based on quantitative chemical and sensory analyses. *LWT-Food Science and Technology* 41, 193–205.

Mazza, G., Shi, J., Le Maguer, M., 2016. *Functional Foods: Biochemical and Processing Aspects.* Vol. 2. Boca Raton, FL, CRC Press.

McDonald, S.T., 2016. Concepts of flavor creation in novel nutraceuticals and functional food formulations, *Developing New Functional Food and Nutraceutical Products.* Amsterdam, the Netherlands, Elsevier, pp. 231–247.

Naim, M., Brand, J.G., Kare, M.R., Carpenter, R.G., 1985. Energy intake, weight gain and fat deposition in rats fed flavored, nutritionally controlled diets in a multichoice ("cafeteria") design. *The Journal of Nutrition* 115, 1447–1458.

Ohloff, G., 1972. Classification and genesis of food flavours. *Flavour Ind* 3, 501–508.

Priyadarshi, S., Shukla, A., 2012. Factors affecting withholding and role of flavour compounds in food vitamins. *International Journal of Biology, Pharmacy and Allied Sciences* 1, 1012–1019.

Ramirez, I., 1985. Oral stimulation alters digestion of intragastric oil meals in rats. *American Journal of Physiology-Regulatory, Integrative and Comparative Physiology* 248, R459–R463.

Rawson, N.E., Li, X., 2004. The cellular basis of flavour perception: Taste and aroma. *Flavor Perception* 57–85.

Reineccius, G., 2013. *Source Book of Flavors*. New York, Springer Science & Business Media.

Sinki, G.S., Gordon, R.J., 2001. *Flavoring agents, Food Additives*. Boca Raton, FL, CRC Press, pp. 354–413.

Threatte, R., Giduck, S., Kling, M., 1986. Oropharyngeal stimulation of glucose-absorption from the small-intestine in conscious, unrestrained rats. *Federation Proceedings*. Federation of American Societies for Experimental Biology. 45, 537–537.

Wesley, F., 1957. Flavor. *Journal of the American Pharmaceutical Association (Practical Pharmacy Edition)* 18, 674–677.

Wikipedia, The Free Encyclopedia, 2019. https://en.wikipedia.org/wiki/Flavor. Webpage. (accessed January 1, 2019.).

Yadav, D., Srikanth, K., Yadav, K., 2018. *Flavors in Probiotics and Prebiotics, Flavors for Nutraceutical and Functional Foods*. Boca Raton, FL, CRC Press, pp. 71–94.

Zhao, J., 2012. Nutraceuticals–Functional foods for improving health and preventing disease. *Pharmaceutical Biotechnology: Drug Discovery and Clinical Applications*, 2nd ed, pp. 599–628.

3

Flavor Manufacturing and Selection Criteria for Functional Food and Nutraceuticals Industries

Jayvadan Patel and Anita Patel

Contents

3.1 Why is flavor important in nutraceuticals and functional foods?

A functional food is one type of food that wields health properties over the conventional nutrients it includes. Foods that naturally contain biologically active, non-nutrient components provide health benefits. These components are *phytochemicals*, which are sometimes called *nutraceuticals*. On the basis of this explanation, every plant foods can be regarded as functional foods seeing as they are all abounding in phytochemicals or nutraceuticals. Functional foods are first and foremost foods, and they contend with traditional foods for a portion of the public's liking. Regardless of the supplementary value of health functionality, users are doubtful to purchase functional foods continuing if the foods don't have good taste (Reineccius 2000). The flavor and odor of functional foods are vital if there is a wish to appeal to the ordinary user. One can predict that a majority of food companies might select to strengthen an existing product line through botanicals, minerals, or vitamins. In such circumstances, great care must take into consideration the flavor quality of the product. It has been repetitively shown that if foods do not have good quality flavor the user will not buy foods in the long run. International Food Information Council conducted a 2011 consumer trending survey on the functional foods, which established that the most imperative features that Americans consider regarding functional foods were cost and flavor (Food Insight 2011).

Functional foods must not be considered as a single group, but rather as merely distinct products within a variety of food groups. Increasing a functional food's functionality, such as increasing bioactive compounds, may cause off-flavors that cause aversive customer reaction due to taste (Siro et al. 2008). A study on off-flavor in juice with specific health claims found that even with knowledge of the health claims, consumers' degree of liking and likelihood of consumption decreased with increasing severity of the flavor defect (Tuorila and Cardello 2002). No support was found for the theory that a slight off-flavor was perceived as a positive marker of health benefits to the consumer.

Also in the medical foods, wherever the user might be taking the product under the prescription of doctor, flavor still is extremely imperative. Flavor has numerous advantages above and beyond creating food taste top quality counting stimulation of digestion in addition to the physiological comfort of the individual (Reineccius 2000).

The function of flavor in a final product is changing the profile (Keskar and Igou 2011), which comprises:

- The choice and compensation of existent of impending flavor factors functioning within the restraints of nutritional requirement, the nature as well as resources of raw materials and additional constituents, and the entire concept of the final product.
- The flavor profile regulation, associated with the technique of processing used to suit specific palates or user expectations.
- Adjustment to conquer any pre-existent or created defects of flavor.
- Impartation of completely new flavor in the products, which are insipid or extremely unpleasant to taste.
- Enlargement of the variety and resilience of the finish products.

All of these proceedings need a personal finding on the part of the product improvement team and entail the facts of existing raw materials, smallest and most favorable processing circumstances, lawful restraints, and expected customer reaction to the end product.

3.2 Criteria for application of flavor to functional foods and nutraceuticals

Flavors can be included into functional foods as well as nutraceuticals formulations for a number of purposes (Keskar and Igou 2011):

1. To contribute flavor to an or/else tasteless product produced to present a little pleasing flavor experience. In this classical function of flavor the final product is attributed to a definite overall flavor.
2. To adapt or balance an available base of flavor, distinctively to please specific necessities of the nutraceuticals base.
3. To mask or cover up objectionable flavor traits or off-notes, either by blocking, disguising, or else trapping the taste buds into not distinguishing the off-flavor.
4. A combination of 1, 2, and/or 3.

3.3 The role of attention in flavor perception

"Flavor" relates to the perceptive incident we experience while we eat and drink (Auvray and Spence 2008; Stevenson 2009; Small 2008). In oral cavity taste, olfaction and somatosensation are the three anatomically distinct sensory systems that provide to flavor (Simon et al. 2006). Taste is detected through receptors mainly positioned on the surface of the tongue, which produces feelings of bitterness, meatiness (umami), saltiness, sugariness, tartness, and perhaps other sensations, as well, pertaining to greasiness and metal like tastes (Schiffman 2002). The volatile compounds liberated by foodstuff as well as drink in the oral cavity, particularly at some stage in chewing, are detected by olfaction. These volatile compounds may possibly be pumped up by way of the nasopharynx to the olfactory receptors situated in the nasal crypt and/or might be transported by exhaled air routed through the nose as soon as the oral cavity is filled (Trelea et al. 2008; Hodgson et al. 2003). Contrary to taste, olfaction has a great variety of feelings connected with it and is a most important contributor to our understanding of flavor (Dravnieks 1985). Somatosensation is the last contributing intellect. Not only is this involved in creating our feeling of food texture through receptors to be found inside the different tissues of the oral cavity (Christensen 1984), but as well it notices feelings in relation to annoyance, pain, and temperature. The somatosensory system presents a series of sense which is almost without doubt more than taste; it is most likely less than for olfaction.

Despite the fact that taste, olfaction, and somatosensation consist of the senses entailed while a food is in the oral cavity, the understanding of flavor can as well be influenced by other food properties that are noticeable at some stage in or just prior to ingestion. For the period of ingestion, the food makes sound when it is being chewed which have an effect on our experience of feel, as may sound before ingestion, even if its influence is most likely quite small (Vickers and Bourne 1976; Zampini and Spence 2004). Far more powerful are the consequences of the appearance of the food and the anticipations that these optical signals can cause (Yeomans et al. 2008). Several investigations have specified that appearance has an effect on both the satisfaction of that food and the manner in which it is apparent when later ingested (Shankar et al. 2010; Spence et al. 2010). This modality-based explanation of senses entailed in the perception of flavor does not appear to be in correspondence with mainly individual's routine experience of flavor, or with modern functionalist hypothesis of flavor insight (Auvray and Spence 2008; Small 2008), which necessitates collating information on a one act (ingestion) into a one percept: a solitary flavor understanding. Flavor, at that moment, gives the impression to be somewhat of a developing property from the person senses like taste, olfaction and somatosensation to create it up.

3.4 Factors influencing the selection process of flavors for functional food and nutraceuticals

The overall quality of the food associated to aroma affected by a number of factors, such as physico chemical factors like temperature, pH, exposure to air and biochemical factors like interactions of volatile fragrance components with food ingredients (Landy et al. 1995). For functional food and nutraceuticals industry experts these variants are very significant to take into account while evaluating and creating new products. Bearing in mind these factors will facilitate precise sensory and bench top tasting consequences as well as the growth of products consumers need.

3.4.1 Physicochemical factors

There are a number of physicochemical factors that can affect selection process for flavors, ranging from temperature of food to exposure of air.

3.4.1.1 Temperature

We are all recognizable with the influence of temperature on the taste—imagine about first to consume beverage or anything when it is lukewarm and finishing it while chilled, or vice versa. Depending on its temperature the same food or drink can taste different. Additionally, about 20%–30% of the populace, heating or cooling little regions of the tongue extends a taste feeling devoid of the existence of food or beverages; these types of people's are called thermal tasters.

On a mixture of "thermal" tasters; "super" tasters, that is, people who are predominantly susceptible to tastes in most cases; and "regular" tasters, the effect of temperature at both 5°C and 35°C on the ephemeral uniqueness of astringency, bitterness, sourness, and sweetness was investigated (Bajec et al. 2012). Over a period of time they were requested to evaluate the taste intensity.

For the entire three categories of tasters, temperature affected the highest sensed intensity from astringent, bitter and sour solutions, however not from the sweet solutions. In particular:

- When the solution was lukewarm astringency was more strong, furthermore with the lukewarm solution the intensity of the flavor lasted longer compared with chilled solution
- When the solution was chilled bitterness was more strong and the intensity of the flavor in the cold solution diminished more rapidly than warm solution
- In the warm solution sourness was more strong and also the intensity of flavor lasted longer compared with chilled solution
- Moreover, amazingly, between the chilled and warmer solutions of sugar there was no distinction in sensed sweetness, nevertheless it acquired more time for the cold solution to achieve its highest intensity of flavor.

At the warmer temperatures both sour and astringent solutions tasted the strongest and also the intensity lasted more than it did with cold solutions. Bitter tastes came through greatest when cooled. Furthermore, temperature is devoid of for sensing sweetness. For the majority individuals, temperature can augment flavors. However for a few, dubbed thermal tasters, temperature alone can be a taste. The perception of taste created via heating or cooling parts of the tongue with no food, determinations that's tough to ingest (Bajec et al. 2012). Effect of temperature on flavor is not consistent. Taste buds can be upset of act by both high and low temperatures. Raising the temperature shows to enhance the reaction to sweetness and decline it to bitterness and saltiness. Declining the temperature shows to enhance the sensitivity to bitterness and decline the sensitivity to sourness (Talavera et al. 2007).

3.4.1.2 Taste medium

The taste buds are capable of detecting those flavors that are only dissolved in a liquid. We cannot taste a dried stuff with a dried tongue. Water is the most excellent medium for sense tests. Taste thresholds are found to be lesser in water compared with tomato juice. Taste sensitivity is reduced by raising the viscosity of the medium. It is easiest to sense tastes in liquid medium, firmer in foams and more complex in gels (Christensen 1980).

3.4.1.3 pH

The majority of fruit comprises natural organic acids like citric acid, malic acid, and tartaric acids that add appreciably to the flavor profile. When simulating a fruit flavor that has an inherently neural taste, the use of acids may be essential; or else, the right flavor feeling is not attained in the final product. A number of flavors and certain spices such as turmeric have components that are responsive to alterations in pH, and it is important that this specific circumstance be cautiously reproduced at some stage in the development of product and shelf-life testing to make sure that unwanted things do not take place. Dependent on what is in the nutraceuticals the pH, either high or low, is case specific (Keskar and Igou 2011).

3.4.1.4 Exposure to air

In the products that are aerated this is of specific interest. Significant volatile losses can be observed in high-speed mixing operations, but more prominently, any closed air creates situations leading to oxidation of any unsaturated lipids present. If an encapsulated flavoring or seasoning is not employed, the pneumatic delivery of crushed flavor components possibly will also result in momentous volatile losses. Any added flavor exposed to the least treatment is the guiding principle. Visibly, addition into the primary mixture cannot be escaped for a few foodstuffs, but wherever feasible, flavoring be supposed to be added at later stage as compatible with consistency of distribution in the final formulation (Keskar and Igou 2011).

3.4.2 Biochemical factors

Biological structure in any food and nutraceuticals formulation interacts with components that can have an effect on the assortment process of that specific flavor for a particular product system. Acetaldehyde, diacetyl, ethyl acetate, and isoamyl acetate are examples of most frequently used flavor components which may interact with non-active components of the nutraceuticals base. At some stage in food processing, the majority of interactions take place between the chief ingredients of the food, for instance, carbohydrates, fats, proteins and vitamins. In the foodstuff, properties of flavor rely on both the nature of the foods as well as the flavor components present. The delivery of flavor is greatly reliant on the accessibility of the flavor components in gaseous phase, and consequently on the affinity of the flavor components for the food matrix. A variety of flavor components properties decide the interactions with food compounds for example functional groups, molecular size, shape and volatility. Properties like boiling point, molecular weight, octanol-water partition coefficient (log P), and vapor pressure have been utilized to envisage the volatility of the components under a stationary state (Van Ruth and Roozen 2002).

In foodstuff, flavor matrix interactions have been extensively studied regarding effects on flavor release as well as sensitivity. Explaining key matrix parameters that have an effect on the release of flavor components from foods would give helpful data to manage the flavor reaction of food formulations and permit for the effectual utilization of flavor resources (Schober and Peterson 2004). When the "binding sites" of food components are still accessible, food matrix compounds are able to bind, entrap or encapsulate volatile as well as nonvolatile flavor ingredients. Accordingly, the flavor release rate declined by the interactions and has an effect on the flavor intensity along with the quality of foods which ultimately effects the user overall acceptance (Naknean and Meenune 2010).

3.4.2.1 Carbohydrate–flavor interactions

The volatility of flavor components can be altered by carbohydrates corresponding to water, but the effect relies on the interaction between the specific flavor and carbohydrate atoms (Godshall 1997). In general, mono- and disaccharides demonstrate a salting out consequence, which brings about an augment in flavor volatility corresponding to water. Simple sugars and/or starches commonly form delivery service for flavor compounds. By reducing vapor pressure or by affecting rate of mass transfer polysaccharides have an effect on the release of aroma. A lot of possible chemical interactions have been observed with complex sugars compared with simple sugars, attributable to the variety of functional groups presented. Study revealed that polysaccharides in general encourage a decrease in release of aroma induced by an augment in the viscosity and/or by molecular interactions with flavor ingredients (Goubet et al. 1998).

3.4.2.2 Protein and free amino acids–flavor interactions

Owing to functional properties of protein, for example emulsifying and stabilizing capabilities in lipid-dispersed food systems, they are frequently used food components. Proteins interact with flavor compounds by reversible or irreversible binding. The binding of volatiles to proteins, particularly soy protein, is the most repeatedly investigated flavor compound interaction, as stated in the literature (Plug and Haring 1994). It was confirmed from binding investigations that heat denaturation of the protein augmented the binding capability of aliphatic aldehydes plus ketones; but for alcohols there are opposing results. Protein binding results can be quite complex, in view of the fact that the binding of volatiles is powerfully contingent on the quantity as well as water content of protein. Propanol has been found to bind to anhydrous zein proteins (Matheis 1993). Sodium caseinate has been found to bind a higher degree of esters and ketones than starch (Le Thanh et al. 1992). Conversely, aldehydes have a tendency to respond chemically with protein amino groups, resulting in irreversible binding (Matheis 1993). Information obtained from model systems demonstrate that a number of factors, e.g., the chemical nature of the flavor ingredient, ionic conditions, presence of ethanol, temperature, and the structure and processing history of the food protein, decide the degree of interactions among proteins and flavors of food (O'Neil 1996; Lubbers et al. 1998).

For determination of protein-aroma interactions various techniques resultant in corresponding information have been employed (Lubbers et al. 1998). In aqueous medium free amino acids are able to bind with numerous flavor components. Through the hydrogen bridges ketones and alcohols are reversibly joined to the amino or carboxyl factions of the amino acids, while the with proteins, a few aldehydes chemically interact with the amino groups to form Schiff's bases. In an aqueous media the amino acid cysteine interacts with aldehyde and ketones to thiazolidine-4-carboxylic acid, this reaction is reversible under heating, mainly under acidic pH. Volatile aldehydes (e.g., hexanal), ketones (e.g., diacetyl), acids, and amines are adsorbed by dry amino acids (Keskar and Igou 2011).

3.4.2.3 Lipid–flavor interactions

Lipid–flavor interactions are generally correlated to partition phenomenon, or else the comparative quantity of flavor dissolved in the lipid and water phases. In foods the principal lipids are fats and oils, which comprise mainly of di- and triglycerides. These triglycerides can attach or solubilize substantial amounts of lipophilic and partly lipophilic flavoring matters. Solid fats have lower binding capacity than oils. The amount of bind flavoring matter relies upon the length of fatty acid chain and extent of unsaturation in the composition of triglyceride (Matheis 1993). Triglycerides having long-chain fatty acids can bind less quantity of ethanol and ethyl acetate compared with those triglycerides having short-chain fatty acids. A triglyceride like triolein, which has unsaturated oleic acid, binds more amount of flavor compared with tripalmitin and trilaurin, both of which have only saturated fatty acids. This entails that the amount in the gas phase decreases as the chain length enlarges (Matheis 1993). Distorted flavor-ingredient interactions create a diverse flavor release behavior in the food system, which exhibits low levels of fat. During the progress of reduced-fat and fat-free foods substantial attention in these phenomena has resulted. A low vapor pressure exhibited in various volatile flavoring matters in lipids, and for that reason a higher odor threshold, than they perform in aqueous media. Decreasing the quantity of fat has the result of increasing the equilibrium vapor pressure of the flavorant, and altering its time-intensity profile of release. Accordingly, in the food matrix volatile flavors, which cannot be maintained and are released instantly, result in a powerful but rapidly dissipating flavor feeling.

3.5 Challenges in flavoring functional foods and nutraceuticals formulation

There has been countless advancement in the nutraceuticals company in terms of improving taste but confrontations still continue. The challenge in flavoring functional foods and nutraceuticals formulation consists of problems with off flavors intrinsic to the functional foods and nutraceutical ingredients, possible change of flavors in the manufacture of the functional foods and nutraceutical formulations, negative interactions between flavoring and the functional element, and any storage issues that may be directly or indirectly related to the functional ingredient. Functional food as a category are relatively new in United States, or may be more properly, have been in recent times rediscovered by the food industry, but the flavoring problems associated with them are not new. Functioning with off flavors in a food system is an ordinary task and a continuing battle.

Functional foods come under food category, therefore flavoring them should be same as flavoring other processed foods. Though, the active ingredients might induce off-flavors that require being reduced or masked. For instance, foods containing soy protein might have grassy or beany flavor that a lot of populace experience unpleasant (Reineccius 2000; Hazen 2003), and the food industry have considerable experience in flavoring soy-based foods in this context for covering beany off flavor (Milo 1999). Vitamins exhibit meaty brothy flavor. Several functional components are acrimonious while some functional components tend to flavor degeneration, for example carotenoids as well as fish oils (oxidation) and thiamin (degradation to meaty or notes like vitamin) (Reineccius 2000). An interaction of base is as well an issue. Fibers added in foods also have the tendency to bind with flavors well, like do proteins and starches. Foods containing full fat flavors work effectively, while foods containing low or no fat may perhaps not taste good. Numerous functional food items have a variety of herbs and spices that hold astringent, bitter, grassy, and other flavor deficiencies. Mineral products have metal like taste (Hazen 2003). Natural flavor chemicals are many times reactive and flavorists are familiar with that and are able to balance in regular food. On the other hand, they might be unknown with the reactive trait of the herbal mix in the food (Reineccius 2000).

The flavor industry also has substantial experience in flavoring pharmaceutical products that many times have very distinct off flavors and necessitate strong flavor systems to make them pleasant. Providentially for pharmaceutical products, the consumer buys and uses them for a long time not by choice but by need. Although the flavoring problems may be alike, the flavoring necessities for functional foods and nutraceutical formulations are much more rigorous. Manufacturers want the consumers to pay for their products not for the instant advantage of decreasing pain but for some distant promising advantages of improved health. This will to a great extent reduce consumer patience for poor sensory quality. However, there is a body of facts in companies that have customarily provided flavorings for pharmaceutical products that can be directly applied to functional foods and nutraceutical formulations (Reineccius 2000). Likewise, other confrontations related to processing, interaction and storage issues are not sole to functional foods and nutraceutical formulations. It is not likely that the processes used to recover or further process a functional foods and nutraceutical formulations will be different considerably from those presently used in the food industry for an existing product. Some of these processes, retorting, for instance, are tremendously insensitive on stability of flavor. The interactions one might imagine are also familiar through either existing food products or pharmaceuticals. While there may be differences in the extent to which each of these problems takes place in a functional foods and nutraceutical formulations that make flavoring more complicated, the solution to the problem will classically follow a customary and similar approach (Reineccius 2000). Nutraceuticals as not food will be expected to contain bitter, metallic, and other off flavors. In addition they might contain distinctive different food, similar to high or low pH or longer shelf lives, and require to be registered with appropriate health authorizations (Wright 2011).

3.6 Strategies for flavoring functional foods and nutraceuticals

3.6.1 Ingredient quality

The types of problem one faces in flavoring functional foods are renowned. For instance, masking the beany note of soy is a common concern. The answer to this problem begins with the soy protein manufacturing process itself. One must process the soybeans in a way that diminishes this off note (early enzyme inactivation). This approach carries over to any functional ingredient-one must produce functional ingredients of the best possible sensory quality. In manufacturing, one should not only focus on the functional quality of the ingredient but also its sensory quality. The range in quality of a given ingredient in the industry drives the point home that large dissimilarities exist and more focus requires to be placed on sensory quality. Psyllium is a product that can be very offensive in taste, texture, and odor or pleasurably bland, depending upon the manufacturing process and the temperature and humidity condition during storage and transport. The point is to insist the best quality ingredient one can acquire. One does not formulate a good flavored product with off-flavored elements (Reineccius 2000).

3.6.2 Flavor selection: Compatible flavorings

If the functional food holds off notes and they cannot be satisfactorily (or cost-effectively) removed, then one has to work with the off notes, either by selecting flavors such that the distasteful notes are positive contributors

Table 3.1 Flavors Most Companionable with Off Tastes in Functional Foods

Taste Sensation	Recommended Flavor Selection
Alkaline	Chocolate; cream; mint combinations; vanilla
Metallic	Burgundy; grape; lemon-lime
Salty	Butterscotch; melon; mixed citrus and fruits; nut; raspberry
Sour	Anise; citrus; cherry; licorice; raspberry; root beer; strawberry
Sweet	Berries; maple and honey; mixed fruits; specific fruits and berries; vanilla

to the flavor or by adding other components that will battle for sensory binding sites and thus lessen the sensitivity of the objectionable notes (masking). The first strategy is most widely used to select the flavor such that offensive sensory notes are positive contributors to the flavoring. For bitter functional components, one might select cheese, coffee, dark chocolate, or tea flavoring. All of these products have bitter notes as a pleasing part of the flavoring, and the bitterness innate to the ingredient may be used to give this feature of the flavoring. Fischer (1998) has suggested flavors that are compatible with some of the common taste effects associated with functional foods (**Table 3.1**).

Cooked protein off notes may be flavored with caramel, butterscotch, or chocolate flavoring, as the cooked notes would be attractive. Functional foods containing thiamin can be delivered in a meat soup or broth form since the thiamin degradation products would be in fact desirable in these products. Milk chocolate is frequently used to hide oxidized notes. Hershey's milk derives its distinctive flavor from the use of carefully oxidized milk. Thus, the American consumer who has grown up with Hershey's chocolate finds an oxidized note to be very satisfactory as a part of chocolate flavor profile. Another example of balancing flavor was suggested by Manley (1997) who recommended the use of citrus and fruit flavors to coat fish oil off notes. The consumer is incredibly tolerant of oxidized citrus oil, and some companies intentionally "age" orange oil to make a characteristic oxidized character that can the consumer finds quite tolerable. The use of berry flavors to mask oxidized oil off notes is less understandable except for in the case of raspberry. Raspberry has a sensory character that can coat oxidized sensory notes quite well (Reineccius 2000).

3.6.3 Masking

There is an incomplete amount one can do to truly mask an off note, that is, to block the sensory receptors so that the offensive notes are not perceived. While there are examples of masking in the sensory literature, those that are well recognized are hardly any. One of the best established examples is the masking of bitterness by sweetness. If one adds sugar to a bitter product, the bitter sensation will be declined since the same taste receptors serve to sense bitter and sweet (Heijden 1993; Walters and Roy 1995). Therefore, if there is a great deal of sugar in a product, the taste receptors are occupied with it only owing to mass action, and the bitter sensation is reduced. Anecdoctal information in the literature recommends that salt masks a given note or that monosodium glutamate augments or masks a given note, but there is small conformity on these cases, and the documentation is often weak (Reineccius 2000).

3.6.4 Flavor–ingredient interactions

In recent times the literature on food/flavor interaction has grown enormously. This knowledge base has been to a great extent enhanced by concerted efforts such as the European COST 96 Action (European Co-operation in the Field of Scientific and Technical Research) on flavor interaction with food matrices (Guichard 2000). As a result, a greater admiration has been gained for how a given food ingredient or component will interact with a flavor component and thus influence flavor sensitivity.

It is appreciated that protein and certain carbohydrate polymers are good binders of flavoring. The physical properties of a food like viscosity or friability will influence how the flavor is released and subsequently sensed. The effect of fat content on food flavor has been well acknowledged and modeled. Consequently, functional foods created that have known ingredients and physical properties, to manage the interactions that occur (Reineccius 2000).

In terms of interactions between flavors and specific components, the flavoring of pharmaceutical products offers considerable information. For instance, the flavor industry has learned how to make a cherry flavor without benzaldehyde when the pharmaceutical has an active ingredient with an amine function, or grape without methylanthranilate when the active ingredient is a carbonyl. This is necessary to get rid of the amine/carbonyl reactions that result in the loss of both the flavor and active ingredient (Millard reaction). The difficulty with functional foods is that often the active ingredient of the food may be unidentified, for example in herbal extracts. Therefore, the creative flavorists may have to test with the flavoring system and instinctively make such changes in flavor formulation (Reineccius 2000).

3.7 Flavor stability

In consumer satisfaction flavor plays a significant role and persuades further eating of foods. The majority of existing aroma ingredients are formed via chemical synthesis or extraction. Food comprising artificial or synthetic flavor are often dodged, as the users believe that these ingredients are harmful or injurious to their health (Teixeira et al. 2004). In accordance with the multifaceted matrix of the food system as well as the structure of flavor themselves, one imperative feature that plays a pivotal role in the quality trait of food is stability of flavor (Weerawatanakorn et al. 2015). Flavor stability may or may not be a problem in functional foods and nutraceuticals. Stability will depend on whether the base product itself is stable or will build up off-notes during storage. This will depend upon numerous factors, for example the functional component being delivered, its concentration, the food matrix itself, the presence of stabilizing ingredients like antioxidants or chelating agents, packaging, and storage conditions (Reineccius 2000).

Among the lot of factors concerned with flavor quality, stability of flavor is the most significant factor. The individual flavor ingredients chemical structure is related with the chemical reaction that is liable for stability of flavors. The chemical reactivity of these ingredients is affected by the existence of active functional groups, like carbonyl, hydroxyl, and thiol functional groups. Either high- or low-volatility flavor ingredients, in spite of whether they are acidic, neutral, or nitrogen- and sulfur-containing elements, can be vulnerable to chemical alterations taking place in different types of interactions, counting hydrolysis, thermal degradation, oxidation, photo-oxidation, polymerization of unsaturated components, and interaction with protein in foodstuff. Aldehydes might be easily oxidized to acid, amines can create a complex with metal ions, and terpenes are able to undergo reorganization and isomerization under acidic media. These susceptible outcomes act on the overall flavor valuc of food (Fisher and Scott 1997).

Flavor stability in diverse foods has been of growing attention due to its connection with the quality and tolerability of foods, but it is not easy to manage. Today one confrontation faced by the flavor industry is to handle the binding of flavors to food matrices, diffusion and interactions with the surroundings wherein they find appliance and the stability of process/reaction flavors to heat, UV-light, oxidation. Seeventer et al. (2001) investigated the stability of thiols produced from model system ribose/cystein, and stated that 2-furfurylthiol, 2-mercapto-3-butanone, 2-methyl-3-furanthiol and furaneol declined at some stage in storage. In brewed coffee, after application of heat, the amount 2-furfurylthiol, methional, and 3-mercapto-3-methylbutyl formate was reduced, and it has been based on interactions with melanoidins (Hofmann et al. 2001; Hofmann and Schieberle 2002; Kumazawa and Masuda 2003). The detriment of 2-furfurylthiol was associated to fenton-type oxidation on heating at 37°C for 1 hour (Blank et al. 2002).

Many times overall flavor alterations induced by the manufacturing and storage processes, packaging materials and ingredients in foods resulted in reduction in intensity of aroma ingredient or generation of off-flavor ingredients (Lubbers et al. 1998). There are lots of variables that make flavors extremely knotty systems. A few of which are water soluble and are steadier carbohydrates while some are more stable in lipid-based coating.

3.8 Flavor manufacturing for functional food and nutraceuticals

Flavors as well as fragrances are commonly employed in food, beverage, cosmetic, and pharmaceutical products. One of the most momentous trends in flavoring functional foods, specifically beverages and candy store products, is the employment of flavors that are linked with health and function. Flavoring preparations are not chemically distinct materials, however multifaceted mixture whose constitution is defined by natural raw materials. Established examples are extracts of different vegetable and fruit, spice and herb as well as yeast. Flavoring preparations are many times the major constituent of citrus, spice and mint flavorings, reflective of distinctive aroma and taste. In recent times because of consumer's augmented attention and health consciousness in natural formulations, more pressure has been given for the use of natural flavors acquired from natural origins. The global market for flavors and fragrances was valued at $26.0 billion in 2015. This market is expected to increase from $27.1 billion in 2016 to $37.0 billion in 2021 at a compound annual growth rate (CAGR) of 6.4% for 2016–2021 (BCC Research 2016). The majority of these flavoring and fragrance ingredients are produced by chemical technology and barely a little portion of the requirement is fulfilled from extraction from plant and animal sources or through microbial origins. Study has revealed that microbes can also be utilized to make flavors. Researchers are using microbial systems to give new flavors to the fermentation formulations like beer, cheese wine, etc from eras. Natural flavoring preparations are obtained from plant, animal or microbiological source materials by means of physical or biotechnological production processes. The first artificial or synthetic fragrance and flavor compounds were vanillin (1874) and coumarin (1868) prepared for use in the food industry (Bomgardner 2012; Gupta et al. 2015). Choice of flavor ingredients not only presents taste options to consumers but may also help mask objectionable flavor notes imparted by functional ingredients. A few flavors are more transparent than others, so manufacturers should consider matching stronger flavors with nutrient profiles that present stronger off-notes. Chocolate, for example, is a relatively stronger taste masker than vanilla.

3.9 Sources of flavoring materials

Near the beginning in the history of flavor formation, merely natural extracts were handy. The most important sources were the extracts of herbs and spices, in addition to several other botanicals, various parts of plant can be used viz. counting bark, buds, exudates, flowers, rhizomes and seeds. A restricted number of animal-sourced flavoring were as well obtainable (hardly ever if in any way in flavor today), for example castoreum from beavers and musk from musk deer.

By utilizing solvents or distillation different plant extracts are alienated from the non-flavorful fraction of the plant material. Ethyl alcohol acts as an excellent solvent and one can use the resulting extract as such (with no elimination of the ethanol) to make a tincture. Hexane or acetone can also be used as solvent, but, these organic solvents have to be eliminated via vacuum to lower than a federally allowable level, and the resultant flavor extracts are called oleoresins. There is no residual solvent left, if super critical carbon dioxide (SC-CO_2) is utilized as a solvent to extract the materials, as the CO_2 is liberated from end product. The complete flavor of the source material has been found in solvent extracts, together with volatile and nonvolatile ingredients. Distillation is used to separate out flavorings from the source materials, to make essential oils that contain the fragrance of the source material but is deficient in the nonvolatile part of flavors. For instance, solvent extracted black pepper oleoresin exhibits the fragrance of black pepper, in addition to the acrid "bite" from piperine, and is pigmented and viscous. On the other hand, black pepper essential oil scents are similar to black pepper but it is clear as well as non-viscous and devoid of piperine. Other processes of distillation extraction comprise spinning cone technology, which produces aqueous extracts.

In the flavor industry, the expansion of organic chemistry brought about the formation of lots of synthetic chemicals for use which contained synthetic edition of naturally found flavor chemicals, for example vanillin resulting from petroleum or pulpwood. Furthermore chemicals not present in food or even in nature were established for flavor exploit, ethyl methyl phenyl glycidate or aldehyde C16 is one such ingredient employed for "red pop" as a flavor similar to raspberry or strawberry.

In the 1980s, natural flavors became more famous while synthetic flavors were less accepted by users. Organic chemists were capable of creating lots of the similar chemicals established in nature with techniques that followed the FDA definition of natural flavor. Thereafter the accessibility of such flavor compounds is greater than before, as a result at this time it is possible to produce the majority of the flavors as natural that was one time just existing synthetically.

A reaction flavor is the one more source of natural flavors which includes the formation of flavorful matters from usually low or non-flavorful precursors. A reasonable chocolate flavor can be produced if the amino acids phenylalanine and leucine are subjected to heating under definite conditions in the existence of a reducing sugar. This method is incredibly well known for producing meat-type flavors with no animal-derived components, ideal for vegetarian foods. For producing smoke and grill flavors pyrolysis is an excellent method.

There continue to be advances in fermentation, enzyme, and extraction technologies to offer natural flavorings. The flavor extracts mentioned earlier are available for flavor creator to make many natural flavors for use in many applications (McDonald 2017).

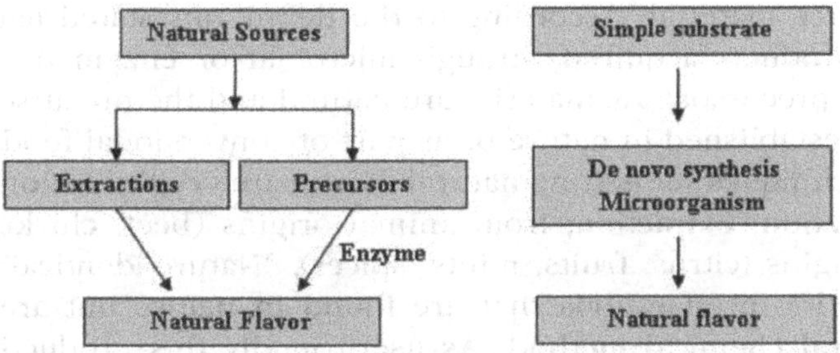

Figure 3.1 Pathway for preparation of natural flavor.

3.10 Flavors and fragrances: Types and synthesis

Flavors and fragrances are generally categorized into two types, natural and nature-identical. **Figure 3.1** shows three pathway for preparation of natural flavor (Dubal et al. 2008). In which the first two involve the extraction of the flavor from plants or precursors from natural sources. The last method is the de novo synthesis of the flavor in microbe's or else in plants or by single-step biotransformation of natural bases with microbes or their enzymes or plant cells, for example creation of nootkatone by citrus cell cultures. Microbes convert carbon or nitrogen components into flavor ingredients by means of enzymes such as esterases, lipases, nucleases, proteases and a few glycosidases in the de novo synthesis. In industry for the synthesis of flavoring ingredients these enzymes are widely utilized as they catalyze single-step 2 transformations of substrates into natural flavor compound. Additionally, where complex substrates like lignin, phenyl-propanoids and phenolic stilbenes are transformed, to prefer flavors microbiological or enzyme based methods have also been urbanized (Priefert et al. 2001).

3.11 Types of natural and nature-identical flavors

Natural flavors are those flavoring preparations that are achieved by suitable physical, enzymatic or microbiological techniques and acquired from plant or animal source material. According to Code of Federal Regulations, USA (1990), in general, natural flavor comprises dairy products, distillate of any product of roasting, essence or extractive, essential oil, fermentation products, oleoresin, protein hydrolysate which includes the flavoring ingredients obtained from a bud, bark, edible yeast, eggs, fruit juice, herb, leaf or comparable plant material, poultry, root, spice, seafood, vegetable or vegetable juice whose important function in food is giving flavor instead of nutrition. And as per European Commission Flavor Directive (88/388/EEC) natural flavors are defined as flavoring preparations that are achieved by suitable physical, enzymatic or microbiological procedures from plant or animal source material. According to the definitions stated natural flavors comprise products acquired through microbial or enzymatic methods as long as the precursor/raw material are natural and the precursor and product can be established in nature or as part of conventional foods. The main physical techniques for getting natural flavors are crystallization, concentration, distillation, extraction; from animal origins (beef, chicken, seafood) or plant origins (citrus, fruits, mints, spices). "Nature-identical" flavors are those flavoring preparations that are found in nature but are formed by a non-natural chemical method. As user-friendly this production mode is no longer popular (Kim 2005). There are some perfume ingredients e.g. esters (ethyl and butyl acetates, ethyl butyrate, caproate, isobutyrate, isovalerate, 2-methyl butyrate, menthyl acetate), aldehydes and ketones, (acetaldehyde, diacetyl), acids (acetic, butyric, caproic, caprylic, isobutyric, isovaleric, 2-methyl butyric) and lactones (gamma-decalactone) which can either be created by microbial fermentation or by using enzymes (**Table 3.2**) (Krings and Berger 1998).

Table 3.2 Flavors and Fragrances Manufactured by Various Microorganisms or Using Enzymes

Flavor Type	Flavor Active Constituent	Microorganism/Enzyme Entailed in Manufacture
Almond	Benzaldehyde	*Ischnoderma benzoinum* (a bracket fungus)
Apple and pine apple	Butyric acid	*Clostridium butyricum*
Citrus type fragrance	(–) Isopulegol	Lipase (*Pseudomonas* sp.)
Chocolate flavors	Thaumatin and monellin	*Kluyveromyces* sp. (yeast)
Coconut-peach	δ-Decalactone	Enzymatic reduction of the α, β-unsaturated compound (massoia lactone)
Expensive fragrance compound	Nor-patchoulenol	*Pithomyces* sp. (mould)
Flavor component of dairy products	2,3-butanedione	*Lactic streptococci*
Flavor component of many essential oils	(+)-Curcumene	*Saccharomyces cerevisiae* (baker's yeast)
Flavor component of many essential oils	(+)-Dehydro-curcumene	*Saccharomyces cerevisiae* (yeast)
Flavor component of many essential oils	(+)-Nuciferal	*Saccharomyces cerevisiae* (yeast)
Flavor component of many essential oils	(+)-Turmerone	*Saccharomyces cerevisiae* (yeast)
Mint	(–)-Menthol	Lipase (*Candida rugosa*)
Peach	γ-Decalactone	*Yarrowia lipolytica* (yeast)
Raspberry	Raspberry ketone	*Beauveria bassiana* (fungus)
Rosary	Phenolethanol	*Kluyveromyces* sp. (yeast)
Rose-like odor	Citronellal	*Rhodotorula minuta* (yeast)
Spearmint flavor	β-pinene	Lipase
Vanilla	Vanillin	*Pycnoporus cinnabarinus* (fungus)

3.12 Flavors and fragrance chemicals

In general, flavor is the result of the existence, inside complex matrices, of numerous volatile and nonvolatile compounds containing different chemical as well as physicochemical properties. Whilst the nonvolatile ingredients mostly affect the taste, the volatile ones affect both taste and fragrance. A vast collection of molecules might be accountable for the fragrance of the foodstuffs, for example alcohols, aldehydes, dicarbonyls, esters, lactones, methyl ketones, phenolic compounds, short- to medium-chain free fatty acids, and sulfur molecules (Urbach 1997; Gatfield 1988). In-depth knowledge on the manufacture of various normally employed food flavor ingredients by different methods is presented below.

3.12.1 Alcohols

Long-chain and complex alcohols formed by the yeasts in alcoholic fermentations, excluding ethanol that compounds as well as their derived esters contain attractive organoleptic properties. In an investigation, on the delignified cellulosic material as well as gluten pellets *Saccharomyces cerevisiae* was immobilized. High amounts of ester are produced from the delignified cellulosic material whereas high amounts of alcohol are produced from gluten pellets (Mallouchos et al. 2003). The 2-phenylethanol, a chemically produced aroma from substrates like benzene, methylphenylacetate, toluene, or styrene, is an important fragrance-related alcohol which has smell like a rose (Nomura et al. 2001). The natural 2-phenylethanol is extracted from petals of the rose, which involves expensive method. Definite yeast strains such as *Kluyveromyces marxianus, Saccharomyces cerevisiae,* and *Hansenula anomala* are also able to create aroma components via bioconversion of 2-phenylalanine into 2-phenylethanol (Stark et al. 2002).

3.12.2 Methyl ketones

Methyl ketones give cheesy flavors and therefore are in large demand. In the spores of mold ripened cheese *Penicillium roqueforti* the existence of methyl ketones ($RCOCH_3$) was first-time detected (Longo and Sanroman 2006). The distinctive flavor and odor of ripened cheese is attributable to the existence of methyl ketones, predominantly methyl n-pentyl ketone in addition to other short chain which adds to fruity-spicy notes to aromas. Fatty acid chains with less than 14 carbon atoms ($RCH_2CH_2CO_2H$) converted into methyl ketones by the fungal mycelium of *P. roqueforti*. The acids are oxidized to methyl ketones with one less carbon atom than the original acid in every reaction. Likewise it is as well feasible to convert vegetable oil and triglycerides to methyl ketones (Longo and Sanroman 2006).

3.12.3 Diacetyl

Diacety ($CH_3COCOCH_3$) is a naturally arising flavoring chemical that has a strong characteristic "butter" odor and taste on dilution, so it is extensively used to imitate the buttery flavor to the dairy or milk products and also employed in perfumes as well as in reconstituting essential oils. Through microbiological oxidation diacetyl was produced from acetoin, and *Lactococcus lactis, Lactobacillus* sp., *Streptococcus thermophilus,* and *Leuconostoc mesenteroides* are the main diacetyl forming bacteria. Greater production of diacetyl can be achieved by the employing humectants like glycerol or sucrose which reduces the water activity of the medium. The production is further improved by employing low pH (specifically less than 5.5), low temperature as well as aeration (Longo and Sanroman 2006).

3.12.4 Lactones

Lactones are cyclic esters of first and foremost g- and d-hydroxy acids, and they are universally found in food, causative to taste and flavor touch like buttery, coconut-like, creamy, fruity, or nutty sweet. Lactones can be produced by a variety of fungi from substrates like lecithin, oleic acid, sebum, triolein, and Tween 80. The γ-lactones are most vital elements for coconut, dairy and fruit flavors, particularly apricot, peach and strawberry. An additional lactone containing a coconut odor is 6-pentyl-2-pyrone that is created by *Trichoderma viridae* fungus. The buttery, coconut-like and milky flavor offered by these lactones is pleasing in dairy and milk products. Nevertheless, the presence of lactones may add to the stale flavor of heated milk, even though to a smaller extent than ketones (Longo and Sanroman 2006).

3.12.5 Butyric acid

Butyric acid ($CH_3CH_2CH_2CO_2H$) is mostly manufactured by oblige anaerobic bacteria be in the genera *Butyri-vibrio, Clostridium, Eubacterium* and *Fusarium.* In low concentration they are commonly used to provide butter-like flavor to diverse food products similar to natural cheese flavors. Butyric acid is naturally found in ester form in the butter ranging from 2% to 4%; however, its isolation process is awfully complex and costly. An eye-catching and lucrative alternating is microbial manufacture via fermentation method. For the making of butyric acid the clostridia, principally *C. acetylbutyricum* which have the capacity to make organic solvents like acetone and butanol, can be adapted and customized (Longo and Sanroman 2006).

3.12.6 Esters

Esters are incredibly valued for the fruity fragrances they supply and most widely used flavoring agents. They are used in baked goods, dairy products, fruit-flavored products, and wines. A number of strains of lactic acid bacteria produce ethyl esters and thioesters. *Lactococcus lactis* have distinctive esterase enzyme that is accountable for the development of these fragrance ester components (Nardi et al. 2002). Acetate ester is produced by a number of non-*Saccharomyces* wine yeasts (Rojas et al. 2001). Ethyl or methyl esters of short-chain fatty acids add to the fruity flavor whereas thioesters resulting from thiols are linked with cabbage or sulfur aromas in cheese production (Liu et al. 2004).

3.12.7 Pyrazines

A nutty and roasted flavor imparted by pyrazines which are chemically heterocyclic, nitrogen-containing composites that are in general created during usual cooking or roasting of food via the Maillard reaction (Seitz 1994). On the other hand, the microwave cooking does not support formation of pyrazine and as a result the natural pyrazines are outwardly added as food additives to give a roasty flavor. Few bacteria for instance *Corynebacterium glutamicum* generate significant amounts of tetramethyl pyrazine from amino acids (Longo and Sanroman 2006).

3.12.8 Vanillin

Vanillin is a distinctive flavor chemical that is found in *Vanilla planifolia* beans and it is chemically 4-hydroxy-3-methoxybenzaldehyde. This flavor is extensively employed in beverages, foods, perfumes, pharmaceuticals and in a variety of medical companies (Priefert et al. 2001). Even if vanillin can be chemically manufactured, however the demand for natural vanillin is increasing (Walton et al. 2000). Natural vanillin can be isolated from vanilla beans, although this source is very inadequate and costly which forms this ingredient a hopeful intention for biotechnological production of flavor. A number of bacterial and fungal strains of *Arthrobacter globiformis, Aspergillus niger, Corynebacterium glutamicum, Pseudomonas putida,* and *Serratia marcescens* are able to convert of natural eugenol and isoeugenol from essential oils into vanillin (Priefert et al. 2001; Shimoni et al. 2003). As a substitute, marketable processes have been urbanized on the basis of the bioconversion of ferulic acid from rice bran into vanillin, this procedure has been explained in some patents, counting in a Givaudan patent using *Streptomyces setonii* (Muheim et al. 1998) and a Symrise patent using *Amycolatopsis* species as the active strain (Rabenhorst and Hopp 1997). Vanillin ex ferulic acid from rice bran is at present completely accepted worldwide by the flavor company and legislative authorities as a natural flavoring matter.

3.12.9 Benzaldehyde

Benzaldehyde is naturally isolated from fruit kernels like apricots and gives flavor like cherry, but this isolation leads to creation of poisonous hydrocyanic acid that is detrimental. Another technique of extraction is microbial fermentation without forming any destructive by-products. Nevertheless, benzaldehyde is deadly toward microbial metabolism and their accumulations in the culture medium inhibit cell growth. As a consequence, merely a small number of microorganisms like *Pseudomonas putida* and the white rot fungi *Bjerkandera adusta, Phanerochaete chrysosporium, Polyporus tuberaster* and *Trametes suaveolens* can be employed as benzaldehyde creators. They are concerned in the biosynthesis of benzaldehyde from phenylalanine (Longo and Sanroman 2006).

3.13 Production of flavor: Microbial and enzymatic methodologies

Currently chemical synthesis of food flavor is under question, because of downsides for example contamination, higher manufacturing expenses, lower yield, poor reaction selectivity conductive to unwanted side reactions and unfeasibility of labeling the resultant products as natural. Further, a lot of ingredients employed for flavoring are present as optical isomers with diverse flavor distinctiveness, and resolution of racemic mixtures is usually difficult to achieve chemically. On the other hand, the employment of enzyme-catalyzed reactions avoids lots of these difficulties, as a result of the regio- and enantio selectivity of these biocatalysts and substrate specificity, which can be exploited at mild reaction circumstances. Enantiomerically pure flavor ingredients can be achieved by selecting the appropriate enzymes, in single step, accordingly rising process effectiveness and reducing downstream expenses. In addition, flavors attained via biocatalysis can be regarded as natural products (Longo and Sanroman 2006). The most well-liked advances entail the utilization of enzyme preparations or microbial cultures, while plant cell cultures have as well be described as proper production systems (**Figure 3.2**).

During fermentation process on nutrients like amino acids and sugars, microorganisms can generate flavors as secondary metabolites. This ability can be utilized in two dissimilar ways:

- Generation of the *in situ* flavor, as a fundamental element of food or beverage manufacture processes (i.e. beer, cheese, wine, yogurt), which establishes the organoleptic characters of the finished product,
- To acquire flavor ingredients microbial cultures purposely designed which can be separated and used afterward as additives in food production. This approach permits the achieved flavors to be labeled as natural.

To facilitate the biosynthesis of definite flavors in both ways, intermediates or precursors can be added to the culture medium. Also, the data achieved during the study of microbial metabolism in food fermentation methods may possibly be used for the development of appropriate manufacture methods for exacting flavor additives. In contrast, enzyme technology presents a very

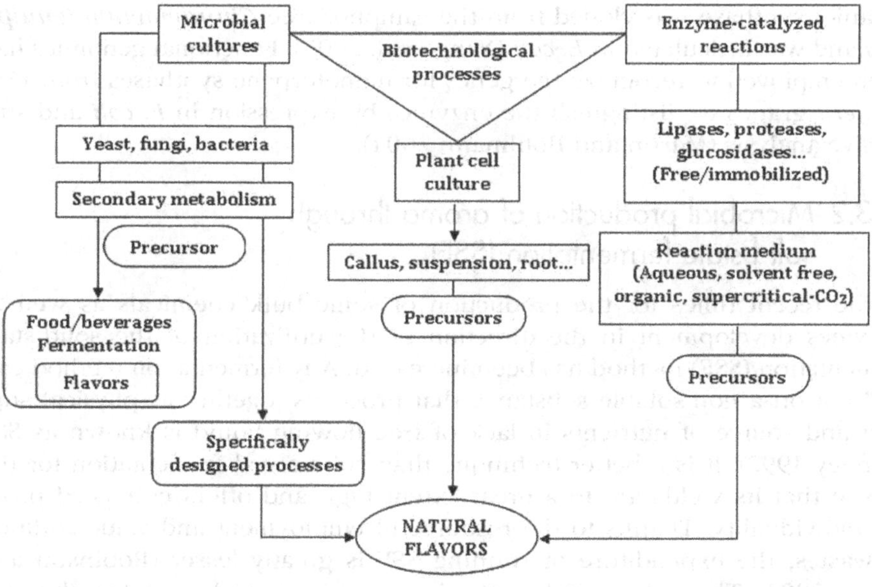

***Figure 3.2** Biotechnological processes for the production of flavor compounds.*

talented alternative for biosynthesis of the natural flavor. Many enzymes such as glucosidases, lipases and proteases catalyze the manufacture of aroma-related ingredients from precursors. The employment of enzyme-catalyzed reactions has the distinguished benefit of providing high stereoselectivity compared with chemical routes (Longo and Sanroman 2006). Microbial biotransformation and biosynthesis of flavor and fragrance chemicals present the prospective advantages of making optically active isomers which regularly have noticeable dissimilarities in flavor and fragrance quality as well as sensory intensity (Kim 2005).

3.13.1 Flavor and fragrance manufacture during microbial oxidation of monoterpenes

Investigation has revealed that flavors can as well be created during microbial oxidation of monoterpenes. A few microorganisms can be metabolically customized to perform particular conversions of substrates of lower marketable worth. Through oxidation of terpene components flavors have been produced by *Aspergillus niger* IOC-3913 and these species are very defiant to noxious monoterpenic substrates (Rozenbaum et al. 2006). Lots of microorganisms comprise an aptitude to break terpenes and to perform precise conversions making value added products. Monoterpene synthase has been sequenced, the enzyme concerned in terpene biosynthesis, and it is also a chief enzyme for the making of geraniol and has been sequenced to make recombinant geraniol synthase (Pichersky et al. 2005). Additionally, one more geraniol synthase was cloned from the camphor tree *Cinnamomum tenuipilum* and was articulated in *E. coli* (Yang et al. 2005). Functional genomics had been employed to recognize the genes for monoterpene synthases from *Vitis vinifera* grapes to distinguish the enzymes by expression in *E. coli* and successive analysis (Martin and Bohlmann 2004).

3.13.2 Microbial production of aroma through solid state fermentation (SSF)

In the recent times for the production of some bulk chemicals as well as enzymes development in the direction of the utilization of the solid-state fermentation (SSF) method has been increased. Any fermentation method carried out on a non-soluble substance that proceeds together as physical support and source of nutrients in lack of free flowing liquid is known as SSF (Pandey 1992). It is a better technique than submerged fermentation for the reason that its yields are to a great extent high and offers enhanced product individuality. Thanks to the resourceful employment and value-addition of wastes, the expenditure of running SSF is greatly lesser (Robinson and Nigam 2003). Through a number of microorganisms, such as *Aspergillus* sp., *Neurospora* sp., and *Zygosaccharomyces rouxii*, a range of investigations have been performed using SSF for the production of flavors on different solid substrates like cellulose fibers, pre-gelatinized rice, and miso fibers correspondingly (Longo and Sanroman 2006).

Some agro-horticultural wastes like amaranth, apple pomace, cassava bagasse and soybean extraordinary fruity fragrance can be created using *Ceratocystis fimbriata* in solid-state cultures. In SSF a strong pineapple fragrance can be formed with *Ceratocystis fimbriata* on coffee husk as a substrate (Soares et al. 2000). Acetaldehyde, ethanol, ethyl acetate and ethyl isobutyrate are the main flavor components identified. By addition of leucine in the substrate for SSF, a powerful banana odor can be formed that additional improved the manufacture of ethyl acetate and isoamyl acetate. In SSF technique, De Alberto et al. (2002) investigated the formation of 6-pentyl-a-pyrone (6-PP), an unsaturated lactone with a powerful coconut-like fragrance, with both solid and liquid substrates. For the development of characteristic dairy flavor ingredients like butyric acid, diacetyl, and lactic acid semisolid maize-based culture can be employed by utilizing *Lactobacillus acidophilus* and *Pediococcus pentosaceus* mixed cultures (Escamilla-Hurtado et al. 2005).

The effectiveness of SSF in the field of food flavor development does not only relate to the direct microbial synthesis of these components, nevertheless also the formation of biocatalysts that may perhaps be utilized in different enzymatic synthesis techniques. The main restrictions for widespread industrial appliance of microbial enzymes are their expenditure, which is calculated through the manufacture yields, downstream processing supplies as well as stability of enzyme (Kanwar et al. 2002). For that reason, SSF is of attention to augment the efficiency of the fermentation procedures by optimization of culture conditions. As the raw materials used in the culture medium make a payment to the total making expenses (Castilho et al. 2000), the lessening in the substrate costs would be an appropriate approach to enlarge the output of the method. Hence, SSF grasps remarkable prospective for the manufacture of enzymes.

3.13.3 Flavors by enzymatic esterification process

For biosynthesis of the natural flavor, enzyme technology presents an incredibly talented alternative approach. From precursors, numerous enzymes catalyzed the manufacture of aroma-related components. Chemically produced flavor esters are very familiar, however not biodegradable in nature. The biotransformation of these components is latent of substantial attention for relevance in the food flavor manufacturing. At the present time, a number of investigators and industries have shifted to biocatalytic flavor production owing to user's preference toward natural flavors over chemically produced flavors. Mild operating conditions are used by these reactions, encompass higher specificity with lesser side reactions, and make flavor components having higher purity by eliminating the costlier separation procedures (Singh et al. 2014).

Most adaptable enzymes for flavor development are lipases and comprise a wide range of industrial appliances because of the variety of reaction they catalyze (Singh et al. 2014). For the formation of flavor esters like isoamyl acetate (Larios et al. 2004; Vulsan 1994), isoamyl butyrate (Abbas and Comeau 2003; Claon and Akoh 1994), geranyl acetate (Gatfield 1995), and octyl acetate (Singh et al. 2014), lipases are used for esterification in the organic solvents. Lipase-mediated production of flavor esters under solvent-free conditions has momentous significance in diverse food and pharmaceutical industries because of the evasion of poisonous solvent and removal of its recovery in the operation (Ghamgui et al. 2006). By esterification reactions, lipase catalyzed formation of flavor esters, which is affected by numerous esterification variables like alcohol molarity, reaction time, addition of water, quantity of immobilized enzyme, speed of agitation, and temperature.

To conquer the troubles connected with free enzyme separation and solvent toxicity lots of workings were carried out for the production of flavor esters by using immobilized lipases and solvent-free conditions. Krishna and Karanth (2001) described solvent-free synthesis of ethyl oleate which resulted in a 78.6% conversion in 7 h using *Candida antarctica B* lipase adsorbed on powder of polypropylene. For the development of butyl acetate (pineapple like flavor), ethyl valerate (green apple like flavor) and isoamyl acetate (banana like flavor) immobilized lipase from *C. rugosa* and porcine pancreatic lipase were used in n-hexane (Güvenç et al. 2002). Some researchers evaluated the immobilized lipases for esterification capacity to make a variety of flavor esters (Ghamgui et al. 2006; Torres et al. 2009).

3.14 Biotechnological production of flavor compounds

The biotechnological approach for production of flavor entails added benefits. Flavors are bioactive components, and the identified effects of chirality on aroma sensitivity propose the exploit of biocatalysts. Additional benefits linked with the biotechnological theory are liberty from agriculture and show potential shortages induced by local conditions of manufacture such as climate, diseases, fertilizers, pesticides, trade restrictions, sociopolitical instabilities and aptitude for scaled-up and industrial-scale manufacturing via engineered trails, up-regulated metabolisms, and moderate product recovery to produce an infinite resource of homogenous, distinct product and liable care of natural resources in emergent countries (Krings and Berger 1998). Interminably essential oils obtained from higher plants, fruit juices, vegetable extracts, and only some products of animal source (e.g., amber, musk, zibet) were the single sources of natural flavors. Biotechnological alternatives consist of one-step de novo synthesis with microorganisms, plant cells, biotransformations, bioconversions and biocatalysis (Krings and Berger 1998).

3.14.1 De novo synthesis

Fermentation process, also called de novo synthesis, which involves the manufacture of flavor components with simple culture media, is exclusive of adding any particular carbon source. Generally, this technique employs the complete metabolic arsenal from the microorganism and makes a blend of some flavor components, which are vital for the development of the bunch of the product (Berger 1995). Even if for a huge number of microorganisms, the metabolic budding for de novo flavor biosynthesis is enormous and a broad range of precious products can be identified in microbial culture media, the amounts created in nature are generally incredibly small for marketable appliances. In the case of some nonvolatile ingredients the flavor components are derived from primary metabolism (Schrader 2007). Carbohydrates, fats, and proteins are further catabolized by whole cell and additionally convert the collapse products to supplementary difficult flavor elements, a property that is conventionally employed at some stage in the making of fermented foods with their astonishing amount of flavor chemicals components (Krings and Berger 1998). Initial cultures make primary metabolites in significant quantities, although more multifaceted flavor chemicals produce trace amount, such as very well-organized lactic acid creators give to dairy flavors. Concomitantly little quantities of chemically somewhat diverse volatile flavors, like aldehydes, ketones, lactones, methyl ketones, short-chain alcohols, and acids, in addition to pyrazines, and thiols, are produced (Cogan 1995; Imhof and Bosset 1994; Krings and Berger 1998).

3.14.2 Biotransformation

The conversion of a molecule into the product by means of enzymes, microorganisms, or living plant cells is known as biotransformation. As said by (Berger 1995) biotransformation is defined as a reaction able to catalyze the transformation of the substrate in only single step. In contrast, bioconversion takes place with two or more biochemical steps, even though there are arguments about these theories being dissimilar in the literature. Anyway, biotransformation or bioconversion is a thought that passes on to the production of one or more flavor components during the adding up precursors in the culture media (Berger 1995). For this approach as a precondition, the precursor must be there in nature, and its separation in adequate quantities from the natural source should be simply practicable in a cost-effectively feasible manner. Reasonably priced, easily available, and renewable natural precursors, like fatty or amino acids, were able to be transformed to more highly appreciated flavors. Vanillin, one of the main aroma compound applied worldwide, may also be produced in high yields through microbial biotransformations (Daugsch and Pastore 2005; Walton et al., 2000).

3.14.3 Biocatalysis

The biocatalytic transformation of a structurally associated precursor particle is again a better approach, which permits the buildup of preferred flavors to be appreciably improved. In the following types of reactions, biocatalysis battles are most excellent one with below chemical catalysis:

- chirality resolution,
- functionalization of chemically inert carbons,
- discriminating alterations of single functional group in multifunctional molecules,
- racemates resolution.

Biocatalysis has an extensive variety of appliances on fragrance making. The enzymes can be used straightforwardly on the food as additives, in addition to offer free scents of the product and to circumvent objectionable flavors induced by a few components (Macedo 1997). In the development of fragrance making in food the enzymes at hand can be endogenous (be in the food) or enzymes from microbial origins. For the production of esthers derived from fatty acids and short-chain alcohols, lipases from *Rhizopus* sp. can be utilized as biocatalyst, both recognized as essential flavor components (Macedo 1997).

3.14.4 Production of flavors by plant cell and tissue culture

The first testing's with plant tissue culture were urbanized over 100 years back (by Haberland in 1902); however, the exploitation of it to the flavor making started just in the 1970s (Hrazdina 2006). This technique has some benefits as compared with the manufacture of perfumed components by customary agriculture. Cultivation of plant relies on the season, whereas cell plant cultivation presents a scheme that employs the entire year and is not reliant on the type of weather, time of year, place, and so on. Despite that, the procedure of flavor formation from plant cell cultivation has its boundaries. A number of species are somewhat complicated to cultivate *in vitro* (Berger 1995). *Coffea arabica* cell culture turns out vanillin -β-D-glucoside from vanillin as precursor (Giri et al. 2001). Some plants showing their distinctive aroma are also good candidates for cell culture. Vanilla is the most used flavor constituent being the major constituent cultured in *Vanilla planifolia*, followed by *p*-hydroxybenzaldehyde, vanillic acid, and *p*-hydroxybenzyl methyl ether. Strawberry fragrance is extremely complex; it is comprised by 278 volatile substances, among them are 33 acids, 39 alcohols, 17 aldehydes, 14 ketones, and 103 esters. Tissue culture for strawberry fragrance was simply achieved for component groups like esters. Other flavors generated by plant cell cultures consist of apple, cinnamic acid, caryophyllen, cocoa, garlic, and onion (Longo and Sanroman 2006).

3.15 Commercial feasibility of natural flavor production by biotechnological/enzymatic/microbial approach

Mainly during the last four decades, numerous microbial and enzymatic processes have been described in the literature for the synthesis of attractive flavors. Though, the described processes have more academic than practical value. The number of industrial application is therefore limited to some fragrance components, for example, vanillin, benzaldehyde, decalactone. A reason for this in the majority cases is the low yield. The microbial flavors are often present only in low concentrations in the fermentation broths, resultant in higher costs for down stream processing. Other problems hindering industrial appliances have a technological character in that the volatility and the low solubility in water of numerous flavors make often their recovery complicated to perform. On the other hand, it is often crucial to maintaining the concentration of the end-product in the fermentation broth below a certain level due to end-product inhibition and toxicity toward the microorganisms themselves. In order to defeat these limitations, the processes for in situ product removal, by coupling bioreactor to the down stream unit, have been developed in many works. These systems permit the fast removal of the products from the fermentation broths. The selective product recovery during downstream processing becomes a major issue for the bio-process development. Although the in situ product removal has been intensively studied, the applications remain limited to the laboratory scale. The developed systems are generally of high costs for a production on a large scale and often lack economic profitability (Shaaban et al. 2016).

Another obstruction to commercialization is the framework of legal regulations. New products have to be examined by regulatory authorities before the status "natural" is awarded. The process required can take years and bring about high costs for biotechnological products. Despite these difficulties, a number of flavor compounds are already produced by micro organisms on an industrial scale. Nearly all important flavor companies declare indeed that they regularly use fermentation techniques for the production of fragrance compounds, yet only a few of them mention the specific products (Van Brunt 1985; Dziczak 1986).

The field of production of natural aroma compounds is still in development and needs highly innovative processes. More academic studies have to be undertaken in order to find novel biocatalytic routes for flavor production. This will rely on genetically superior biocatalysts and on process engineering features. Biotechnological progress, including genetic engineering techniques and modern techniques of molecular biology and process engineering, are now increasingly practical to improve the efficiency of the biocatalysts. Because of the rising demand from the consumer's side for "natural" additives for food, feed, and cosmetics, the commercial significance of biotechnologically produced flavors will certainly grow further in the near future (Shaaban et al. 2016).

3.16 Conclusion

Functional foods are first and foremost foods, and they contend with traditional foods for a divide up the public's liking. The flavor and odor of functional foods are vital if there is a wish to create these products plea to the ordinary user. In the foodstuff, properties of flavor rely on both the nature of the foods as well as the flavor components present. Functional foods are foods, therefore flavor used for them should be like a flavor used for other processed foodstuffs. There has been lots of advancement in the nutraceuticals company in terms of enhancing taste. Nutraceutical formulations are typically pungent and to be effectual, and therefore, they have to be at an offensive level. Accordingly confrontation is converting unpalatable active ingredients to palatable level. The flavor industry also has considerable experience in flavoring pharmaceutical products that often have very pronounced off flavors and require intense flavor systems to make them palatable. At present the majority of the fragrance ingredients are extracted from synthetic origins or from natural origins of plants. However, a latest progress in metabolic engineering has made an enormous attention for natural products mainly the fragrance company to search for novel techniques to achieve aroma and flavor ingredients naturally. Industries are still struggling with finding a joyful means between flavoring and masking, as others are concentrating on fundamental factors for example mouth-feel and fragrance. The thought of just adding up flavors is not a reply. We have to appreciate the functionality of the components and how components act in response with one another as flavor, which is just a solitary part of how to formulate a high-quality tasting end product.

References

Abbas H, Comeau L. Aroma synthesis by immobilized lipase from *Mucor* sp. *Enzyme Microb Technol.* 2003; 32: 589–595.

Auvray M, Spence C. The multisensory perception of flavor. *Conscious Cogn.* 2008; 17:1016–1031.

Bajec MR, Pickering GJ, DeCourville N. Influence of stimulus temperature on orosensory perception and variation with taste phenotype. *Chem Percep.* 2012; 5:243–265.

BCC Research. Global Markets for Flavors and Fragrances, 2016. https://www.bccresearch.com. Accessed 23 October 2018.

Berger RG. *Aroma Biotechnology*, 1st ed. Springer—Verlag, New York, 1995.

Blank I, Pascual EC, Devaud S, Fay LB, Stadler RH, Yeretzian C, Goodman B. Degradation of the coffee flavor compound furfuryl mercaptan in model Fenton-type reaction systems. *J Agric Food Chem.* 2002; 50:2356–2364.

Bomgardner MM. The sweet smell of microbes. *Chemical and Engineering News.* 2012; 90(29):25–29.

Castilho LR, Polato CMS, Baruque EA, Sant' Anna GL, Freire DMG. Economic analysis of lipase production by *Penicillium restrictum* in solid-state and submerged fermentations. *Biochem Eng J.* 2000; 4:239–247.

Christensen CM. Food texture perception. In: Chichester CO, Orlando FL, editors. *Advances in Food Research.* Volume 29. Academic Press; Cambridge, MA, 1984. pp. 159–199.

Christensen, CM. Effects of taste quality and intensity on oral perception of viscosity. *Percept Psycho.* 1980; 28(4):315–320.

Claon PA, Akoh CC. Enzymatic synthesis of geranyl acetate in n-hexane with *Candida antarctica* lipases. *J Am Oil Chem Soc.* 1994; 71(6):575–578.

Cogan TM. Flavor production by dairy starter cultures. *J Appl Bacteriol.* 1995; 79:49–64.

Daugsch A, Pastore GM. Obtenção de vanilina: oportunidade biotecnológica. *Química Nova.* 2005; 28(4):642–645.

De Alberto AA, Pastore GM, Berger RG. Production of coconut aroma by fungi cultivation in solid-state fermentation. *Appl Biochem Biotechnol.* 2002; 98–100:747–751.

Dravnieks A. *Atlas of Odor Character Profiles*, American Society for Testing and Materials: West Conshohocken, PA, 1985.

Dubal SA, Tilkari YP, Momin SA, Borkar IV. Biotechnological routes in flavor industries. *Advanced Biotech.* 2008; 20–31.

Dziczak JD. Biotechnology and flavor development: An industrial research perspective. *Food Technol.* 1986; 40:108–113.

Escamilla-Hurtado ML, Valdes-Martinez SE, Soriano-Santos J, Gomez- Pliego R, Verde-Calvo JR, Reyes-Dorantes A, Tomasini-Campocosio A. Effect of culture conditions on production of butter flavor compounds by *Pediococcus pentosaceus* and *Lactobacillus acidophilus* in semisolid maize-based cultures. *Int J Food Microbiol.* 2005; 105(3):305–316.

Fischer L. Flavoring nutraceuticals and functional foods. *Presented at IFT Workshop on Flavor/Food Interactions.* New Jersey; 1998.

Fisher C, Scott TR. *Introduction-Problems in Flavor Research.* Royal Society of Chemistry: London, UK; 1997.

Food Insight. IFIC (International Food Information Council) Foundation Releases. Functional foods/foods for health consumer trending survey. 2011. http://www.foodinsight.org/2011_functional_Foods_Foods_For_Health_Consumer_Trending_Survey. Accessed 11 April 2018.

Gatfield IL. Production of flavor and aroma compounds by biotechnology. *Food Technol.* 1988; 10:110–122.

Gatfield IL. Enzymatic and microbial generation of flavors. *Perfumer and Flavorist.* 1995; 20(5):5–14.

Ghamgui H, Karra-Chaâbouni M, Bezzine S, Miled N, Gargouri Y. Production of isoamyl acetate with immobilized *Staphylococcus simulans* lipase in a solvent-free system. *Enzyme Microb Technol.* 2006; 38(6):788–794.

Giri A, Dhingra V, Giri CC, Singh A, Ward OP, Narasu ML. Biotransformations using plant cells, organ cultures and enzyme systems: Current trends and future prospects. *Biotechnol Adv.* 2001; 19:175–199.

Godshall MA. How carbohydrates influence food flavor. *Food Technology.* 1997; 51:63–67.

Goubet I, Le Quere LJ, Voilley JA. Retention of aroma compounds by carbohydrates: Influence of their physiochemical characteristics and of their physical state. A review. *J Agric Food Chem.* 1998; 46:1981–1990.

Guichard E. COST Action 96: Interaction of food matrix with small ligands influencing flavour and texture. *Food Res Int.* 2000; 33:187–190.

Gupta C, Prakash D, Gupta S. A biotechnological approach to microbial based perfumes and flavors. *J Microbiol Exp.* 2015; 3(1).

Güvenç A, Kapucu N, Mehmetoğlu U. The production of isoamyl acetate using immobilized lipases in a solvent-free system. *Process Biochem.* 2002; 38(3):379–386.

Hazen C. Secrets of Masking Flavors, In: *Food Product Design.* Virgo Publishing LLC. 2003. http://www.foodproductdesign.com/articles/2003/11/secrets-of-masking-flavors.aspx?pg=3. Accessed 8 April 2018.

Heijden A. van der. Sweet and bitter tastes. In: Acree TE, Teranishi R, editors. *Flavor Science: Sensible Principles and Techniques.* ACS Publishers, Washington, DC; 1993. pp. 67–116.

Hodgson M, Linforth RST, Taylor A. Simultaneous real-time measurements of mastication, swallowing, nasal airflow and aroma release. *J Agric Food Chem.* 2003; 51:5052–5057.

Hofmann T, Czerny M, Calligaris S, Schieberle P. Model Studies on the influence of coffee melanoidins on flavor volatiles of coffee beverages. *J Agric Food Chem.* 2001; 49:2382–2386.

Hofmann T, Schieberle P. Chemical interactions between odor-active thiols and melanoidins involved in the aroma staling of coffee beverages. *J Agric Food Chem.* 2002; 50(2):319–326.

Hrazdina G. Aroma production by tissue cultures. *J Agric Food Chem.* 2006; 54:1116–1123.

Imhof R, Bosset JO. Quantitative GC—MS analysis of volatile flavour compounds in pasteurized milk and fermented milk products applying a standard addition method. *Lebensmittel Wissenschaft und Technologie.* 1994; 27:265–269.

Kanwar L, Gogoi BK, Goswami P. Production of a *Pseudomonas* lipase in *n*-alkane substrate and its isolation using an improved ammonium sulfate precipitation technique. *Bioresour Technol.* 2002; 84:207–211.

Keskar A, Igou Jr. W. Flavoring of nutraceuticals. In: Pathak Y, editor. *Handbook of Nutraceuticals: Vol-II Scale Up, Processing and Automation.* CRC Press, Boca Raton, FL, 2011. pp. 485–512.

Kim AY. Application of biotechnology to the production of natural flavor and fragrance chemicals. *Natural Flavors and Fragrances ACS Symp Ser.* 2005; 908:60–75.

Krings U, Berger RG. Biotechnological production of flavors and fragrances. *Appl Microbiol Biotechnol.* 1998; 49(1):1–8.

Krishna SH, Karanth NG. Lipase-catalyzed synthesis of isoamyl butyrate: A kinetic study. *Biochim Biophys Acta.* 2001; 1547(2):262–267.

Kumazawa K, Masuda H. Investigation of the change in the flavor of a coffee drink during heat processing. *J Agric Food Chem.* 2003; 51:2674–2678.

Landy P, Druaux C, Voilley A. Retention of aroma compounds by proteins in aqueous solution. *Food Chem.* 1995; 54:387–392.

Larios A, García HS, Oliart RM, Valerio-Alfaro G. Synthesis of flavor and fragrance esters using *Candida antarctica* lipase. *Appl Microbiol Biotechnol.* 2004; 65:373–376.

Le Thanh M, Thibeaudeau P, Thibaut MA, Voilley A. Interactions between volatile and non-volatile compounds in the presence of water. *Food Chem.* 1992; 43:129–135.

Liu SQ, Holland R, Crow VL. Esters and their biosynthesis in fermented dairy products: A review. *Int Dairy J.* 2004; 14(11):923–945.

Longo MA, Sanroman MA. Production of food aroma compounds. *Food Technol Biotechnol.* 2006; 44(3):335–353.

Lubbers S, Landy P, Voilley A. Retention and release of aroma compounds in foods containing proteins. *Food Technol.* 1998; 52(5):68–74, 208–214.

Macedo GA. S í ntese de é steres de aroma por lipases microbianas em meio livre de solvente org â nico. Doctor of Science Thesis (Portuguese), Univ. Estadual de Campinas, Brazil; 1997.

Mallouchos A, Komaitis M, Koutinas A, Kanellaki M. Wine fermentations by immobilized and free cells at different temperatures. Effect of immobilization and temperature on volatile by-products. *Food Chem.* 2003; 80(1):109–113.

Manley C. The flavoring of nutraceuticals, dietary supplements, functional foods and medical foods, Presented at Nutracon 97, Las Vegas; NV, 1997.

Martin DM, Bohlmann J. Identification of Vitis vinifera (–)-alpha-terpineol synthase by in silico screening of full-length cDNA ESTs and functional characterization of recombinant terpene synthase. *Phytochemistry.* 2004; 65(9):1223–1229.

Matheis, G. Interactions between volatile flavoring substances and food constituents. Part 1: Carbohydrates, proteins and amino acids. *Dragco Report 3.* 1993; 38(3):98–114.

McDonald ST. concepts of flavor creation in novel nutraceuticals and functional food formulation. In: Bagchi D, Nair S, Editors. *Developing New Functional Foods and Nutraceuticals Products.* Academic Press Publications, Elsevier, US; 2017. pp. 235–236.

Milo OL. Functional foods put the flavor up front. *Prepared Foods.* 1999; 168(2):59–64.

Muheim A, Mueller B, Muench T, Wetli M. (Givaudan Roure Int.). Process for the production of vanillin. EP0885968 (A1), 1998-12-23.

Naknean P, Meenune M. Factors affecting retention and release of flavor compounds in food carbohydrates. *Int Food ResJ.* 2010; 17:23–34.

Nardi M, Fiez-Vandal C, Tailliez P, Monnet V. The EstA esterase is responsible for the main capacity of *Lactococcus lactis* to synthesize short chain fatty acid esters *in vitro. J Appl Microbiol.* 2002; 93(6):994–1002.

Nomura K, Ogura H, Imanishi Y. Direct synthesis of 2-phenylethanol by hydrogenation of methyl phenyl acetate using homogeneous ruthenium-phosphine catalysis under low hydrogen pressure. *J Mol Catal A-Chem.* 2001; 166(2):345–349.

O'Neil TE. Flavor Binding by food proteins: An overview. In: McGorrin RG, Leland JV, Editors. *Flavor–Food Interaction.* American Chemical Society: Washington, DC; 1996. pp. 59–74.

Pandey A. Recent process developments in solid-state fermentation. *Process Biochem.* 1992; 27:109–117.

Pichersky E, Iijima Y, Lewinsohn E, Gang DR, Simon J. Cloning and sequence of geraniol synthase from sweet basil and use in the production of geraniol and metabolites. International patent PCT WO 2005060553, 2005.

Plug H, Haring P. The influence of flavor-ingredient interactions on flavor perception. *Food Qual Prefer.* 1994; 5:95–102.

Priefert H, Babenhorst J, Steinbuchel A. Biotechnological production of vanillin. *Appl Microbiol Biotechnol.* 2001; 56(3–4):296–314.

Rabenhorst J, Hopp R. (Haarmann & Reimer). Process for the preparation of vanillin and suitable microorganisms. EP0761817 (A2), 1997-03-12.

Reineccius GA. Flavoring system for functional foods. In: Schmidl MK, Labuza TP, editors. *Essentials of Functional Foods.* Aspen Publishers, Inc. Gaithersburg, Maryland, MD; 2000. pp. 89–98.

Robinson T, Nigam P. Bioreactor design for protein enrichment of agricultural residues by solid state fermentation. *Biochem Eng J.* 2003; 13(2–3):197–203.

Rojas V, Gil JV, Pinaga F, Manzanares P. Studies on acetate ester production by non-*Saccharomyces* wine yeasts. *Int J Food Microbiol.* 2001; 70(3): 283–289.
Rozenbaum HF, Patitucci ML, Antunes OAC, Pereira N. Production of aromas and fragrances through microbial oxidation of monoterpenes. *Braz J Chem Eng.* 2006; 23(3):273–279.
Schiffman SS. Taste quality and neural coding: implications from psychophysics and neurophysiology. *Physiol Behav.* 2002; 69:147–159.
Schober LA, Peterson GD. Flavor release and perception in hard candy: Influence flavor compound-compound interactions. *J Agric Food Chem.* 2004; 52(9):2623–2627.
Schrader J. Microbial flavor production. In Berger RG, editor. *Flavors and Fragrances.* Berlin, Germany: Springer, 2007. pp. 507.
Seeventer PB, Weenen H, Winkel C, Kerler JJ. Stability of thiols in an aqueous process flavoring. *Agric Food Chem.* 2001; 49:4292–4295.
Seitz EW. Fermentation Production of Pyrazines and Terpenoids for Flavor and Fragrances. In: Gabeliman A, editor. *Bioprocess Production of Flavor, Fragrance, and Colour.* Wiley, New York; 1994. pp. 95–134.
Shaaban HA, Mahmoud KF, Amin AA, EL Banna HA. Application of biotechnology to the production of natural flavor and fragrance chemicals. *Res J Pharm Biol Chem Sci.* 2016; 7:2670–2717.
Shankar MU, Levitan CA, Spence C. Grape expectations: The role of cognitive influences in color-flavor interactions. *Conscious Cogn.* 2010; 19:380–390.
Shimoni E, Baasov T, Ravid U, Shoham Y. Biotransformations of propenylbenzenes by an *Arthrobacter* sp. and its t-anethole blocked mutants. *J Biotechnol.* 2003; 105(1–2):61–70.
Simon SA, de Araujo IE, Gutierrez R, Nicolelis MAL. The neural mechanisms of gustation: A distributed processing code. *Nat Rev Neurosci.* 2006; 7:890–901.
Singh P, Saxena DK, Naik SN. Synthesis of food flavors by enzymatic esterification process. *Int J Sci Res.* 2014; 3(9):2113–2116.
Siro I, Kapolna E, Kapolna B, Lugasi A. Functional food. Product development, marketing and consumer acceptance—A review. *Appetite.* 2008; 51(3):456–467.
Small DM. Flavor and the formation of category-specific processing in olfaction. *Chemosens Percept.* 2008; 1:136–146.
Soares M, Christen P, Pandey A, Soccol CR. Fruity flavor production by *Ceratocystis fimbriata* grown on coffee husk in solid state fermentation. *Process Biochem.* 2000; 35(8):857–861.
Spence C, Levitan C, Shankar MU, Zampini M. Does food color influence taste and flavor perception in humans? *Chemosens Percept.* 2010; 3:68–84.
Stark D, Munch T, Sonnleitner B, Marison IW, von Stockar U. Extractive bioconversion of 2-phenylethanol from L-phenylalanine by *Saccharomyces cerevisiae. Biotechnol Prog.* 2002; 18(3):514–523.
Stevenson RJ. *The Psychology of Flavor* Oxford, UK: Oxford University Press, 2009.
Talavera K, NinomiyaY, Winkel C, Voets T, Nilius B. Influence of temperature on taste perception. *Cell Mol Life Sci.* 2007; 64:377–381.
Teixeira MI, Andrade LR, Farina M, Rocha-Leão MHM. Characterization of short chain fatty acid microcapsules produced by spray drying. *Mat Sci Eng C.* 2004; 24:653–658.
Torres S, Baigorí MD, Swathy SL, Pandey A, Castro GR. Enzymatic synthesis of banana flavor (isoamyl acetate) by *Bacillus licheniformis* S-86 esterase. *Food Res Int.* 2009; 42(4):454–460.
Trelea IC, Atlan S, Déléris I, Saint-Eve A, Marin M, Souchon I. Mechanistic mathematical model for *in vivo* aroma release during eating of semi liquid foods. *Chem Senses.* 2008; 33:181–192.
Tuorila H, Cardello AV. Consumer responses to an off-flavor in juice in the presence of specific health claims. *Food Qual Prefer.* 2002; 13(7–8):561–569.

Urbach G. The flavor of milk and dairy products: II. Cheese: Contribution of volatile compounds. *Int J Dairy Technol.* 1997; 50:79–89.

Van Brunt J. Nibbling at the flavor market. *Biotechnology,* 1985; 3:525–530.

Van Ruth SM, Roozen J P. Delivery of flavours from food matrices. In: Taylor AJ, Editor. *Food Flavour Technology.* Sheffield Academic Press: Sheffield, UK, 2002. pp. 167–184.

Vickers ZM, Bourne MC. A psychoacoustical theory of crispness. *J Food Sci.* 1976; 41:1158–1164.

Vulsan EN. Industrial application of lipase. In: Wooley P, Peterson SB, editors. *Lipase—Their Structure, Biochemistry and Application.* Cambridge University Press: Cambridge, UK, 1994. pp. 271–288.

Walters DE, Roy G. Taste interactions of sweet and bitter compounds. *Institute of Food Technology Annual Meeting Abstracts*; 1995:171.

Walton NJ, Narbad A, Faulds C, Williamson G. Novel approaches to biosynthesis of vanillin. *Curr Opin Biotechnol.* 2000; 11(5):490–496.

Weerawatanakorn M, Wu JC, Pan MH, Ho CT. Reactivity and stability of selected flavor compounds. *J Food Drug Anal.* 2015; 23(2):176–190.

Wright J. Applications (Chapter 11). In: *Flavor Creation.* 2nd ed. Allured Publishing Corp: Carol Stream, IL, 2011.

Yang T, Li J, Wang HX, Zeng Y. A geraniol-synthase gene from *Cinnamomum tenuipilum. Phytochemistry.* 2005; 66(3):285–293.

Yeomans MR, Chambers L, Blumenthal H, Blake A. The role of expectancy in sensory and hedonic evaluation: The case of smoked salmon ice cream. *Food Qual Prefer.* 2008; 19:565–573.

Zampini M, Spence C. The role of auditory cues in modulating the perceived crispness and staleness of potato chips. *J Sens Stud.* 2004; 19:347–363.

4

Flavor Challenges in Legume-Based Functional Food Products

M. Selvamuthukumaran and Yashwant V. Pathak

Contents

4.1 Introduction

Legumes contain several off-flavor components, which can limit their use in processing various legume-based functional food products. The majority of labels on functional food products say, for instance, whether they are flour, beverage, or even whole legumes, and if they will give a bitter taste, astringency and/or off odor. Oxidative deterioration of legumes leads to the production of several objectionable off-flavor components. These off flavors were developed by default in the plant or sometimes during cultivation and value addition. The majority of the legume off-flavor components belong to a group of alcohols, aldehydes, ketones, and so on. Off flavors were also formed during heat processing, such as the maillard reaction, when sugar and proteins are subjected to high heat treatment during legume processing. Heat degradation of vitamins like thiamine and phenolic acids and lipid oxidation will also lead to the formation of off-flavor components in legumes (**Table 4.1**). The various off-flavor components found in legumes and technical approaches essential to eradicate such off-flavor aspects in legumes are discussed in this chapter.

4.2 Off-flavor components in legumes

4.2.1 Peas

The development of objectionable aromas and tastes for peas starts from the field itself, such as harvesting followed by processing and storage. Alcohols, ketones, and aldehydes were mainly responsible for these objectionable flavors (**Figure 4.1**). The oxidation of unsaturated fatty acids leads to the formation of such objectionable off flavors in these legumes. Heng (2005) reported that Hexanal is one of the predominant flavoring constituents, which are responsible for contributing off flavor in processed products like pea flour and pea protein isolates. The 2-ethyl-1-hexanol is the off-flavor constituent resulting from the separation of pea proteins, from legumin and vicilin. Heng's study shows that protein purity also significantly affects the retention of off-flavor constituents. Like proteins, the lipids significantly contribute to retain the off-flavor volatile compounds during processing, which will further limit its end use for consumption. The saponin, which also contributes to the bitter taste of peas, even at lesser threshold levels ranging from 2 to 8 mg (Heng et al., 2006). There are two different types of saponins identified from peas, which

Table 4.1 Types of Off-Flavor Components Formation in Legumes

Off flavor (Inherent)	Off Flavor (Chemically Derived)
1. Volatiles of flavors derived by mode of enzymatic action	1. Oxidation of lipids
2. Off flavor volatile release as a result of protein degradation	2. Degradation of vitamins
	3. Maillard reaction

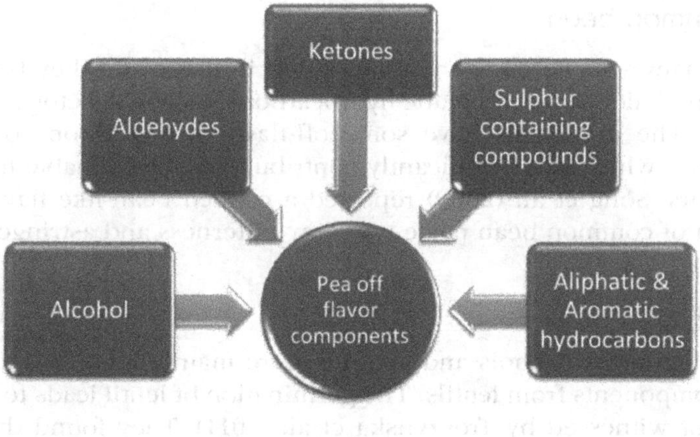

Figure 4.1 *Off-flavor components in peas.*

include saponin bg and saponin Bb (Daveby et al. 1998). Therefore, it is mandatory to prepare the flour so that it is free or devoid of saponin content to avoid metallic or bitter taste in the final product.

4.2.2 Chickpea

The chickpea contains meager off-flavor constituents compared to other legume crops. However, Simons (2011) and Troszynska et al. (2011) reported that phenolic components have brought both bitter and astringent tastes. The other constituents like isoflavones from formononetin and biochanin may also contribute to bitterness to a certain extent. Sanchez-Vioque et al. (1998) reported one bitter component in defatted chickpea flour, i.e. phosphatidylcholine, which can alter the taste of the products prepared from defatted flour. The processed products such as flour obtained from the chickpea is found to contain various free fatty acids that may react with lipoxygenase enzymes (LOP) to produce several volatile off-flavor components.

4.2.3 Broad beans

Oomah et al. (2011) reported that there are several volatile components present in Broad beans like alcohols, furans, ketones, alkanes, aldehydes, and so on. The off-flavor constituents for beans at an acidic pH give an unacceptable or unpleasant fruity flavor, while at base pH it gives a pea-like flavor. The processed broad bean products have some volatile organic acids, which lead to the formation of an off flavor. In addition to that, the presence of 8% tannin in the beans leads to a bitter taste in the final processed products. Baysal and Aslihan (2007) reported that the bean contains one enzyme, i.e. lipoxygenase, which will oxidize the fatty acids to give hydroperoxides. Therefore, the processed products made from this bean is quiet a problematic issue because of the presence of lipoxygenase enzymes.

4.2.4 Common bean

There are many volatile components reported in this bean; they belong to a major group of aldehydes, aromatic hydrocarbons, alcohols, ketones, alkanes, and so on. The dry beans have some off-flavor hydrocarbon constituents like Geosmin, which will significantly contribute to objectionable moldy and earthy flavors. Song et al. (2009) reported a cooked bean-like flavor during preparation of common bean paste with low bitterness and astringency.

4.2.5 Lentils

Like other legumes, alcohols and ketones were mainly identified as the off-flavoring components from lentils. The germination of lentil leads to off-flavor development witnessed by Troszynska et al. (2011). They found that germinated lentils give a pronounced bitter taste, astringency, and off odor after seven days of germination. This may be due to the presence of phenolic substances like catechin gallate and kaempferolglycosides. The lentils also contain one kind of saponin i.e. saponin bg, which drastically contributes to an increased bitter taste in the final product. It is found that the cooking time significantly enhances the concentration of saponin content. During cooking, the saponin bg form is converted into the saponin Bb form. Both the forms lead to more bitterness, with bg form found to exhibit higher bitterness attributes.

4.2.6 Lupin

Schindler et al. (2011) reported that protein extracts obtained from lupin seeds contains several high-intensity odor compounds like *n*-pentanal, 1-pyrroline, *n*-hexanal, and dimethyl trisulfide. The flour obtained from lupin legumes showed various undesirable flavors like metallic, fruity, fatty, grassy, and meat-like. It contains a higher content of polyunsaturated fatty acids (PUFA) like linoleic and linolenic acids (Chiofalo et al. 2012). Belitz et al. (2001) found that oxidation of these PUFAs leads to the creation of various off-odorous components. During storage, the degradation of amino acids in lupin seed hulls by microorganisms leads to the formation of carboxylic acid resulting in the occurrence of off odor (Bader et al., 2009). Therefore, processing of products from lupin seeds is found to exhibit a variety of off flavors. Lupins also possess one alkaloid, i.e. quinolizidine, which contributes to a higher degree of bitterness. Therefore, the presence of such alkaloids makes the lupin seed inedible to a greater extent (Cristian et al. 2010).

4.2.7 Peanut

The development of off flavor in peanuts initiates during the storage period (Neta et al., 2010). The lipids get oxidized to give objectionable off flavor. Hydroperoxides are the result of the initial oxidation process. The later oxidation process leads to the formation of chemical components like hydrocarbon, aldehydes, ketones, furan, alcohol, and so on. An initial period of oxidation results in a cardboard-like flavor; the compounds responsible for this flavor is *trans*, *trans*,-2,6-Nonadienal. In the later stages, the oxidation leads to the formation of a fishy flavor, from *trans*, *cis*, *trans*-2,4,7-decatrienol and pent-2-enal.

4.2.8 Soybean

The raw soybean and processed products found to contain major volatile off-flavor constituent from lipoxygenase enzyme. The lipids of soybeans gets oxidized to produce a variety of off flavors like beany, painty and unpleasant flavors. Wszelaki et al. (2005) reported that during the processing of soybeans during roasting, germination, or fermentation, the enzymes present in soybeans (i.e. lipoxygenase) degrade the lipids present in soybeans into several off-flavor constituents, from hexanal and methanethiol. Wang et al. (2008) reported that the PUFA present in soybeans get oxidized to give fatty acid peroxides, which produces off flavor and a bitter taste.

Lozano et al. (2007) reported that lipoxygenase will speed up the PUFA oxidation and further result in the formation of methanethiol, aldehydes, alcohols, and so forth. The peroxides, lipoxygenase, and phospholipids significantly contribute to off flavor in various soya products. The grassy and beany flavor leads to the production of volatile acids, and non-volatile components contribute to bitter and astringent flavors. Volatile compounds, which are responsible for an off flavor, are created because of an enzymatic mode of action or by oxidation of PUFA. Soybean oil produces a beany or grassy flavor due to component furan, i.e. 2(l-pentenyl), which is formed as a result of autooxidation of linolenic acid.

4.3 Techniques applied for removing, modifying, or masking off flavor in legumes

4.3.1 Soaking and heat treatment

Soaking is a process of covering or dipping the legumes in water for a stipulated period of time under controlled conditions. During this time, several bitter components diffuse in water and get leached out to make the legumes acceptable and edible to a certain extent. Heat treatment can eliminate various antinutritional factors and control enzyme activation in legumes (Chango et al. 1995). For lupin, the bitter components were removed by leaching via boiling in hot water. Santana and Empis (2001) found that this process results in some loss of protein, which is water soluble along with a bitter alkaloid component i.e. lupanine. Antinutritional factors were also removed to a certain extent. Cristian et al. (2010) observed that treating lupin flour with alkali exhibited good results by removing the bitter alkaloids to a greater extent with good protein quality. Roasting can remove the beany flavor from lupin seeds. Lupin seeds were roasted at a temperature of 80°C–90°C at different periods of time ranging from 10 to 40 minutes. A heating time of 10 minutes further helps to eliminate beany flavor with fine protein quality (Yannez et al., 1986).

Soaking soybeans at 100°C in hot water at a pH of 8 prior to soyamilk processing helps to reduce the undesirable flavor, i.e. beany flavor, and enhances color with greater solubilization properties of protein (Rehman et al., 2007). Heat treatment in soybeans helps to alter soybean flavor without loss of nutritional aspects. It can also reduce lipoxygenase enzyme activity, which can cause off flavor in soybean products.

4.3.2 Germination

The main advantage of exposing legumes to germination is to reduce antinutritional factors, thereby enhancing the nutritional quality (Simons, 2011). It helps to remove objectionable odorous and taste components caused due to the action of lipoxygenase enzymes in legumes. Simultaneously, it also enhances the nutrient content, especially vitamins, along with phenols and saponins. Therefore, compared to soaking and heat treatment, germination is quiet inefficient because of enhancing the contents of saponin and phenols in legumes, which may lead to the creation of off-flavor bitter components even though the beany flavor is removed. Troszynska et al. (2011) found that germination of lentil seeds leads to a reduction in beany and green aroma with enhanced bitterness and astringency. They observed that this germination process won't enhance the sensory attributes in lentil legumes. This is quite different with chickpeas, because in case of the chickpea germination process, the results show that germination had enhanced the intensity of sweetness, thereby masking .or modifying the perception of bitterness (Bellaio et al. 2013).

4.3.3 Solvent extraction

The lipids need to be extracted prior to the pea protein manufacturing process to remove the off flavor. The use of hexane alone for removing lipids is not enough for removing the off flavor. Other solvents can also give good results like ethanol, methanol, or chloroform (Heng et al. 2004). However, using solvents such as hexane, other than methanol, chloroform, or ethanol, may further limit its use for food consumption because of its safety aspects, further hindering the functionality properties of protein.

Legumes phenolic constituents can be best removed by adopting organic solvents. The various aqueous extraction of solvents, from acetone, ethanol, and methanol, can effectively remove phenolic components, but one has to guard the functional properties of protein as well as off-flavor removal aspects (Xu and Chang, 2007). Schultz et al. (1988) reported that the isoelectric washing and defatting process helps to minimize the dried pea off flavor during the preparation of protein isolates from fava beans.

The extraction of soyaflour with alcohol helps to reduce the intensity of its off flavor. The addition of alcohol will leach out the off-flavor constituents during processing of soybean products.

4.3.4 Fermentation

Debittering of legumes can be done by the fermentation process. In lupins, the bacteria can be effectively utilized to remove a bitter alkaloid component, i.e. quinolizidine. Santana and Empis (2001) reported that some bacterial strains can be successfully employed to debitter components like lupanine and lupin alkaloids. In the case of pea and lupin, the lactic acid fermentation will improve the flavor and protein quality. A variety of objectionable flavors can be minimized to a greater extent as a result of the fermentation process. Schindler et al. (2012) found that pea flavor can be modified by the lactic acid fermentation process. Heng (2005) found that saponin Bb can also be modified to a greater extent by this fermentation process in legumes. Therefore, the aroma profile can be significantly modified with these fermentations.

4.3.5 Enzyme treatment

The use of enzymatic treatment for debittering or minimizing the off-flavor compounds is also possible to some extent (Song et al., 2009). Takenawa et al. (1990) utilized enzymes to the wet soybeans, which helps to remove the oxidative flavor after extraction of soyamilk.

4.3.6 Microwave treatment

Using broad beans, as such, leads to a beany aroma and taste; therefore, it needs some kind of treatment prior to application in foods. To solve this problem, Jiang et al. (2016) subjected broad beans to microwave and conventional oven heating. They found that exposing the seed to a microwave at 950 W for a time of 1.5 minutes deactivated the off-flavor components from lipoxygenase and peroxidase. In addition, they also observed after microwave cooking that certain quality parameters for this bean are also improved to a greater extent like milling quality, viscosity of flour pasting, and a reduction in the hardness of the seed. The solubility properties of protein were also preserved to a greater extent. The main advantage of microwave heating is that higher heating rates can be achieved in a shorter period of time with efficient results compared to the conventional cooking method of heating the legumes (Chandrasekaran et al., 2013).

4.3.7 Breeding approach

In legumes, genetic modification for removing off flavor is a possible way to solve this objectionable flavor problem. In soybeans, the transformation of protein and lipids leads to removing off-flavor components to a greater extent, which further helps to minimize the intensity of a beany flavor and other off-flavor compounds (Heywood et al., 2002). Changing the gene code for lipoxygenase enzymes during its synthesis can help prevent fatty acid peroxidation and eradicate creation of hexanal.

4.3.8 Blanching

Blanching is one of the effective methods for leaching undesirable flavor components, especially for soybeans. During soya milk processing, i.e. before milk extraction, beans are blanched under stipulated conditions, which can deactivate lipoxygenase enzymes. As a result, bland and oxidized flavors can be significantly removed from beans.

4.3.9 Extrusion cooking

Extrusion cooking helps to remove undesirable off-flavor components by inactivating soybean enzymes during soya flour processing. The flour obtained by extrusion cooking, as a preliminary treatment before milling, helps to modify the flavor components of the processed products.

4.3.10 Modification of pH

pH also plays a prominent role in flavor modification of legumes. Exposing beans—especially soybeans—to acidic pH condition results in preventing enzyme activity with reduced off-flavor components during processing.

4.3.11 Use of food additives

In the food industry, removing astringent peptides, polysaccharides like cyclodextrin, can modify and improve the flavor. Several masking agents were also used to modify such flavors. These agents will solve issues like off odor, astringency, and after taste in functional food products. These masking agents will neutralize the various off-flavor components and won't impart any kind of new or unacceptable flavor when being added to the food. The various masking agents and approaches used for removing off flavor during processing of various legume-based functional food products were given in **Table 4.2** and **Figure 4.2**. In legume-based functional foods and nutraceutical products, mainly sucrose, salt was added in the base mix before a desirable specific flavor was incorporated, which further helps to remove the off-flavor components and further enhance the overall

Table 4.2 Masking Agents Used in Functional Food Products

Bitter Masking Components	Flavor Enhancer	Flavor Modulator	Flavor Modifier
Sucrose	Monosodium Glutamate	Polydextrose	Pectin
Sodium			
High-intensity synthetic sweeteners			
Phosphatidic acid			

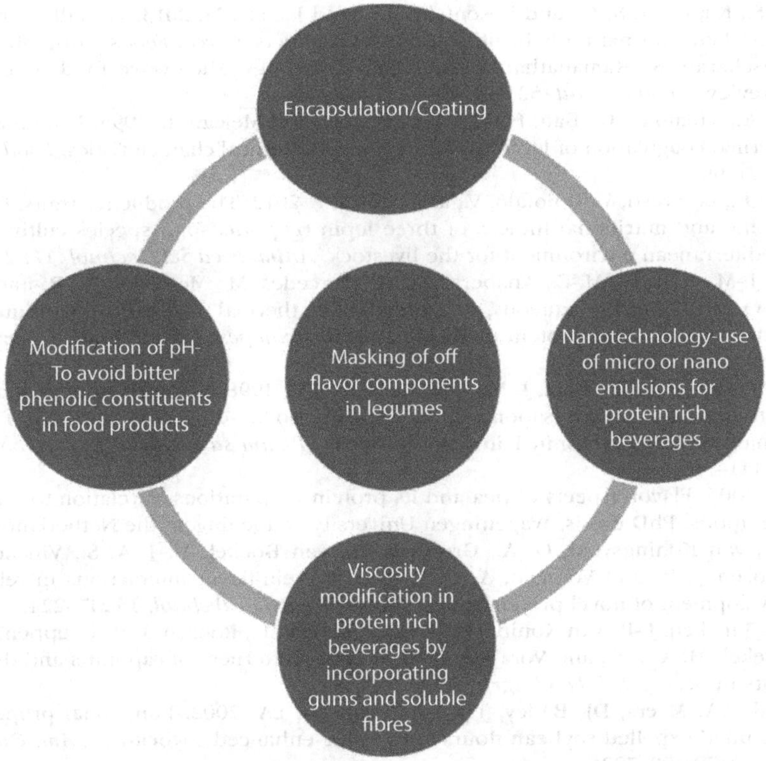

***Figure 4.2** Approaches for masking off-flavor components in legumes.*

consumer acceptability to a greater extent because of the masking effect or properties of added materials.

4.4 Conclusion

Off-flavor components can be removed during processing, either by adopting individual or integrated approaches, so that the beany odor, bitter taste, and after taste can be eliminated with a higher degree of consumer acceptability in the final-processed, functional legume food product.

References

Bader, S., Czerny, M., Eisner, P., and Buettner, A. 2009. Characterisation of odor-active compounds in lupin flour. *J. Sci. Food Agric.* 89:2421–2427.

Baysal, T., and Demirdoven, A. 2007. Lipoxygenase in fruits and vegetables: A review. *Enzyme Microb. Technol.* 40:491–496.

Belitz, H.-D., Grosch, W., and Schieberle, P. 2001. Lipids. Pages 157–244 in: *Food Chemistry*, 3rd ed. Springer: Berlin, Germany.

Bellaio, S., Kappeler, S. R., and Rosenfeld, EZ., and Jacobs, M. 2013. Partially germinated ingredients for naturally healthy and tasty products. *Cereal Foods World* 58:55–59.

Chandrasekaran, S., Ramanathan S., and Basak, T. 2013. Microwave food processing—A review. *Food Res. Int.* 52:243–261.

Chango, A., Villaume, C., Bau, H. M., Nicolas, J. P., and Mejean, L. 1995. Fractionation by thermal coagulation of lupin proteins: Physicochemical characteristics. *Food Res. Int.* 28:91–99.

Chiofalo, B., Lo Presti, V., Chiofalo, V., and Gresta, F. 2012. The productive traits, fatty acid profile and nutritional indices of three lupin (*Lupinus* spp.) species cultivated in a Mediterranean environment for the livestock. *Anim. Feed Sci. Technol.* 171:230–239.

Cristian, J.-M., Rosalva, M.-E., Anaberta, C. M., Mercedes, M., Mercedes, M. P., and Gloria, D.-O. 2010. Effect of aqueous, acid and alkaline thermal treatments on antinutritional factors content and protein quality in *Lupinus campestris* seed flour. *J. Agric. Food Chem.* 58:1741–1745.

Daveby, Y. D., Aman, P., Betz, J. M., and Musser, S. M. 1998. Effect of storage and extraction on ratio of soyasaponin I to 2,3-dihydro-2,5-dihydroxy-6-methyl-4-pyrone-conjugated soya saponin I in dehulled peas (*Pisum sativum* L). *J. Sci. Food Agric.* 78:141–146.

Heng, L. 2005. Flavor aspects of pea and its protein preparations in relation to novel protein foods. PhD thesis. Wageningen University: Wageningen, the Netherlands.

Heng, L., van Koningsveld, G. A., Gruppen, H., van Boekel, M. J. A. S., Vincken, J.-P., Roozen, J. P., and Voragen, A. G. J. 2004. Protein-flavor interactions in relation to development of novel protein foods. *Trends Food Sci. Technol.* 15:217–224.

Heng, L., Vincken, J.-P., van Koningsveld, G. A., Legger, L., Roozen, J. P., Gruppen, H., van Boekel, M. A. J. S., and Voragen, A. G. J. 2006. Bitterness of saponins and their contents in peas. *J. Sci. Food Agric.* 86:1225–1231.

Heywood, AA, Myers, DJ, Bailey, TB., and Johnson, LA. 2002. Functional properties of extruded-expelled soybean flours from value-enhanced soybeans. *J. Am. Oil. Chem. Soc.* 79(7):699–702.

Jiang, Z., Pulkkinen, M., Wang, Y., Lampi, A., Stoddard, F. L., Salovaara, H., Piironen, V., and Sontag-Strohm, T. 2016. Faba bean flavour and technological property improvement by thermal pre-treatments. *LWT—Food Sci. Technol.* 68:295–305.

Lozano PR, Drake, M., Benitez, D., and Cadwallader, K. 2007. Instrumental and sensory characterization of heat-induced odorants in aseptically packaged soy milk. *J. Agric. Food Chem.* 55(8):3018–3026.

Neta, E. R., Sanders, T., and Drake, M. A. 2010. Understanding peanut flavor: A current review. In: *Handbook of Fruit and Vegetable Flavors*, edited by Hui et al. CRC Press. pp. 985–1022.

Oomah, B. D., Luc, G., Leprelle, C., Drover, J. C. G., Harrison, J. E., and Olson, M. 2011. Phenolics, phytic acid, and phytase in Canadian-grown low-tannin fava bean (*Vicia faba* L.) genotypes. *J. Agric. Food Chem.* 59:3763–3771.

Rehman, S., Nawaz, H., Ahmad M. M., Hussain, S., Murtaza, A., and Shahid, S. H. 2007. Physico-chemical and sensory evaluation of ready to drink soy-cow milk blend. *Pak. J. Nutr.* 6(3):283–285.

Sanchez-Vioque, R., Clemente, A., Vioque, J., Bautista, J., and Millan, F. 1998. Polar lipids of defatted chickpea (*Cicer arietinum* L.) flour and protein isolates. *Food Chem.* 63:357–361.

Santana, F. C., and Empis, J. 2001. Bacterial removal of quinolizidine alkaloids from *Lupinus albus* flours. *Eur. Food Res. Technol.* 212:217–224.

Schindler, S., Wittig, M., Zelena, K., Krings, U., Bez, J., Eisner, P., and Berger, R. G. 2011. Lactic fermentation to improve the aroma of protein extracts of sweet lupin (*Lupinus angustifolius*). *Food Chem.* 128:330–337.

Schindler, S., Zelena, K., Krings, U., Bez, J., Eisner, P., and Berger, R. G. 2012. Improvement of the aroma of pea (*Pisum sativum*) protein extracts by lactic acid fermentation. *Food Biotechnol.* 26:58–74.

Schultz, M., Hoppe, K., and Schmandke, H. 1988. Off-flavour reduction in *Vicia faba* bean protein isolate. *Food Chem.* 30:129–135.

Simons, R. 2011. Prenylated Isoflavonoids from Soya and Licorice: Analysis, Induction and *In Vitro* Estrogenicity. PhD thesis. Wageningen University: Wageningen, the Netherlands.

Song, D., Chang, S. K. C., and Ibrahim, S. A. 2009. Descriptive sensory characteristics of no-flatulence pinto bean. *J. Food Qual.* 32:775–792.

Takenawa, S., Takeda, H. and Honkoshi, M. 1990. Process for eliminating disagreeable odor from soya milk. US Patent. 4,929,451.

Troszynska, A., Estrella, I., Lamparski, G., Hernandez, T., Amarowicz, R., and Pegg, R. B. 2011. Relationship between the sensory quality of lentil (*Lens culinaris*) sprouts and their phenolic constituents. *Food Res. Int.* 44:3195–3201.

Wang, X. S., Tang, C. H., Li, B. S., Yang, X. Q., Li, L., and Ma, C. Y. 2008. Effects of high-pressure treatment on some physicochemical and functional properties of soy protein isolates. *Food Hydrocoll.* 22(4):560–567.

Wszelaki, A. L., Delwiche, J. F., Walker, S. D., Liggett, R. E., Miller, S. A., and Kleinhenz, M. D. 2005. Consumer liking and descriptive analysis of six varieties of organically grown edamame-type soybean. *Food Qual. Prefer.* 16:651–658.

Xu B. J., and Chang, S. K. 2007. A comparative study on phenolic profiles and antioxidant activities of legumes as affected by extraction solvents. *J Food Sci.* 72(2):S159–S166.

Yannez, E., Lobos, P., Diaz, G., and Ballester, D. 1986. Effect of roasting on the chemical composition and protein quality of lupin seeds (*Lupinus albus* cv Multolupa). *J. Food Sci.* 51:1235–1238.

5

Flavor Challenges in Cereal-Based Functional Food Products

M. Selvamuthukumaran and Yashwant V. Pathak

Contents

5.1 Introduction

The flavor is one of the important sensory attributes for product acceptability. Cereal products become more functional if the bran is used for formulation of products. Bran has many health benefits, which make the ultimate end product to create more functions for the consumers, when compared to normal cereal-based food products. The major disadvantage of using bran is the prominent flavor, which will further limit its usage during preparation of several bran-fortified cereal products. Therefore, in this chapter the significance of bran fortification, off-flavor components in cereal-based food products and techniques employed to increase the palatability of the bran-fortified cereal-based functional food products were discussed in detail.

5.2 Functional aspects of cereal bran

The bran obtained from various cereals was found to contain various functional benefits (**Figure 5.1**). They were the richest source of fiber, B-complex vitamins and several minerals. The presence of fibers and bioactive components helps to prevent lifestyle-related diseases like diabetes and heart diseases (Balandrán-Quintana et al., 2015). Wheat bran was found to exhibit higher antioxidant activity, which can be ascribed to presence of several phenolic components. Stevenson et al. (2012) reported that the bran obtained from wheat as a result of milling was found to contain antioxidant i.e. alkyl resorcinol, which can exhibit biological activity by reducing the formation of several triglycerides and also induce thromboxane production. It can also prevent constipation problems. Wheat bran contains more amount of mineral i.e. magnesium, which improves immunity and further strengthens the bones with reduced incidence of heart-related ailments. It also contains an enormous amount of folate, which can help to synthesize DNA to achieve cell growth and to reduce Alzheimer's disease. Wheat bran possesses cancer-reduction

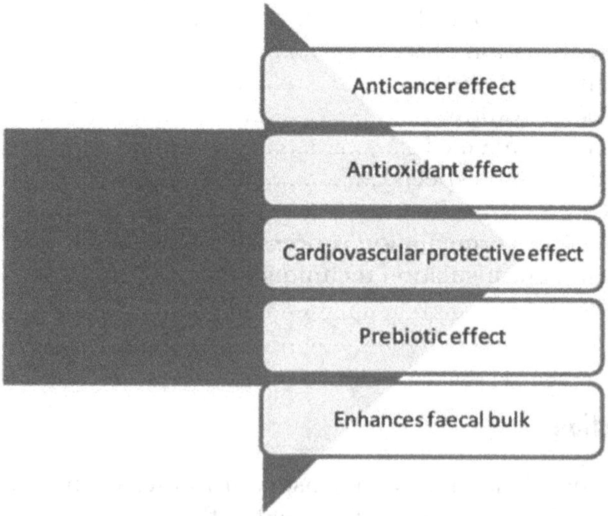

Figure 5.1 *Functional properties of cereal bran.*

properties, which may be due to presence of bioactive constituents like lignans, flavonoids and phenolic acids. The consumption of wheat bran leads to the dilution of carcinogen components, thereby forming a very bulky stool, which further stops the entry of such carcinogen components into the cells lining the colon (Lupton and Turner, 1999). Intestinal transit time will be reduced to a greater extent as a result of consumption of wheat bran. It also prevents tumor growth by fermenting the bran to give short-chain fatty acids by several colonic beneficial bacteria's, which can significantly prevent tumor formation tumor and enhance the prebiotic activity.

Wang et al. (2009) reported that feruloyl oligosaccharides, a kind of phenolic constituents in wheat bran, were found to exhibit antioxidant activity by preventing free radical mediated oxidative damage in erythrocytes of human.

Like wheat, oats were also found to possess a significant amount of dietary fiber, which can reduce LDL cholesterol and also cardiovascular diseases. The oats soluble dietary fiber β-glucan helps in controlling the glucose level in blood, especially for diabetic people.

Consumption of oat-based products for a time period of 42 days helps to reduce the cholesterol and blood pressure among consumers (Saltzman et al., 2001). Different types of cancers from prostate, breast and colon can also be minimized by regularly consuming oats as part of a daily diet.

Chen et al. (1998) portrayed that the consumption of wheat bran @ 28–30 g per day results in stool weight enhancement of 52 g/day. Costabile et al. (2008) reported prominent microbiota changes in gut as a result of consuming wheat bran compared to consuming whole grain as such. Freeland et al. (2009) reported that consumption of fiber from wheat bran will significantly reduce the intake of energy and enhance satiety. These fibers play a prominent role in regulating the appetite.

Costabile et al. (2008) found that regular intake of wheat bran-based fortified breakfast foods containing 13 g fiber can reduce cholesterol from @ 5.5 mmol/L.

5.3 Off-flavor components in cereal bran

Cereal brans were the rich source of several bioactive constituents, but the presence of several non-volatile components contributes to bitter flavor. The components solely responsible for contributing to off notes in cereals were fatty acids, peptides, phenolic constituents and amino acids (Heiniö et al., 2008). The dark-colored rice varieties produce bitter flavor (Bett-Garber et al., 2012).

The milling fractions of various cereals produce a variety of unacceptable off notes ascribed to phenolic constituents, which are found in the external layer of grain. The sensory and nutritional properties may differ for various rye fractions, from whole kernel, floury endosperm and bran processed by different methods (Heiniö et al., 2008). The flour produces only mild flavor, but bran produce sharp bitter flavor, which further enhances during storage. Heiniö et al. (2008) reported that the acceptable taste components were found in between endosperm and the bran layer of rye grain.

Tannins present in sorghum grain leads to astringent characteristics (Kobue-Lekalake, 2008).

The consumption of whole-grain cereal-based processed foods are reduced due to presence of several phenolic components in outer layer of different cereal grains from syringic acid, phenolic acid, ferulic acid, lignans, benzoxazinoids etc. (Andersson et al., 2014) (**Figure 5.2**). Challacombe et al. (2012) reported that products, especially crackers and breads, made from whole-wheat grain of red- and white-colored varieties exhibited bitter taste. The intensity of taste may vary according to forms of phenolic components present in grains, the free form showed a higher degree of bitterness when compared to bound form.

Bitter peptides were formed during milling of whole cereal grains because of proteolytic enzymes, which are present in the outer surface of the grain (Brijs et al., 1999). Jiang and Peterson (2013) and Bin et al. (2012) reported that bitter note in bread made out of whole wheat grain seems to be taste bitter due to proteolysis and Maillard reaction in flour. The amino acids like albumin, globulin, tryptophan and various end products of Maillard reaction played a prominent role in bringing a bitter note to the bread made from whole wheat grain. It is reported that peptide amino acid constituents significantly contribute to a more bitter note when compared to the size of the peptide molecule (Lemieux & Simard, 1992).

The bitter taste can be noticed more in bran-fortified processed products. The bran contains more lipids, which can undergo oxidation under favorable condition with the help of lipase enzymes, which ultimately produces bitter components and make the products unacceptable. Off-flavor components were derived as a result of lipid hydrolysis. In case of wheat, corn and barley, the germ portion contains high amount of such lipids; for oats, it is distributed

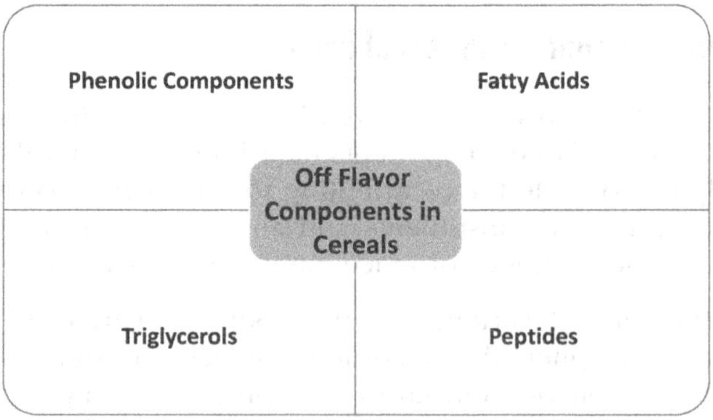

Figure 5.2 *The various off-flavor component in cereal grains.*

in whole grain. Therefore, lipid oxidation occurrence will be more for the germ portion in these types of grains.

Among all cereal grains, oat grain needs special processing care to devoid off-flavor taste formation during product development. Lipid hydrolysis may occur in this grain as a result of enzymatic reaction or by chemical reaction, which will add up the off note to the end products.

Break down of lipids is progressive during storage of whole-grain flours; as a result, the processed products made from stored whole grain yield bitter taste and further reduce the level of acceptability. Free fatty acids will be formed during a storage period and lead the resultant flour in rancid taste and putrid odor. Triacylglycerols upon hydrolysis will lead to form several fatty acids, which are responsible for causing rancidity in stored flours especially whole-grain flours. The enzyme lipase plays a prominent role in converting triacylglycerols into fatty acids (Heiniö et al., 2002). The enzyme lipoxygenase will also oxidize free fatty acids, which are found in the bran and germ portions of wheat grain to bring a rancid taste. During rye grain fraction storage, intensity of bitterness can be increased because of triacylglycerol and free fatty acid formation (Nordlund et al., 2013). The cause of bitter note may be either due to presence of either phenolic constituents or protein or lipids.

The oat cereal has very low shelf life, which may be due to occurrence of rancid formation during storage. This problem limits the use of processed products made out of oats (Heiniö et al., 2011). Bitter taste also occurs because of the presence of phenolic components (Molteberg et al., 1996), and it is mandatory to prohibit the lipase enzyme activity during the processing of oats in order to decrease the off-flavor and at the same time increase the sensory profile of the products.

5.4 Off-flavor in processed cereal products

5.4.1 Off-flavor in bread

Fortification of wheat flour with bran produces bread with a bitter taste, further affecting the color and texture (Della Valle et al., 2014). The processed bread made out of either bran fortified wheat flour or from whole wheat atta flour yields bread with dark color, unacceptable bran odor and dense crumb texture with drastically reduced sensory acceptability. Gan et al. (1992) reported that quality wise the processed bread made out of refined wheat flour is much superior than whole wheat flour ascribed to protein dilution in kernel endosperm.

5.4.2 Off-flavor in biscuit

Spreadability is one of the important quality parameter for biscuit. The incorporation of higher amount of bran enhances product chewiness and crumbliness with unacceptable darker color. Incorporation of wheat bran in the range of 25%–30% in the flour results the product in dry mouthfeel nature (Stanyon & Costello, 1990; Sudha et al., 2007; Sozer et al., 2014). Similar results were observed during preparation of biscuits from composite flours (Omoba et al., 2015). Sensory profile of biscuits seems to be acceptable for biscuits made or prepared by incorporating brans of oat (30%) and wheat or barley (20%).

Sozer et al. (2014) reported that biscuit-fresh flavor got diminished as a result of bran addition and ultimately resulted in an aftertaste of the product (Sudha et al., 2007).

5.4.3 Off-flavor in pasta

Many researchers reported that bran-fortified pasta products were not liked by several consumers because of typical flavor, color and texture (Edward et al., 1995; Sahlström et al., 1993; Sozer et al., 2007). The incorporation of bran during processing affects the quality of the final products, as it produces product with bitter taste, branny aroma with hard textural properties (West et al., 2013a). A flavor interaction study was performed by West et al. (2013b). They had selected 12 trained panellist, who analyzed the product for interaction of flavor of this product with cheese sauce of lower sodium content. The product was developed by incorporating whole cereal grain @ 0%–100%. Their results show that enhancing the whole-grain cereal gives bitter notes with branny flavor. The flavor intensity was reduced for 75%–100% whole-grain incorporated pasta product compared to lower incorporation levels (West et al., 2013c).

Similarly, the pasta was prepared by blending wheat semolina of durum variety with wheat bran and wheat germ @ 5%–17% (Aravind et al., 2012). Their sensory results showed that incorporation of 20% wheat bran and germ @ of 30% accorded poor acceptability score. The product tastes to be bitter with poor texture. West et al. (2013c) reported that pasta developed from whole-grain cereal accorded bitter taste with hard texture compared to the product made from refined flours.

5.4.4 Off-flavor in porridge

The porridge prepared from sorghum exhibited different sensory characteristics ascribed to various milling conditions (Kebakile et al. 2008). Astringent product with branny odor was pronounced for product made from roller-milled sorghum grain fractions (**Table 5.1**). Colorwise, the product is very dark with unpleasant flavor. The product prepared from hand-pounded fractions also yielded similar sensory characteristic as like roller-milled sorghum fractions. In addition to that the hand-pounded fractions exhibited rancid flavor. Porridges prepared by abrasive-action and hammer-milling process from light-colored sorghum grains was highly acceptable and recommended. Porridge prepared from pigmented pericarp sorghum varieties pronounced bitter porridge with off aroma.

Table 5.1 Sensory Characteristics of Porridges Prepared from Different Milled Fractions of Sorghum

Milling Conditions	Color	Odor	Flavor	Overall Acceptability
Roller milled sorghum porridge	Dark	Branny	Unpleasant	Slightly acceptable
Hand pounded sorghum porridge	Dark	Branny	Unpleasant & rancid	Slightly acceptable
Hammer milled sorghum porridge	Light	Less branny	Pleasant	Highly acceptable

5.5 Techniques for increasing the palatability of bran-fortified, cereal-based functional food products

5.5.1 Selection of raw materials

Raw material plays a prominent role in contributing flavor aspect to the processed product. There are several varieties, which are available, which can significantly contribute to flavor intensity of the derived products. The wide array of wheat varieties, which are available globally, and their color may vary from red, white, golden, yellow and so on. The bread processed from red-colored wheat variety produces darker-colored product with unacceptable sensory characteristics, which may be because of presence of reddish-brown particles in bran, and its fortification can further reduce the overall product acceptability level to a greater extent. The bitter taste of product may be due to presence of constituents like tannins and phenolic acids. The bread processed from white-colored varieties produces product of acceptable lighter-color with less bran flavor (Doblado-Maldonado et al., 2012). Starr et al. (2013) reported that some ancient species and old varieties, from spelt, emmer, etc., produced bread of an acceptable flavor when compared to recently available varieties.

The bread manufactured by using two types of different wheat varieties showed varied flavor content, which can be ascribed to formation of volatile constituents as a result of fermentation of damaged starch, and also occurrence of a Maillard reaction during the baking process (Galey et al., 1994).

Kobue-Lekalake et al. (2009) reported that some sorghum varieties contain low content of tannins, detectable even below threshold level, such varieties will produce product with a high degree of overall acceptability, even though the bitter components were present. Foods with greater antioxidant activity can be produced by use of sorghum with varieties of lesser tannin content.

5.5.2 Biotechnological approach

Flavor of cereal-based products can be modified by using wide variety of microbes as well as enzymes. The age-old sourdough fermentation process is said to alter sensory characteristics of flour obtained from whole-grain flour

(Poutanen et al., 2009). During fermentation process the enzyme proteases get activated under acidic conditions, induces proteolytic activity and during fermentation the bitter phenolic substances were released (Liukkonen et al., 2003), and further degraded by microorganisms during fermentation process, thereby altering the flavor of the products.

Bread flavor is solely dependent upon biochemical changes, which occur during the fermentation process. Sensory characteristics of bread can still be further improved by optimizing the fermentation process by incorporating a flavor modifier like sourdough, which can yield product with improved sensory attributes. Among the microorganisms used, lactic acid bacteria produce acceptable quality of bread by altering the flavor profile. The sour dough, which contains lactic acid bacteria that can be used as a fermentation substrate during the production of bread, can further significantly modify the flavor of the product when compared to yeast. Roasted crust flavor cannot be modified by using yeast as a sour dough improver. Zehentbauer and Grosch (1998) found that effective use of yeast preferment can drastically reduce the roast flavor. Heiniö (2014) reported that flavor can be successfully modified or altered by using various enzymes during the processing of several bakery products. Use of hydrolytic enzymes specifically will break open cell wall of the grains and cause flavor changes. Katina et al. (2005) found that use of the microorganism from *Lactobacillus brevis* for fermenting whole-grain cereal flour for 20 hrs at room temperature enhances the flavor to a greater extent.

5.5.3 Germination process

Grain flavor can also be altered by effective germination process. It is evident that during malting of barley grains, the germination process will modify flavor characteristics as like or similar to oats (Heiniö et al., 2011).

5.5.4 Milling process

The milling of several cereal grains release variety of enzymes from amylase, peroxidase, polyphenoloxidase, which are responsible for development of off-flavors. Compared to the baking process, the milling conditions determine the degree of bitterness to a greater extent. Kihlberg (2004) compared that the bread made out of flour by adopting two milling techniques. First one is bread processed from roller-milled wheat flour, which was found to be highly acceptable rather than bread obtained from traditional-milled wheat flour. The traditional-milled wheat flour yield bread with roasty flavor. Sharp off-flavor intensity can be reduced by separating first outer portion of the grains.

Flavor constituents were widely found in the whole kernel; therefore, milling process plays a significant effect in contributing flavor to the final processed food products. Heiniö et al. (2008) reported that outer pericarp layer i.e. bran layer composes of bitter taste compared to inner endosperm layer; therefore, performing operations like high degree of polishing may improve the flavor of milled products, but most of the nutrients were highly removed to a greater extent. The flavor of bran fortified products like bread can be enhanced by fermenting bran during its production process, thereby the flavor modification may concurrently occur. The use of various enzymes like amylases, proteases, xylanases, etc., can degrade or disintegrate the cell wall constituents of grains thereby altering the off-flavor and also avoiding the retro gradation of starch, which may cause bread staling (Heiniö, 2014).

5.5.5 Use of additives

Several foods contain bittering components and it can be successfully leached through debittering approach or techniques. This technique is highly useful for the wide application in liquid food products, e.g., debittering of lime juice. With respect to solid grains, i.e., cereals that contain dietary fibers, which can mask the off-flavor by acting as an off-flavor adsorber. The other possible way of reducing the bitter taste is to block the bitter taste receptor hTAS2R39 (Roland, 2014). The bitter block receptors activation can be suppressed by component 6-methoxyflavanones. Protein complexation with Epigallocatechin Gallate can also reduce the bitter block receptors. Sweetening agents were widely employed to mask the off-flavor and to reduce the degree of unacceptability. In case of bran-fortified breakfast cereals, they were used @ 10% which will highly improve the sensory perception of the final product, for whole wheat products, honey is extensively used. The bitter taste can also be nullified by incorporating salt.

Bett-Garber et al. (2013) found that addition of salt alone or salt with oil and water during cooking of rice drastically reduced the bittering effects, presence of sodium helps in masking the bitter constituents. Incorporation of additives like gluten, enzymes, emulsifying agents and surfactants during processing can improve the sensory parameters of bran-fortified cereal functional food products.

The incorporation of ingredients like fat will lead to development of aromatic carbonyl compounds during baking process (Johnson & EI-Dash, 1969). The use of ascorbic acid will diminish aldehydes formation (Feillet, 2000). However, Nishibori and Kawakishi (1990) reported that enhancement of ingredients from sugar and fat during product development will simultaneously enhance the hydroxyl methyl furfural content, which can produce an unacceptable odor.

The use of various yeast strains other than baker's yeast, from *Saccharomyces exiguus, Candida* spp., *and Pichia* spp., can greatly contribute to better flavor compared to *Saccharomyces cerevisiae.* Tomer et al. (1992). reported that use of such yeast species will lead to produce several aromatic compounds. Bread with enhanced aroma is produced by use of sour dough containing yeast i.e. *Saccharomyces cerevisiae* in the fermentation process. The volatile aromatic components like 2-methylpropanoic, 3-methylbutanoic acids are being produced in a greater extent. The flavor can also be modified or improved by using by-products obtained from distillery unit. Spent grains, a by-product of distillery unit, which can be successfully utilized as an ingredient to modify the flavor of the final product, thereby enhancing the texture, especially the soft quality of the final baked food products. Spent grains are a rich source of fiber, which can increase the intensity of the flavor, when it is incorporated for ready to eat cereal products @ 10%–15%.

The bread aroma can also be improved by adding acid whey obtained from cheese industries (Shenkenberg et al., 1972). Similarly, other products from puddings, cakes, etc. of their aroma profile can be enhanced by adding such acid whey obtained from cottage cheese manufacturing industries (Guy, 1978).

5.5.6 Use of enzymes

The incorporation of enzymes during food processing helps to improve flavor, texture and enhance the shelf life of the processed products (Martfnez-Anaya, 1996). Use of amylase enzymes during bread production helps to significantly improve the bread sensory properties especially flavor and color. Additionally dough characteristics like consistency and its fermentation capability are greatly enhanced. These enzymes can be adopted from various sources, like the addition of malted cereals or from the microbial fungal source. The protease enzyme obtained from *Aspergillus oryzae* can modulate bread quality, thereby significantly improving the physical as well as sensory qualities from bread dough consistency and its aroma thereby enhancing the amount of peptide units (Kieffer et al., 1990; Rashed et al., 1990).

The soybean or broad bean flour is incorporated with bread flour to increase the lipoxygenase enzyme activity. The purpose of incorporation is to increase the intensity of whiteness and also to enhance dough raising capacity during fermentation process. One drawback of such incorporation, which leads to development of free radicals that can produce hydroperoxides, that will have greater negative impact over aroma of the final baked product (Van Dam and Hille, 1992). Martfnez-Anaya (1996) reported that use of lipase enzymes can lead to form several short chain fatty acids, which have influence over flavor characteristics and use of oxidase kind of enzymes results in oxidation, which lead to creation of unacceptable aroma components. Other than amylase, lipase and oxidase, several other new enzymes were recently used, which includes maltogenic a-amylase, xylanase, etc., which are widely employed to prevent staling of bread and also to enhance the dough handling nature of the bread (Si, 1997; Pszczola, 2001).

5.5.7 Processing aspects

5.5.7.1 Kneading

Kneading is a prominent step for getting the product with fine flavor and good texture especially for the baked food products. During this process, enzymes present in flour help in creating pleasant aroma. In case of manufacturing bread, formation of aldehyde flavor compound in bread crumbs especially hexenal, which solely depend upon the process of kneading. Kneading helps to enhance such flavor component with good product (Richard-Molard, 1994).

5.5.7.2 Fermentation process

It is the crucial step for manufacturing bakery products. Extending the fermentation time before baking process has enhanced the intensity of 2-phenyl-ethano1 and 3-metyl-butanol (Hansen and Hansen, 1996). The proteolysis can effectively occur if the fermentation time is increased, which results in production of several free amino acids that will react as strecker aldehyde precursors, which can else cause malty flavor. The greater amount of yeast incorporation to dough during fermentation process will increase the intensity of the bread odor components from 2-acetyl-2-pyroline (Zehentbauer and Grosch, 1998).

5.5.8 Microencapsulation techniques

It's a widely used technique, where flavor is being trapped into a carrier material, which may be made up of either carbohydrates or lipids or proteins, which further helps to mask the unacceptable flavor and it also, protect the core component against various oxidative changes that occur during processing and storage. Polarity of core, volatile properties of core, chemical properties of core decides the flavor retention in the final product (Goubert et al., 1998). The stabilized bran fractions can be encapsulated by using spray drying techniques to yield fine powder form, which can be further incorporated into several cereals based functional food products.

5.6 Conclusion

Therefore bran palatability can be increased by adopting techniques like proper selection of raw materials, germination process, biotechnological approach, use of food additives, manufacturing processes optimization, microencapsulation techniques. These several techniques can modify the flavor of bran, so that they can be fortified successfully to develop and formulate several cereals based functional food products.

References

Aravind, N., Sissons, M., & Fellows, C. M. (2012). Effect of soluble fibre (guar gum and carboxymethylcellulose) addition on technological, sensory and structural properties of durum wheat spaghetti. *Food Chemistry*, 131(3), 893–900.

Andersson, A. A. M., Gimberg, L., Åman, P., & Landberg, R. (2014). Recent findings on certain bioactive compounds in whole grain wheat and rye. *Journal of Cereal Science*, 59, 294–311.

Balandrán-Quintana, R. R., Mercado-Ruiz, J. N., & Mendoza-Wilson, A. M. (2015). Wheat bran proteins: A review of their uses and potential. *Food Reviews International*, 31, 279–293.

Bett-Garber, K., Lea, J., Champagne, E., & McClung, A. (2012). Whole-grain rice flavor associated with assorted bran colors. *Journal of Sensory Studies*, 27(2), 78–86.

Bett-Garber, K. L., Lea, J. M., Watson, M. A., & Champagne, E. T. (2013). Influence of cooking formulation on flavor and hydrophilic oxygen radical absorption capacity values of whole grain colored rice. *Cereal Chemistry*, 90(1), 58–64.

Bin, Q., Jiang, D., Cho, I. H., & Peterson, D. G. (2012). Chemical markers for bitterness in wheat bread. *Flavour and Fragrance Journal*, 27(6), 454–458.

Brijs, K., Bleukx, W. & Delcour, J. A. (1999). Proteolytic activities in dormant rye (Secale cereale L.) grain. *Journal of Agricultural and Food Chemistry*, 47(9), 3572–3578.

Challacombe, C. A., Abdel-Aal, E. M., Seetharaman, K., & Duizer, L. M. (2012). Influence of phenolic acid content on sensory perception of bread and crackers made from red or white wheat. *Journal of Cereal Science*, 56(2), 181–188.

Chen, H-L., Haack, V. S., Janecky, C. W., Vollendorf, N. W., and Marlett J. A. (1998). Mechanisms by which wheat bran and oat bran increase stool weight in humans. *American Journal of Clinical Nutrition*, 68, 711–719.

Costabile, A., Klinder, A., Fava, F., Napolitano, A., Fogliano, V., Leonard, C., Gibson, G. R., and Tuohy, K. M. (2008). Whole-grain wheat breakfast cereal has a prebiotic effect on the human gut microbiota: A double-blind, placebo-controlled, crossover study. *British Journal of Nutrition*, 99, 110–120.

Doblado-Maldonado, A. F., Pike, O. A., Sweley, J. C., & Rose, D. J. (2012). Key issues and challenges in whole wheat flour milling and storage. *Journal of Cereal Science*, 56(2), 119–126.

Della Valle, G., Chiron, H., Cicerelli, L., Kansou, K., Katina, K., Ndiaye, A., Whitworth, M., & Poutanen, K. (2014). Basic knowledge models for the design of bread texture. *Trends in Food Science & Technology*, 36(1), 5–14.

Edwards, N., Biliaderis, C., & Dexter, J. (1995). Textural characteristics of wholewheat pasta and pasta containing non-starch polysaccharides. *Journal of Food Science*, 60(6), 1321–1324.

Feillet, P. (2000). De la farine à la pâte et d4 la pâte au pain. ln Le grain de blé. Composition et utilisation, INRA Editions; Paris, 137–173.

Freeland, K. R., Anderson G. H., Wolever, T.M. S. (2009). Acute effects of dietary fibre and glycaemic carbohydrate on appetite and food intake in healthy males. *Appetite*, 52:58–64.

Galey, C., Potus, J., Drapron, R., Poiffait, A., Bar, C., Fischer, J., Giampaoli, P. (1994). Crumb bread flavour: Influence of wheat variety and bread making process. *Sciences Des Aliments*, 14 (5), 643–653.

Gan, Z., Galliard, T., Ellis, P., Angold, R., & Vaughan, J. (1992). Effect of the outer bran layers on the loaf volume of wheat bread. *Journal of Cereal Science*, 15(2), 151–163.

Goubert, I., Le Quere, J. L., Voilley, A. J. (1998). Retention of aroma compounds by carbohydrates: Influence of their physicochemical characteristics and of their physical state. *Journal of Agricultural and Food Chemistry*, 46, 1981–1990.

Guy, E. J. (1978). Dried Cottage cheese whey: Review of its composition and use in bakery products. *Bakers Digest*, 52(5), 10–12,14,16.

Hansen, A., and Hansen, B. (1996). Flavour of sourdough wheat bread crumb. *Zeitschrift für Lebensmittel-Untersuchung und Forschung*, 202, 244–249.

Heiniö, R-L. (2014). Sensory attributes of bakery products. In W. Zhou, Y. H. Hui, I. De Leyn, M. A. Pagani, C. M. Rosell, J. D. Selman & N. Therdthai (Eds.), *Bakery Products Science and Technology* (pp. 391–407). West Sussex, UK: John Wiley & Sons.

Heiniö, R-L., Kaukovirta-Norja, A., & Poutanen, K. (2011). Flavor in processing new oat foods. *Cereal Food Worlds*, 56(1), 21–26.

Heiniö, R-L., Lehtinen, P., Oksman-Caldentey, K., & Poutanen, K. (2002). Differences between sensory profiles and development of rancidity during long-term storage of native and processed oat. *Cereal Chemistry*, 79(3), 367–375.

Heiniö, R-L., Liukkonen, K., Myllymäki, O., Pihlava, J., Adlercreutz, H., Heinonen, S., and Poutanen, K. (2008). Quantities of phenolic compounds and their impacts on the perceived flavour attributes of rye grain. *Journal of Cereal Science*, 47(3), 566–575.

Johnson, J. A., EI-Dash, A. A. (1969). Role of nonvolatile compounds in bread flavor. *Journal of Agricultural and Food Chemistry*,17(4), 740–746.

Jiang, D., & Peterson, D. G. (2013). Identification of bitter compounds in whole wheat bread. *Food Chemistry*, 141(2), 1345–1353.

Katina, K., Arendt, E., Liukkonen, K., Autio, K., Flander, L., & Poutanen, K. (2005). Potential of sourdough for healthier cereal products. *Trends in Food Science & Technology*, 16(1), 104–112.

Kebakile, M. M., Rooney, L. W., de Kock, H. L., & Taylor, J. R. (2008). Effects of sorghum type and milling process on the sensory characteristics of sorghum porridge. *Cereal Chemistry*, 85(3), 307–313.

Kieffer, R., Rashed, M. M., Belitz, H. D. (1990). Modification of the functional properties of wheat flour by proteolytic enzymes. Part 1. Activity against substrates in solution and rheological effects. *Zeitschrift Fur Lebensmittel-Untersuchung Und-Forschung*, 191, 104–109.

Kihlberg, I. (2004). Sensory quality and consumer perception of wheat bread: Towards sustainable production and consumption (Doctoral dissertation). University Dissertations.

Kobue-Lekalake, R. I. (2008). Sensory perception of bitterness and astringency in sorghum (Doctoral dissertation). University of Pretoria Dissertations and theses [9088].

Kobue-Lekalake, R. I., Taylor, J. R. N., & DeKock, R. (2009). Influence of PROP taster status on the consumer acceptability of food made from tannin sorghums. *Journal of the Science of Food and Agriculture*, 89, 1809–1814.

Lemieux, L., & Simard, R. (1992). Bitter flavour in dairy products. II. A review of bitter peptides from caseins: Their formation, isolation and identification, structure masking and inhibition. *Le Lait*, 72(4), 335–385.

Liukkonen, K-H., Katina, K., Wilhelmsson, A., Myllymäki, O., Lampi, A. M., Kariluoto, S., et al. (2003). Process-induced changes on bioactive compounds in whole grain rye. *Proceedings of the Nutrition Society of London*, 62, 117–122.

Lupton, J. R. and, Turner, N. D. (1999). Potential protective mechanisms of wheat bran fiber. *The American Journal of Medicine*, 106(1A), 24S–27S.

Martfnez-Anaya, M. A. (1996). Enzymes and bread flavor. *Journal of Agricultural and Food Chemistry*, 44(9), 2469–2480.

Molteberg, E., Solheim, R., Dimberg, L., & Frølich, W. (1996). Variation in oat groats due to variety, storage and heat treatment. II: Sensory quality. *Journal of Cereal Science*, 24(3), 273–282.

Nishibori, S., Kawakishi, S. (1990). Effects of dough materials on flavor formation in baked cookies. *Journal of Food Science*, 55, 409–412.

Nordlund, E., Heiniö, R.-L., Viljanen, K., Pihlava, J., Lehtinen, P., & Poutanen, K. (2013). Flavour and stability of rye grain fractions in relation to their chemical composition. *Food Research International*, 54(1), 48–56.

Omoba, O. S., Taylor, J. R. N. & de Kock, H. L. (2015). Sensory and nutritive profiles of biscuits from whole grain sorghum and pearl millet plus soya flour with and without sourdough fermentation. *International Journal of Food Science and Technology*, 1–8.

Pszczola, D. E. (2001). From soybeans to spaghetti: The broadening use of enzymes. *Food Technology*, 55 (11), 54–64.

Poutanen, K., Flander, L., & Katina, K. (2009). Sourdough and cereal fermentation in a nutritional perspective. *Food Microbiology*, 26(7), 693–699.

Rashed, M. M., Kieffer, R., Belitz, H. D. (1990). Modification of the functional properties of wheat flour by proteolytic enzymes. Part 2. Activity against substrates in 10w moisture systems, rheological effects, and baking behavior. *Zeitschrift Fur Lebensmittel-Untersuchung Und-Forschung*, 191, 110–115.

Richard-Molard, D. (1994). Le goût du pain. ln La panification Française; Guinet, R., Godon, B.S., Ed.; Collection Sciences et techniques agro-alimentaires; Tec & Doc Editions: Paris, 454–476.

Roland, W. S. U. (2014). Intrinsic bitterness of flavonoids and isoflavonoids and masking their taste activity. (Doctoral dissertation). Wageningen University dissertation [5723].

Sahlström, S., Mosleth, E., Bævre, A. B., Gloria, H. & Fayard, G. (1993). Influence of starch, gluten proteins and extraction rate on bread and pasta quality. *Carbohydrate Polymers*, 21(2), 169–175.

Saltzman, E., Das, S. K., Lichtenstein, A. H., Dallal, G. E., Corrales, A., Schaefer E. J., Greenberg, A. S., and Roberts, S. B. (2001). An oat-containing hypocaloric diet reduces systolic blood pressure and improves lipid profile beyond effects of weight loss in men and women. *Journal of Nutrition*, 131(5), 1465–1470.

Shenkenberg, D. R., Barnes, F. G., Guy, E. J. (1972). New process for sourdough bread improves uniformity and reduces process time. *Food Product Development*, 6 (1), 29–30, 32.

Si, QJ. (1997). Synergistic effect of enzymes for breadbaking. *Cereal Food World*, 42, 802–807.

Sozer, N., Dalgic, A. C., Kaya, A. (2007). Thermal, extural and cooking properties of spaghetti enriched with resistant starch. *Journal of Food Engineering*, 81, 476–484.

Sozer, N., Juvonen, R., Katina, K., & Poutanen, K. (2014). Patent application FI 20145873. Method for modifying brans, and products comprising modified bran. EU Patent and Trademark Office.

Stanyon, P., & Costello, C. (1990). Effects of wheat bran and polydextrose on the sensory characteristics of biscuits. *Cereal Chemistry*, 67(6), 545–547.

Starr, G., Bredie, W., & Hansen, Å. (2012). Sensory profiles of cooked grains from wheat species and varieties. *Journal of Cereal Science*, 57, 295–303.

Stevenson, L., Phillips, F., O'sullivan, K., & Walton, J. (2012). Wheat bran: Its composition and benefits to health, a European perspective. *International Journal of Food Sciences and Nutrition*, 63(8), 1001–1013.

Sudha, M., Vetrimani, R., & Leelavathi, K. (2007). Influence of fibre from different cereals on the rheological characteristics of wheat flour dough and on biscuit quality. *Food Chemistry*, 100(4), 1365–1370.

Tomer, M. J., Martfnez-Anaya, M. A., Antuna, B., Benedito de Barber, C. (1992). Headspace flavour compounds produced by yeasts and lactobacilli during fermentation of preferments and bread doughs. *International Journal of Food Microbiology*, 15, 145–152.

Van Dam, H. W., Hille, J. D. R. (1992). Yeast and enzymes in bread making. *Cereal Food Word*, 37, 245–250.

Wang, J., Sun, B., Cao, Y., & Tian, Y. (2009). Protection of wheat bran feruloyl oligosaccharides against free radical-induced oxidative damage in normal human erythrocytes. *Food and Chemical Toxicology*, 47, 1591–1599.

West, R., Duizer, L. & Seetharaman, K. (2013a). The effect of drying and whole grain content on the pasting, physicochemical and qualitative properties of pasta. *Starch-Stärke*, 65(7–8), 645–652.

West, R., Seetharaman, K. & Duizer, L. M. (2013b). Whole grain macaroni: Flavour interactions with sodium-reduced cheese sauce. *Food Research International*, 53(1), 149–155.

West, R., Seetharaman, K., & Duizer, L.M. (2013c). Effect of drying profile and whole grain content on flavour and texture of pasta. *Journal of Cereal Science*, 58, 82–88.

Zehentbauer, G., & Grosch, W. (1998). Crust aroma of baguettes II. Dependence of the concentrations of key odorants on yeast level and dough processing. *Journal of Cereal Science*, 28(1), 93–96.

6

Encapsulation of Flavoring Compounds in Functional Foods and Dairy Nutraceuticals

Kanchan Kashyap, Kiran Yadav, and Deepak Yadav

Contents

6.1 Introduction

Encapsulation entraps active agents within a carrier material (wall material) and is a useful tool for delivery of functional as well as bioactive molecules (Naknean and Meenune, 2010) in medicine and food products (Dias et al., 2015). Encapsulating materials must be of food grade, biodegradable and must be capable of forming a barrier between the internal phase and its surroundings. Flavoring agents are of enormous interest in food products currently because consumers are attracted to such products with improved taste. Encapsulation of flavoring compounds leads to flavor retention for longer duration (Bakker et al., 1996; Pandey et al., 2010).

Functional foods are those similar in function to conventional foods in appearance and consumed as part of usual diet, but these are modified to offer health benefits (Bech-Larsen and Grunert, 2003). They may reduce the risks of diseases beyond the stipulation of normal nutrient requirements (Siro et al., 2008). They may also include the food products that are genetically customized, food ingredients with health promoting additives such as vitamin-enriched products (Chawda et al., 2017). Functional foods are the pyramids of the nutrition (Milner, 2000) because they may help prevent cancer, diabetes, hypertension and/or other diseases (Bech-Larsen and Grunert, 2003). Functional dairy products are healthy food products that include probiotics, foods enriched with vitamins and minerals (Chawda et al., 2017).

Basically, functional foods are rich in soluble fiber such as oat bran—psyllium, which reduces the incidence of coronary heart disease. Other examples of functional food components include soya proteins and sterols which reduce body cholesterol and thus reduce the chances of coronary heart disease (NDRI, 2015). Omega-3 is an essential nutrient that cannot be produced by the human body but needs to be taken from dietary sources. Omega 3 fatty acids are alpha linolenic acid obtained from the flax oil, which are added to milk in the dairy industry. Its consumption has increased from last few years in the form of omega 3 milks, cheese, yogurts, etc., and it generally promotes heart health. Conjugated linoleic acid (CLA) is present in the dairy products. So these dairy products provide essential fatty acid (CLA) health benefits to consumers in addition to that of the natural dairy products. CLA-containing products are tremendously emerging as nutraceutical and functional foods.

Probiotics are small micro organisms or bacteria that can help keep the natural balance of organisms (microflora) in the gut. The largest group of probiotics is lactic acid bacteria—*Lactobacillus acidophilus* found in yogurt. It offers various health remuneration including combating intestinal infections. These probiotic cultures are delivered through dairy products and have been marketed with different flavors (Yadav et al., 2018). It may help prevent food-borne infections. Curd is also a dairy nutraceutical that act as a probiotic functional food. It has preventive and therapeutic effect on pathologies such as acute diarrhea, antibiotic associated diarrhea and allergy pathology (Helkar et al., 2016).

Whey is a by-product of dairy nutraceuticals that contains various commercial constituents like beta lactoglobulin, immunoglobulin, lactoferrin and lacto peroxidase. They are widely used as food ingredients in the dry form. They have anticarcinogenic, antimicrobial and immune modulatory properties (Helkar et al., 2016). Composition of milk determines its nutritive quality. It contains vitamins, minerals, proteins, water, carbohydrates (lactose) and fats. Sweeteners are added as flavoring agent not only for the reason that it imparts sweetness but it also acts as a preservative. It also aids in the fermentation process. Sweeteners that supply energy are called nutritive sweeteners and provide 4 calories per gram of consumption. Sweeteners that do not provide calories are known as non-nutritive sweeteners. There are various artificial sweeteners that may be nutritive or non-nutritive. These are used in functional foods and dairy nutraceuticals e.g. neotame, saccharin, aspartame and so on (NDRI, 2015).

6.2 General principles for use of flavoring agents

The purity of flavoring compounds must be high. Impurities should not be present in final food at levels that would pose an unacceptable risk to the health. Use of flavoring compounds in food should not lead to unsafe levels. No health hazard should arise from any amount of flavoring compounds and are justified only where they impart or modify the flavor to food. Flavoring compounds may contain nonflavoring food ingredients, including food additives necessary for their storage, handling and use. These ingredients may be useful for dilution, dissolution or dispersion of flavoring agents in food. Nonflavoring food ingredients should be limited to the lowest level required to ensure the safety and quality of the flavoring agents.

Mostly flavoring agents used in liquid foods are volatile, and these are chemically unstable in the presence of air, light, moisture, temperature and so on. But currently, encapsulation has become an attractive approach to transform these liquid food flavorings into stable and free-flowing powders flavor that improves their handling and assimilation into the dry food system (Bhandari et al., 1992).

The coating material must be capable of forming a film that is cohesive with the core material and should be physically and chemically compatible. Many terms can be given to the wall material including carrier, coating membrane, shell and so on. Semi-synthetic cellulose derivatives, proteins, resins or polysaccharides can be used in the coating of flavoring compounds in the functional foods. These materials provide sufficient strength, flexibility, impermeability as well as stability to flavoring agents.

Coating materials should have adequate properties like cohesiveness, permeability, solubility, stability and clarity (**Table 6.1**). These properties must be considered in selection of proper encapsulating or coating material. The coating material must be inert toward the active ingredient, other ingredients, and should provide stabilization of the core material. It must be non-hygroscopic, less viscous and economical. Under specific conditions, it should provide the ability of controlled release of flavoring agents and desired degree of protection. It must be of food grade and legally approved (Gharsallaoui et al., 2007; Nazzaro et al., 2012). It should hold the core material within its structure during the process of manufacturing and storage.

Table 6.1 Classification of Coating Materials Used for Encapsulation of Flavoring Compounds

S. No.	Water Soluble Resins	Water Insoluble Resins	Enteric Resins	Waxes/Lipids
1	Gelatin	Silicones	Shellac	Paraffin
2	Gum Arabic	Poly lactide-co-glycolide	Cellulose acetate phthalate	Carnauba
3	Starch	Nylon (polyamide)	Zein	Beeswax
4	Hydroxylpropyl methyl cellulose	Cellulose nitrate		Stearic acid
5	Polyvinylpyrollidone	Polyethylene		Stearyl alcohol
6	Carboxymethyl cellulose	Polymethacrylate		Glyceryl stearates
7	Polyvinyl alcohol	Polyethylene vinyl acetate		

6.3 Encapsulation of flavoring compounds

Encapsulation of flavoring agents provides protection from interaction of sensitive ingredients and protection from moisture. During processing ingredients can be protected from the effects of temperature and pH. Flavor losses can be prevented through encapsulation. It can speed up or limit the release of flavors (Meiners, 2012) and can also mask the odor. Encapsulation of flavoring compounds separates the incompatible flavor constituents to avoid adverse reactions. Stabilization of citrus flavors for long shelf-life performance can be provided through encapsulation (Makinen-Aakula, 2006). It can modify the physical characteristics of the material to allow easier handling as well as for the separation of the components of the mixture so that they would not react with each other (Nedovic et al., 2011). Encapsulation of flavoring agents offers other advantages, such as prevention from hygroscopicity, burst release and/or premature leaching.

6.4 Current encapsulation techniques employed

Encapsulation of active compounds into coating materials can be achieved by several methods. There are various methods available for encapsulation of flavoring compounds including freeze-drying, super-critical fluid-based processes, spray chilling, extrusion, co-crystallization, coacervation and so on. Every method has its own advantages and limitations in ways of encapsulation efficiency, degree of protection to flavoring compounds, biodegradability and biocompatibility. They provide long-term health remuneration beyond the nutritional requirements. The type of encapsulation technique also influences stability, bioavailability of active ingredients present inside the core. The selection of the process is based on the physical and chemical properties of the core and the coating materials. There are many microencapsulation processes available. Some techniques are based on the physical phenomenon,

Table 6.2 Different Methods Used for Encapsulation

Physical Methods	Particle Size (µm)	Physiochemical Methods	Particle Size (µm)	Chemical Methods	Particle Size (µm)
Melt extrusion	300–5000	Spray cooling	50–600	Polymerization	80–200
Spray cooling	20–200	Solvent evaporation	5–5000	Interfacial polymerization	0.5–1000
Fluidized bed coating	5–5000	Complex Coacervation	10–800	In situ polymerization	0.5–1100
Freeze drying	20–5000	Emulsification	0.2–5000	Matrix polymerization	10–800
Spray drying	10–400				

Source: Zuidam, N.J. and Shimoni, E., *Encapsulation Technologies for Active Food Ingredients and Food Processing*, Springer, New York, pp. 3–29, 2010.

and some are based on both physical and chemical phenomenon (Desai and Jin Park, 2005). **Table 6.2** enlists some of the physical and chemical methods currently being employed for encapsulation of flavoring compounds.

Different types of microcapsules have been classified based on their morphology (**Figure 6.1**). Mononuclear consists of a continuous core region and a continuous shell with spherical geometry.

Polynuclear contains number of particles or small droplets with irregular geometry. In matrix, the entrapped substance (core material) is dispersed homogeneously in the shell material. In addition to these three basic morphologies, a microcapsule can also be mononuclear with multiple shell or they may form cluster. A number of techniques have been used commercially for encapsulation of flavoring compounds (**Table 6.3**) which are briefly discussed in the following sections.

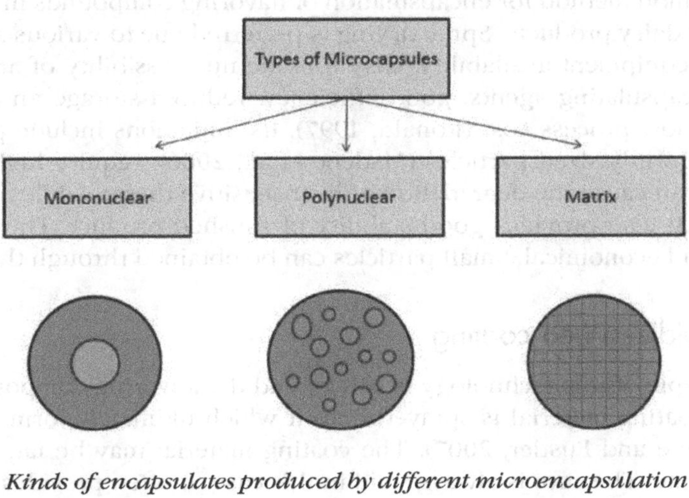

Figure 6.1 *Kinds of encapsulates produced by different microencapsulation techniques.*

Table 6.3 Various Technologies Used Commercially for Encapsulation of Flavoring Compounds

S. No.	Technology	Matrix/Core	Particle Size	Cost	Production	Use
1	Melt extrusion	Matrix	50–100	Low	Continuous	++
2	Spray drying	Matrix	2–100	Low	Continuous	+++
3	Fluidized bed coating	Core	>50	Medium	Batch/continuous	++
4	Spray cooling	Core	20–200	Low	Batch	++
5	Extrusion	Matrix	>50	Low	Continuous	++
6	Coacervation	Core	1–500	High	Batch	++
7	Lyophilization	Core	1–500	High	Batch	++

Source: Papen-Botterhuis, N., *Coating Technology and Encapsulation–Closely Connected*, 2014.

6.4.1 Melt extrusion

It is a process that involves the entrapment of the core material in a molten shell formulation at temperatures of 85°C–125°C. It can encapsulate flavoring compounds with or without the aid of surfactant. The materials are extruded as filaments into a relatively cooled environment that solidifies the extruded mass. Gas phase can be used as receiving environment or a tank containing solvent, which can also act as same. Flavor-loaded particles produced in this manner have an excellent ability to prevent oxidation during storage (Desai and Jin Park, 2005).

6.4.2 Spray drying

Spray drying involves using an emulsion, solution or suspension of core material and wall material for nebulization in a drying chamber with circulating hot air. Water evaporates when it comes in the contact of hot air and the material encapsulates the core (Laohasongkram et al., 2011). Spray drying is the most common method for encapsulation of flavoring compounds in functional foods and dairy products. Spray drying is preferred due to various advantages like easy equipment availability, easy to scale up, possibility of adding variety of encapsulating agents, good efficiency, reduced storage and transport costs and low process cost (Ronald, 1997). Its limitations include production of nonuniformly sized particles (Madene et al., 2006), require high temperature that can cause the degradation of heat-sensitive thermolabile and volatile products. It also provides good stability of finished product. The process is flexible and economical. Small particles can be obtained through this process.

6.4.3 Fluidized bed coating

The principle of this technology is to suspend the flavoring compounds in air and the coating material is sprayed onto it which ultimately forms a capsule (Champagne and Fustier, 2007). The coating material may be fat, protein or carbohydrate. The core is always solid in this case. It often provides advantage of providing an additional layer of molecules for the purpose of controlled release of the flavoring compounds (de Vos et al., 2010).

The process offers advantages like tuneable shell thickness, suitable and compatible with wide range of coating materials. It can be used for much larger food ingredients. It is useful in pharmaceuticals, nutraceuticals and cosmeceuticals. Disadvantages include that it can be used only for solid-core materials and generally not suitable for the particles less then 50mm. This technique provides moderate payloads, especially for small particles. Also chances of agglomeration may be there.

6.4.4 Spray cooling

It involves nebulization process in which the mixture is nebulized by an atomizer; the mixture then enters in a chamber of low temperature air flow. So there is encapsulation of the core material inside the coating material from the solidification of the wall or encapsulating material due to the low temperature (Madene et al., 2006). It is considered as the cheapest method for encapsulation with low temperature treatment with high potential for scale up. However, there is some disadvantage of this technique in that it has low encapsulation capacity, and there can be expulsion of the core material during the storage. This technique can also be used for the encapsulation of vitamins and minerals apart from flavoring agents (Silva et al., 2014).

6.4.5 Extrusion

In this method, the core material is immobilized within a polysaccharide gel. The core material flavoring compound is incorporated into sodium alginate solution. This mixture is drop-wise extruded via a pipette or syringe into a hardening solution (Swarbrick et al., 1978). It provides a barrier against oxygen, hence provides very long shelf life of the flavoring compounds. The major drawback of this technology is that due to the extrusion, usually large particles are formed, which limits this application where good mouth feel is desired (Gouin, 2004). Microencapsulated *Lactobacillus acidophilus* in calcium alginate gel made by the extrusion method increases its shelf life up to 6 months of storage (Mirzaei et al., 2012).

6.4.6 Coacervation

Coacervation is a technique in which the polymer material is deposited around the core by altering the physicochemical characteristics of the medium such as pH or ionic strength (Gonçalves et al., 2016). The process is called simple coacervation where only a single macromolecule/polymer is present. When two or more macromolecules/polymers of opposite charges are present, it is called complex coacervation (Oliveira et al., 2007). Coacervation is a low-cost process, and there is no requirement of organic solvent or high temperature. This method is used to encapsulate flavor oils. Coacervation occurs only within a limited range of pH, colloidal concentration or electrolyte concentration, which is a major disadvantage of this technique (Comunian et al., 2013). Sweet orange oil has been microencapsulated in soybean protein using the coacervation technique, which provides protection for the core material. Coacervation provides good core shell morphology. High payloads are possible. Sustained release can be obtained. Flavors, fish oils, vitamins can be encapsulated. Disadvantages include evaporation of volatile flavoring compounds and instability of complex coacervates. It is a complex batch process providing polydispersed capsules. Limited choice of shell materials is available for this method.

6.4.7 Lyophilization

Lyophilization is a method in which the frozen material is dehydrated under a vacuum sublimation process (Chen and Wang, 2007). It is a low-temperature treatment that provides excellent-quality products because it minimizes the changes associated with a high temperature. Lyophilization has long processing times and is a costly process. It is beneficial for heat-sensitive flavoring compounds (Augustin and Hemar, 2009) and also increases their shelf life (da Silva et al., 2014).

Flavors should be encapsulated in such a manner so that they can be confined for the duration of food processing and released during consumption. All the food ingredients should be protected from deterioration like oxidation, hydrolysis and so on.

6.5 Characterization of flavoring compounds

There are various methods to detect the flavoring compounds in dairy nutraceuticals such as liquid/liquid extraction, solid phase detection or distillation techniques with the drawback of the organic solvent use. Headspace purge-and-trap methods do not use organic solvents, but the range of their analyte is restricted to volatile components so it donates aroma or smell but not taste. In addition to all these methods, the heating of a sample must be avoided since it would lead to reaction in products, which can change the flavor or taste of the functional food (Hoffmann and Heiden, 2000).

6.6 Mechanism of release

It may involve external pressure, which breaks the microcapsule wall and releases the liquid from the core. The microcapsule wall breaks because of inner pressure in some cases. The mechanism of release can also be dissolution based. The encapsulated substance may also be released by abrasion of the wall of microcapsule. Heat can also enhance the release of core material. If the wall of the core material is permeable then it slowly releases the content of the core. The mechanism can be a controlled release, so the shell may release the flavoring compounds by various mechanisms like diffusion, melting, swelling or degradation by pressure/stress.

Diffusion: It is the process of movement of solute from higher concentration to lower concentration which is controlled by the solubility and permeability of compound in the matrix (Crank, 1975). The vapor pressure of a volatile substance in the matrix is a driving force that influences the diffusion (Dumay et al., 1999). It is called burst release if the release of flavor is very rapid. It can be observed initially.

Melting: It involves the melting of coating material to release the core floating compounds. In such cases, the functional food is stored at temperatures well below the melting point of coating material (F. Gibbs, 1999; Sparks et al., 1995).

Swelling: In some cases flavor is dissolved and dispersed in a polymeric matrix and unable to diffuse up to a significant extent. When the matrix comes in contact with the medium in which it dissolves, the polymer swells by absorbing liquid from the medium, and the flavoring compounds slowly diffuse out (F. Gibbs, 1999).

Degradation: The release of the core material may be controlled by diffusion, erosion of matrix or both. When the degradation is confined to a thin layer at surface, it is called heterogeneous erosion where as in homogeneous erosion the degradation occurs at uniform rate through the polymeric matrix (Pothakamury and Barbosa-Cánovas, 1995).

6.7 Food and flavor interactions

Flavor-food matrix interactions in food products need to be investigated because they can influence the flavor release from the food products. The components of the food matrix can bind or entrap the flavoring compounds, which can be volatile or nonvolatile in nature if the binding sites of food components are still available. If there is interaction between a food component and flavor, the rate of flavor release is reduced. It may also influence flavor intensity, quality of food and consumer overall acceptance (Naknean and Meenune, 2010).

The physicochemical properties of the flavor compounds and food components may predict this type of interaction. With time and processing conditions, flavor also changes. There can be four different types of interactions of flavoring compounds and food matrices (**Figure 6.2**). Covalent bonding, which is irreversible binding, e.g. interaction between aldehyde or ketone groups with amino group of proteins. Hydrogen bonding may occur between polar alcohol of flavor compounds and hetero atom of food components for example nitrogen, sulfur or oxygen. Hydrophobic interactions include weak and reversible binding such as van der Waals forces between apolar

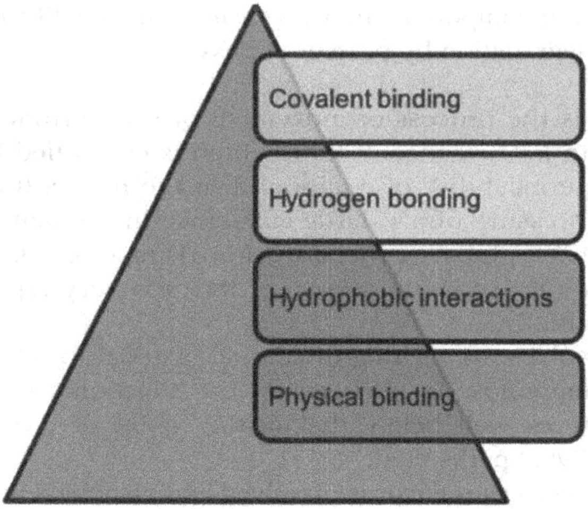

Figure 6.2 *Types of interactions between flavors and food materials.*

compounds and fat molecules. Physical binding may include formation of inclusion complexes. Such binding includes adsorption, absorption, inclusion and retention of flavor compounds into nonvolatile substrate. Afterwards partitioning occurs, which involves distribution of flavor among oil, water and air. The last step is the release of the flavor ingredient, which means the availability of the flavor compound from the bulk food to air.

6.8 Conclusion

Today functional foods and nutraceuticals comprise one of the most promising and dynamically developing segments of food industry. The demand for new functional foods is steadily increasing because of their disease-curing properties. Using different available industrial techniques various encapsulates for flavoring compounds have been designed to incorporate them into functional food matrices to provide them desired aesthetic and flavor appeal. The composition and structure of the encapsulated flavors can be carefully tailored to ensure that the complete functional benefits of the nutraceuticals can be obtained. The final developed products must be tested completely before releasing them to the market, to establish desired physicochemical stability, dispersibility, bioavailability, and desired release profile of the encapsulated flavors.

References

Augustin, M.A., Hemar, Y., 2009. Nano-and micro-structured assemblies for encapsulation of food ingredients. *Chemical Society Reviews* 38, 902–912.

Bakker, J., Brown, W., Hills, B., Boudaud, N., Wilson, C., Harrison, M., 1996. Effect of the food matrix on flavour release and perception. *Special Publications of the Royal Society of Chemistry* 197, 369–374.

Bech-Larsen, T., Grunert, K.G., 2003. The perceived healthiness of functional foods: A conjoint study of Danish, Finnish and American consumers' perception of functional foods. *Appetite* 40, 9–14.
Bhandari, B., Dumoulin, E., Richard, H., Noleau, I., Lebert, A., 1992. Flavor encapsulation by spray drying: Application to citral and linalyl acetate. *Journal of Food Science* 57, 217–221.
Champagne, C.P., Fustier, P., 2007. Microencapsulation for the improved delivery of bioactive compounds into foods. *Current Opinion in Biotechnology* 18, 184–190.
Chawda, P.J., Shi, J., Xue, S., Young Quek, S., 2017. Co-encapsulation of bioactives for food applications. *Food Quality and Safety* 1, 302–309.
Chen, G., Wang, W., 2007. Role of freeze drying in nanotechnology. *Drying Technology* 25, 29–35.
Comunian, T.A., Thomazini, M., Alves, A.J.G., de Matos Junior, F.E., de Carvalho Balieiro, J.C., Favaro-Trindade, C.S., 2013. Microencapsulation of ascorbic acid by complex coacervation: Protection and controlled release. *Food Research International* 52, 373–379.
Crank, J., 1975. Diffusion in a cylinder. *The Mathematics of Diffusion* 69–88.
da Silva, P.T., Fries, L.L.M., de Menezes, C.R., Holkem, A.T., Schwan, C.L., Wigmann, É.F., de Oliveira Bastos, J., da Silva, C.d. B., 2014. Microencapsulação: conceitos, mecanismos, métodos e algumas aplicações em tecnologia de alimentos. *Ciência Rural* 44, 1304–1311.
de Vos, P., Faas, M.M., Spasojevic, M., Sikkema, J., 2010. Encapsulation for preservation of functionality and targeted delivery of bioactive food components. *International Dairy Journal* 20, 292–302.
Desai, K.G.H., Jin Park, H., 2005. Recent developments in microencapsulation of food ingredients. *Drying Technology* 23, 1361–1394.
Dias, M.I., Ferreira, I.C., Barreiro, M.F., 2015. Microencapsulation of bioactives for food applications. *Food & Function* 6, 1035–1052.
Dumay, E., Laligant, A., Zasypkin, D., Cheftel, J., 1999. Pressure-and heat-induced gelation of mixed β-lactoglobulin/polysaccharide solutions: Scanning electron microscopy of gels. *Food Hydrocolloids* 13, 339–351.
Gibbs, B.F., Kermasha, S., Alli, I., Mulligan, C.N., 1999. Encapsulation in the food industry: A review. *International Journal of Food Sciences and Nutrition* 50, 213–224.
Gharsallaoui, A., Roudaut, G., Chambin, O., Voilley, A., Saurel, R., 2007. Applications of spray-drying in microencapsulation of food ingredients: An overview. *Food Research International* 40, 1107–1121.
Gonçalves, A., Estevinho, B.N., Rocha, F., 2016. Microencapsulation of vitamin A: A review. *Trends in Food Science & Technology* 51, 76–87.
Gouin, S., 2004. Microencapsulation: Industrial appraisal of existing technologies and trends. *Trends in Food Science & Technology* 15, 330–347.
Helkar, P., Sahoo, A., Patil, N., 2016. Review: Food industry by-products used as a functional food ingredients. *International Journal of Waste Resources* 6, 1–6.
Hoffmann, A., Heiden, A., 2000. Determination of flavor and off flavor compounds in dairy products using stir bar sorptive extraction (SBSE) and thermal desorption GC/MSD/PFPD, *SANDRA, P. AND RACKSTRAW, AJ Proceeding of the 23rd International Symposium on Capillary Chromatography*, pp. 5–10.
Laohasongkram, K., Mahamaktudsanee, T., Chaiwanichsiri, S., 2011. Microencapsulation of Macadamia oil by spray drying. *Procedia Food Science* 1, 1660–1665.
Madene, A., Jacquot, M., Scher, J., Desobry, S., 2006. Flavour encapsulation and controlled release—A review. *International Journal of Food Science & Technology* 41, 1–21.
Makinen-Aakula, M., 2006. Trends in functional foods dairy market. *Proceedings of the Third Functional Food Net Meeting.*
Meiners, J., 2012. Fluid bed microencapsulation and other coating methods for food ingredient and nutraceutical bioactive compounds, *Encapsulation Technologies and Delivery Systems for Food Ingredients and Nutraceuticals.* Amsterdam, the Netherlands, Elsevier, pp. 151–176.

Milner, J.A., 2000. Functional foods: The US perspective. *The American Journal of Clinical Nutrition* 71, 1654S–1659S.
Mirzaei, H., Pourjafar, H., Homayouni, A., 2012. Effect of calcium alginate and resistant starch microencapsulation on the survival rate of Lactobacillus acidophilus La5 and sensory properties in Iranian white brined cheese. *Food Chemistry* 132, 1966–1970.
Naknean, P., Meenune, M., 2010. Factors affecting retention and release of flavour compounds in food carbohydrates. *International Food Research Journal* 17, e34.
National Dairy Research Institute, 2015. *Vision 2050*, Indian Council of Agricultural Research, New Delhi.
Nazzaro, F., Orlando, P., Fratianni, F., Coppola, R., 2012. Microencapsulation in food science and biotechnology. *Current Opinion in Biotechnology* 23, 182–186.
Nedovic, V., Kalusevic, A., Manojlovic, V., Levic, S., Bugarski, B., 2011. An overview of encapsulation technologies for food applications. *Procedia Food Science* 1, 1806–1815.
Oliveira, A., Moretti, T., Boschini, C., Baliero, J., Freitas, O., Favaro-Trindade, C., 2007. Stability of microencapsulated B. lactis (BI 01) and L. acidophilus (LAC 4) by complex coacervation followed by spray drying. *Journal of Microencapsulation* 24, 685–693.
Pandey, M., Verma, R.K., Saraf, S.A., 2010. Nutraceuticals: New era of medicine and health. *Asian Journal of Pharmaceutical and Clinical Research* 3, 11–15.
Papen-Botterhuis, N., 2014. Coating Technology and Encapsulation–Closely Connected.
Pothakamury, U.R., Barbosa-Cánovas, G.V., 1995. Fundamental aspects of controlled release in foods. *Trends in Food Science & Technology* 6, 397–406.
Ronald, C., 1997. Spray drying innovative use of an old process. *Design Elements* 7, 97–113.
Silva, P.T.D., Fries, L.L.M., Menezes, C.R.D., Holkem, A.T., Schwan, C.L., Wigmann, É.F., Bastos, J.D.O., Silva, C.D.B.D., 2014. Microencapsulation: Concepts, mechanisms, methods and some applications in food technology. *Ciência Rural* 44, 1304–1311.
Siro, I., Kapolna, E., Kapolna, B., Lugasi, A., 2008. Functional food. Product development, marketing and consumer acceptance—A review. *Appetite* 51, 456–467.
Sparks, R., Jacobs, I., Mason, N., 1995. Centrifugal suspension—Separation for coating food ingredients. In: *Encapsulation and Controlled Release of Food Ingredients*, ACS Publications. doi:10.1021/bk-1995-0590.ch008.
Swarbrick, P., Green, W.J., Maillefer, C., 1978. Manufacture of extruded products. Google Patents, US4117195.
Yadav, D., Srikanth, K., Yadav, K., 2018. *Flavors in Probiotics and Prebiotics, Flavors for Nutraceutical and Functional Foods*. Boca Raton, FL, CRC Press, pp. 71–94.
Zuidam, N.J., Shimoni, E., 2010. Overview of microencapsulates for use in food products or processes and methods to make them, in: Zuidam, N.J., Nedovic, V. (Eds.), *Encapsulation Technologies for Active Food Ingredients and Food Processing*. New York, Springer, pp. 3–29.

7

Natural Flavors in Various Nutraceutical Product Applications

Komal Parmar and Jayvadan Patel

Contents

7.1 Introduction

Among the components that are frequently thought to be significant to the consumer in his/her utilization of healthy supplements are the taste of the supplement and the health benefit associated with it. However, taste remains the fundamental component of oral products including food, beverages and nutritive supplements. Flavor is comprehended as an amalgamation of taste, smell, and chemical sensibility of the skin and mucus membranes, such as raciness. Both volatile and non-volatile food components can actuate the real food recognition process (**Figure 7.1**). Various factors affect perception of taste by sensory organs of living beings (Ulla et al. 2016). It is found that consumers have complex and sophisticated choices about what they consume orally. Many parameters add up in the criteria of selection, for instance flavor, color, other additives, exterior look, brand, perception, mood, end cost, serving size, accessibility, nutritional values and others. Regardless of how sound and critical the nutrition product is, it doesn't achieve its objective if the individuals stay away from it. Gone are the days where consumers associated poor taste as an indicator of health benefits to be obtained. Now consumers' believe that healthy eating must be associated with acceptable taste.

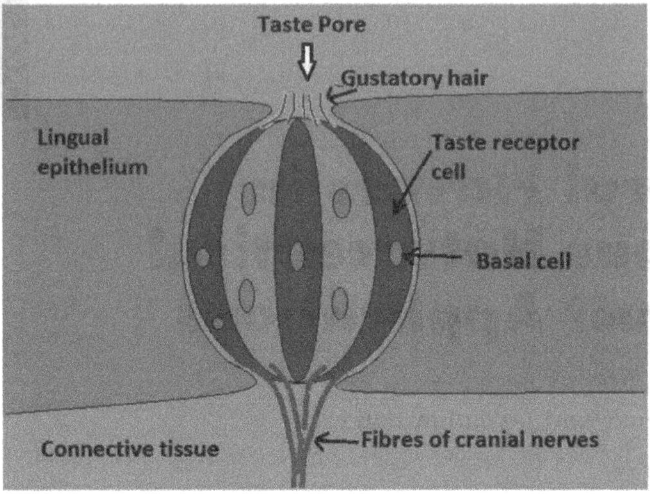

Figure 7.1 *Anatomical features of sensory perception for taste.*

In addition, dietary flavor makes an important factor for quality assessment of oral supplements. Besides it is found that many nutritionally important compounds have bitter, astringent and other off flavors, which may outdistance the consumers (Drewnowski and Gomez-Carneros 2000; LeClair 2000). In one study of Tuorila and Cardello, it was found that disagreeable taste in a juice refreshment had a hindering impact on coveted utilization of the item, regardless of whether the product was exhibited as one that enhances "physical stamina and vitality," "mental readiness and memory power" or "mind-set and emotional prosperity" (Tuorila and Cardello 2002). Further consumers live in their very own tangible universes, which may result in their day by day practices as individualized nutrient orientations. Thus, nutraceuticals need to be formulated in with greater importance to sensory standards, perchance even to the grade given to mainstream foodstuffs. Therefore, companies are constantly researching in developing oral products with satisfactory and appealing flavors.

This chapter focuses on the association of flavors and nutraceuticals. It reviews and takes into account the natural flavors and their applications in nutraceutical products.

7.2 Natural flavors

Consumers bear the notion that though there is no functional difference between natural and artificial flavors; natural flavors are healthier than artificial flavors. Thus across all product categories, consumers and manufacturers are trending toward developing products made up of natural flavors. According to the United States Food and Drug Administration (CFR Title 21): "The term

Table 7.1 Natural Flavor Classification

Principle Source of Flavor	Examples
Fruits	Orange, strawberry, raspberry
Vegetables	Carrot, tomato
Plants and botanicals	Herbs, blossoms, tea
Spices	Cinnamon, garlic, pepper, ginger

natural flavor or natural flavoring means the essential oil, oleoresin, essence or extractive, protein hydrolysate, distillate, or any product of roasting, heating or enzymolysis, which contains the flavoring constituents derived from a spice, fruit or fruit juice, vegetable or vegetable juice, edible yeast, herb, bark, bud, root, leaf or similar plant material, meat, seafood, poultry, eggs, dairy products, or fermentation products thereof, whose significant function in food is flavoring rather than nutritional."

UK Food Law defines a natural flavor as a flavoring substance (or flavoring substances) which is (or are) obtained, by physical, enzymatic, or microbiological processes, from material of vegetable or animal origin which material is either raw or has been subjected to a process normally used in preparing food for human consumption and to no process other than one normally so used.

Many natural flavoring substances are obtained from plant based material after processing to extract their aromatic components. Plant material contains both volatile and non-volatile components that affect sensory organs by providing distinct odor and by influencing gustatory reaction. Flavors can be classified on the basis of their principle natural source as shown in **Table 7.1**.

Fruit flavors are the most acceptable among all flavors that are used in nutraceuticals as well as other therapeutic products. Many industries are manufacturing fruit extracts as fruit flavors to be used in various products, for example allspice berry flavor extract, almond cola flavor extract.

7.3 Nutraceuticals

Functional food refers to natural or processed foods containing components that provide health benefits beyond primary nutrition. Nutraceutical refers to a food, dietary supplement, or biologically active component that provides health benefits. Other than health benefits consumers' may also utilize nutraceuticals to delay aging, prevent chronic disease, or support the body functions. Majorly nutraceutical products are consumed in public without a medical prescription and/or medical supervision. The effect of consumers trust in the development and survival of the nutraceutical industry is, consequently, immediate and huge. Consumers are gaining more awareness regarding consumption of nutritional components which has brought about expanded interest for nutraceuticals globally. The European market is researching on high-performance, natural ingredients used in nutraceuticals. According to new report by Grand View Research, Inc. (2017) global market of nutraceuticals is going to reach USD 578.23 billion by 2025. Increase in awareness of consumer and rise in e-commerce are going to be major contributors to increase the demand of nutraceuticals over the estimated period. The market has seen increased centralization on the ingredients used in manufacturing of nutraceuticals. Consumer finds natural products more beneficial than synthetic products and thus demand is expected to grow apace. Few of the prevalent nutraceuticals incorporate ginseng, green tea, omega-3 fatty acid and folic acid.

7.4 Flavors in nutraceuticals

In the present nutraceutical industry, flavor is never again a bit of second thought. While nutrition is the best need, consumers won't forfeit on taste. Now in dietary supplements industry, unique flavor represents the brand. Flavor remains a main thrust inside the nutraceuticals classification stretching out from useful dietary foods, beverages and refreshments to candies, gummies, powders and so on. In nutraceutical formulations, flavor plays an important role as it helps to mask the unacceptable flavor associated with the dietary supplements and thereby enhances the product acceptance by the consumer.

A conjoint analytical study was carried out by Cardello and Schutz (2003) with military and civilian consumers to evaluate the significance of taste to the proposed use of nutraceuticals. Results revealed among other product characteristics, taste was found to be the most important factor in the consumption of nutraceuticals.

7.5 Application of natural flavors in nutraceuticals

Few botanical constituents have fundamentally pleasant flavors and can upgrade the essence and usefulness of the product. For instance Bigelow® is a tea blender company in United States founded by R.C. Bigelow in 1945; they produce wide variety of herbal teas with real fruit juice added as a flavoring ingredient. List of natural flavors used includes apple, lemon grass, blueberry, cranberry, mango, orange, peach, pomegranate, raspberry, caramel, chamomile, chocolate, cinnamon, ginger, hibiscus, lemon, mint, rose hips, pumpkin, turmeric and vanilla.

Align Prebiotic plus Probiotic gummies supplement, a digestive supplement of Procter and Gamble Company, is made with natural fruit flavors. These probiotics add more good bacteria into the digestive system whereas prebiotics help nourish the good bacteria that live in the digestive system.

Immune support gummies, a health supplement comprising of vitamin C, zinc, and *Echinacea*, is a product of Sundown Naturals, USA. The nutraceutical gummy is available with natural flavors of cran-raspberry and black cherry.

Elite Omega-3 is a pure fish oil supplement in capsule dosage form by Qualia Nutraceuticals. The supplement capsule is available in natural lemon flavor.

Thayers Dry Mouth Spray by Mellen Nutraceuticals, USA, is a natural solution that aids in the temporary relief of dry mouth (Xerostomia) caused by medications, Sjogren's syndrome, lupus, stress, AIDS, radiation therapy, mouth breathing, or aging. The spray comes with natural peppermint flavor.

Vanilla is the most commonly used natural flavor in nutraceuticals and other edible products. It is present in the orchids of genus vanilla, primarily Mexican species, vanilla planifolia. Natural vanilla is used as flavor in many nutraceuticals, few names to mention includes: EsophaSoothe™ of NutraBiogenesis to present nutritive support for healthy functioning of gastric and esophageal mucosa, The Original, PH50 Protein Greens Advanced by Doctors Nutra Nutraceuticals, Whey Isolate Protein Powder by BioChem, Metabolic Cleanse a Nutritional detoxification support by Douglas Laboratories, UltraClear a liver detoxification support by Metagenics and Animal Parade Calcium children's chewable tablets.

Natural Nutra Methyl-B12 Lozenges by Natural Nutraceuticals, USA, is a dietary supplement that supports cardiovascular and brain health contains natural berry flavor.

Enhanced sports drink by Nth Degree, a nutraceutical company, come with various natural flavors including fruit punch, orange, raspberry and lemon citrus.

Infinity, a 100% all-natural pre-workout powder with premium clinically dosed ingredients by RARI Nutrition's, USA, contains natural strawberry lemonade.

ABC natural contains amino acids, beta alanine and citrulline malate by CORE Nutritionals. The dietary supplement contains cherry lemonade.

Rx Select Nutraceuticals Beets, The Circulation Superfood Remolacha Polvo Powder contains natural flavors. The supplement promotes energy and endurance that supports cardiovascular health. It activates essential nitric oxide.

7.6 Future perspective

Among all natural flavors, vanilla, chocolate, coffee, orange, lemon, lime, strawberry, apple, raspberry, etc., will lead the other flavors at all time. Many companies will intermingle these flavors with something more exotic flavors like mango or passion fruit. Overall consumers are becoming more and more curious to try new and different flavors especially natural flavors for the sake of health and have less complexity in terms of chemical additives in the nutraceutical products. Consumers are more exploratory than ever in recent times, and they desire distinctive tastes. A significant boost in botanical-inspired flavors will be seen in almost all nutraceutical products in coming years.

References

ABC natural containing amino acids, beta alanine, citrulline malate by CORE Nutritionals, https://www.corenutritionals.com/products/core-abc. Accessed on 1 November 2018.

Align Prebiotic plus Probiotic gummies supplement, Procter and Gamble, https://www.aligngi.com/en-us/shop-probiotics/align-prebiotic-probiotic-gummies-supplement. Accessed on 5 November 2018.

Animal Parade Calcium children chewable tablets by Naturesplus, https://naturesplus.com/products/productdetail.php?productNumber=29996. Accessed on 10 November 2018.

Bigelow Tea, USA, https://www.bigelowtea.com/. Accessed on 12 November 2018.

Cardello AV, Schutz HG. The importance of taste and other product factors to consumer interest in nutraceutical products: Civilian and military comparisons. *J Food Sci.* 2003; 68(4): 1519–1524.

Drewnowski A, Gomez-Carneros C. Bitter taste, phytonutrients, and the consumer: A review. *Am J Clin Nutr.* 2000; 72(6): 1424–1435.

EsophaSoothe™ of NutraBiogenesis, https://nutrabiogenesis.com/esophasoothe.html.

Immune support gummies supplement, Sundown Naturals, USA, https://www.sundownnaturals.com/products/immunesupportgummies/. Accessed on 15 November 2018.

Infinity, a 100% all natural pre-workout powder by RARI Nutrition's, https://www.rarinutrition.com/products/infinity. Accessed on 16 November 2018.

LeClair K. Breaking the sensory barrier for functional foods. *Food Prod Design.* 2000; 7: 59–63.

Metabolic Cleanse a Nutritional detoxification support by Douglas Laboratories, https://www.douglaslabs.com/metabolic-cleanse-trade-102600.html. Accessed on 16 November 2018.

Natural Nutra Methyl-B12 Lozenges by Natural Nutraceuticals, https://www.natnutra.com/products/b-12-vitamin. Accessed on 26 November 2018.

Nutraceuticals Market Analysis by Product (Dietary Supplements, Functional Food, Functional Beverage), by Region (North America, Asia Pacific, Europe, CSA, MEA), and Segment Forecasts, 2018–2025. Grand View Research, December 2017. Report ID: GVR-2-68038-059-0.

Thayers Dry Mouth Spray by Mellen Nutraceuticals, http://www.mellennutraceuticals.com/shop/thayers-dry-mouth-spray-natural-peppermint-flavor/. Accessed on 30 November 2018.

The Flavourings in Food (Amendment) Regulations 1994, http://www.legislation.gov.uk/uksi/1994/1486/regulation/3/made. Accessed on 30 November 2018.

The Original, PH50 Protein Greens Advanced by Doctors Nutra Nutraceuticals, https://www.yourdoctorsnutra.com/SearchResults.asp?Search = vanilla&Submit = Search. Accessed on 2 December 2018.

Tuorila H, Cardello AV. Consumer responses to an off-flavor in juice in the presence of specific health claims. *Food Qual Pref.* 2002; 13: 561–569.

Ulla H, Antti K, Oskar L, Mari S. Genetic basis of flavor sensitivity and food preferences. In *Flavor, From Food to Behaviors, Wellbeing and Health.* Woodhead Publishing Series in Food Science, Technology and Nutrition, Amsterdam, the Netherlands, Elsevier, 2016; 203–227.

UltraClear a liver detoxification support by Metagenics, https://www.metagenics.com/ultraclear. Accessed on 3 December 2018.

USFDA-United States Food and Drug Administration, CFR - Code of Federal Regulations, Title 21: https://www.accessdata.fda.gov/scripts/cdrh/cfdocs/cfcfr/cfrsearch.cfm?fr = 501.22. Accessed on 7 December 2018.

Whey Isolate Protein Powder by BioChem, http://www.biochem-fitness.com/. Accessed on 8 December 2018.

8

Natural Flavors in Functional Food Supplements

Rahul Shukla, J. Lalrengpuii, W. Abhay, K. Vignesh, and A. Prabakaran

Contents

8.1 History of flavor

Prehistoric periods reported the use of flavors while cooking vegetables, meat or fruits. Many herbs and spices were used not only to enhance the flavor and quality of food but also to preserve meat prior to the use of refrigeration. This promotes the development of culinary techniques used to extract the concentrated flavoring material (Kollmannsberger et al., 2007).

In 5000 B.C., Ayurveda claimed basic tastes were comprised of sweet, salty, sour, pungent, bitter and astringent. Aristotle, in 350 B.C., hypothesized that sweet and bitter are two basic taste. Afterwards, ancient Chinese postulate spiciness as a basic taste. Early Egyptians are known for flavoring food with dill, cumin, marjoram, coriander, sesame, thyme and even cinnamon. They also used honey as a sweetening agent. People in Europe and the Middle East discovered how to distill and extract essential oils and resins. Even in medieval times, monks used to collect these essences and used them as food flavors (Van der Schaft, 2015).

German and Swiss businesses were the first to expand the flavor market significantly. Generally, through the new area of flavor chemistry and technology bulk production of synthetic aroma flavor and derivation of flavoring substances occurring naturally in foods (Sibinovska et al., 2016; Smith, 2015).

The United States emerged as one of the major importers of European essential oils and extracts. They soon expanded to meet local economic and market needs by formulating and manufacturing ingredients domestically. There is a growing number of food and beverage companies driven demand for commercial flavors. For formalizing their association and ensuring a safe supply of flavor, the Pure Food and Drugs Act of 1906 created Flavor and Extract Manufacturers Association of the United States (FEMA) in 1909 (Smith et al., 2018).

Flavor is the sensory impression of functional food and comprises one of the most important classes of food additives. Aspects of flavor are described in **Figure 8.1**. Flavors are used to replace the bad taste of pharmaceutical drugs as well as improving the taste of foods (Attokaran, 2011).

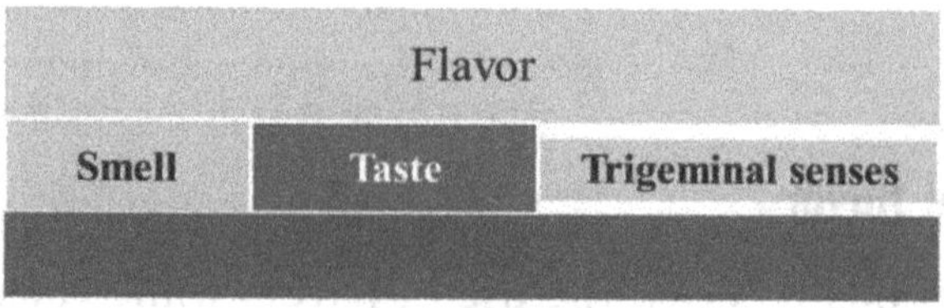

Figure 8.1 *Diagram describing three aspects of flavor.*

8.2 Regulatory definition

1. The US Food and Drug Administration (FDA) defines the term natural flavor or natural flavoring as an essential oil, oleoresin, essence or extractive, protein hydrolysate, distillate or any product of roasting, heating or enzymolysis. It contains the flavoring constituents derived from a spice, fruit or fruit juice, vegetable or vegetable juice, edible yeast, herb, bark, bud, root, leaf or similar plant material, meat, seafood, poultry, eggs, dairy products, or fermentation products thereof, whose significant function in food is flavoring rather than nutritional.
2. According to European law, "A 'natural flavoring substance' shall mean a compound obtained by appropriate physical, enzymatic or microbiological processes from material of vegetable, animal or microbiological origin."
3. International Organization of the Flavor Industry defines, natural flavoring substances are flavoring substances obtained by physical processes that may result in unavoidable but unintentional changes in the chemical structure of the components of the flavoring (e.g., distillation and solvent extraction) or by enzymatic or microbiological processes, from material of plant or animal origin. Such material may be unprocessed or processed for human consumption by traditional food-preparation processes (e.g., drying, torrefaction (roasting) and fermentation). This means substances that have been identified/detected in a natural material of animal or vegetable origin.

8.2.1 Introduction of functional foods

Functional foods are defined as products resembling traditional foods but possess demonstrated physiological benefits (Arihara, 2014). The term "functional food" itself was first used in Japan in the 1980s for food products fortified with special constituents that possess advantageous physiological effects (Doyon and Labrecque, 2008; Nedovic et al., 2011). Functional food is food prepared using scientific intelligence that endows the body with vital nutrition which is required for a healthy lifestyle (Bech-Larsen and Grunert, 2003; Schieber, 2012). Functional foods with example are represented in **Figure 8.2**.

1. *Fortified products*: A food fortified with additional nutrients. Example includes fruit juices fortified with vitamin C.
2. *Enriched products*: A food with added new nutrients or components not normally found in a particular food. Example is of margarine with plant sterol ester, probiotics and prebiotics (Chauhan and Chorawala, 2014).
3. *Altered products*: A food from which a deleterious component has been removed, reduced or replaced with another substance with beneficial effects. Best example is fibers as fat releasers in meat or ice cream products.

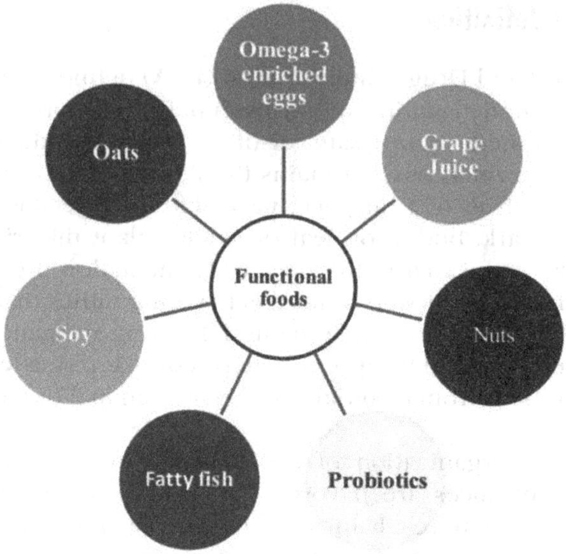

Figure 8.2 *Diagram representing examples of functional foods.*

4. *Enhanced commodities*: A food in which one of the components has been naturally enhanced through special growing conditions, new feed composition, genetic manipulation, or otherwise eggs with increased omega-3 content achieved by altered chicken feed.
5. *Dietary supplements*: These supplements may include nutrients to food or used to decline the risk of health problems such as osteoporosis, arthritis, etc. These are available in the form of pills, tablets, gels, etc.

Functional foods are modified to increase the nutritional value. Functional foods are also termed as nutraceuticals. Foods are now marketed to benefit everything from your intestines to your heart and also to your memory. Functional foods are foods that have health benefits beyond normal nutrition (Holzmeister, 2009). Functional foods with benefits are depicted in **Table 8.1**.

Often, a food is termed as functional because it contains a higher amount of phytochemicals. These natural, active plant chemicals have been found to boost health. Phytochemicals are more abundant in fruits, vegetables, whole grains and spices (garlic and ginger). The presence of antioxidants in a food is another reason it can be termed "functional." Antioxidants, such as vitamin A, C and E, and the mineral selenium, work to destroy harmful particles in the blood that can lead to heart problems and other complications.

Table 8.1 Functional Foods, Their Components, and Their Benefits

S. No.	Name of Functional Food	Functional Components	Benefits	References
1	**Omega-3 enriched eggs**	Omega-3 fatty acids (DHA)	Lower triglycerides, cholesterol, and reduce the risk for coronary heart disease	Volpe (2015)
2	**Oats**	ß-glucan and saponins	Reduce total and LDL cholesterol	Tosh and Miller (2015)
3	**Fatty fish**	Omega-3 fatty acids (DHA and EPA)	Reduce triglycerides, reduce coronary heart disease	Jain et al. (2015)
4	**Fortified margarines**	Plant sterol and stanol esters	Reduce total and LDL cholesterol	De Pee and Bloem (2009)
5	**Soy**	Isoflavones and genistein, and soy protein	Reduce total and LDL cholesterol	Deng (2009)
6	**Tomatoes and tomato products**	Lycopene	Reduction of prostate cancer	Borguini and da Silva Torres (2009)
7	**Probiotics**	Bacteria (lactobacillus)	Support gastrointestinal (GI) health	Parvez et al. (2006)
8	**Nuts**	Monounsaturated fatty acids (healthy fats), vitamin E (antioxidant)	Reduce the risk of coronary heart disease	Kornsteiner et al. (2006)
9	**Grape Juice or Red Wine**	Resveratrol	Exhibit heart-healthy effects	Ortuño et al. (2010)

Examples of foods with antioxidants (Grajek et al., 2005; Victoria Urquiza-Martínez and Fenton Navarro, 2016):

Tea (catechins)
Berries (flavonoids such as quercetin)
Citrus foods (flavonoids/limonoids)
Wine or grape juice (resveratrol)

8.3 Flavors used in functional foods

Flavors are additives used to enhance, modify, and create the taste and the aroma in food products. Odors, tastes, oral irritations, thermal sensations that arise small non-polar to medium polar molecules from a particular food, which are sensory amalgamation of percept flavor. The volatiles give the food its unique, personal identity and character, and the nonvolatile carry primary signal. This allows consumers to identify food stuffs edibility that create psychological–sensory foundation of flavor on which the volatiles

Table 8.2 Properties and Description of Flavors

S. No.	Property	Description
1	APPEARANCE	The size, shape, color, temperature and surface texture imparts the appearance whether the food to be eaten or not.
2	ODOR	The aroma arise from the food felts by human nose is the characteristic odor of that particular food. Odor and taste work together to produce flavor (Richardson, 2013).
3	TASTE	Sweet, sour, salt, bitter, umami are the 5 basic tastes detect by the tongue that is the characteristic of the food (Mouritsen, 2015; Tournier et al., 2007; Vilgis, 2013).
4	TEXTURE	The surface of the food is felt when food is placed in the surface of the tongue, palate or teeth This is called as mouth-feel. Likewise it can be felt with our finger. It include crispy, crunchy, chewy, creamy.

build (Breslin, 2001; Kennedy and Cosnett, 2001). Certain properties with description of flavors are given in **Table 8.2**.

While eating all senses are used, including sight, smell, hearing, taste, and touch thus enabling us to evaluate the food.

Flavor is unique, in having a "dual nature," i.e., it can sense signals originating outside (orthonasal) and inside (retronasal) the body (Mishellany-Dutour et al., 2012).

Orthonasal stimulation: It involves sense odors in an environment by sniffing in through the external nares of the nose to activate the sensory cells in the olfactory epithelium. The resultant odor images in the olfactory bulb processed by the primary olfactory cortex, a part of the prefrontal lobe (Hummel and Seo, 2016).

Retronasal stimulation: It involves release of volatile molecule from food during ingestion and sensed through the nasopharynx-olfactory epithelium. Olfactory pathway combined with the pathways for taste, touch, sound and active sensing by proprioception form a flavor system (Breslin, 2001).

They are flavoring additives (Marnett et al., 2013) that include (as depicted in **Figure 8.3**):

- Flavoring enhancers
- Sweeteners (non-nutritive)
- Flavoring agents

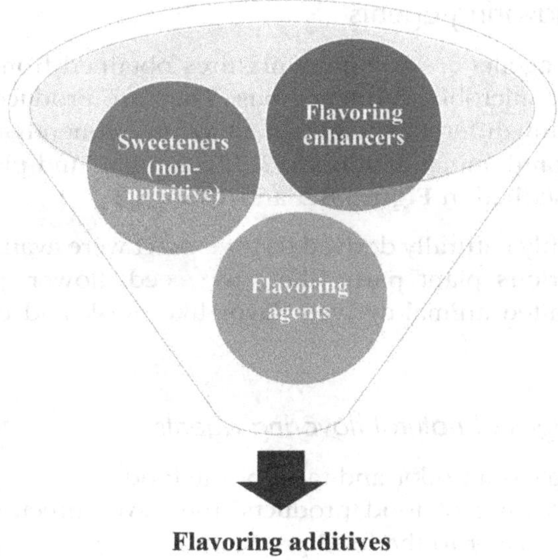

Figure 8.3 *Schematic presentation describing flavoring additives.*

8.3.1 Flavoring agents

Flavoring agents impart flavor that are:

1. Natural—extracted from plants and animal
2. Artificial—chemical synthesis or chemical modification of natural substance
3. Natural identical—chemically processed but identical to natural.

The flavoring agents may contain permitted anti-oxidants, emulsifying anti-caking agents and food preservatives (Kaitano, 2013).

Single component flavor: It is chemically defined as single component of natural or synthetic source usually compounded to form the flavor. Example: benzaldehyde, citral.

Complex component flavor: It is a complex compound naturally obtained from essential oil, oleoresin, extracts. Example: lemon-grass, bitter almond oil. To satisfy the consumer expectation, flavors are added to food products that do not provide aesthetic value but do provide nutritional value to the food.

8.3.2 Natural flavoring agents

Natural flavoring agents are complex mixtures obtained from plants, herbs, spices, animals or microbial fermentations. They are produced by the combined effect of many different chemicals, although frequently the taste of one or two chemical predominates (Bloom, 2017). Animal- and plant-based natural flavors are described in **Figure 8.4** and **Table 8.3**.

In ancient times only naturally derived flavor extract were available. It includes spices, herbs, various plant parts like bark, seed, flower, plant exudates, etc., and only limited animal-derived flavor like musk and castoreum were available.

8.3.2.1 Advantages of natural flavoring agents

- It masks unpleasant odor and taste of the food.
- During processing of food products, the flavor precursors give its characteristic flavor to the food product.
- It helps in marketing by its label claim.
- It acts as code of identification of particular food.
- It enhance the desire of consumer to eat that food.
- It increase and improve stability, texture, appearance of the food (Lanciotti et al., 2004).

8.3.2.2 Disadvantages of natural flavoring agents

Although natural flavors must meet safety criteria, individual allergic reactions might occur due to chemical that contain in natural flavor. For example, Safrole

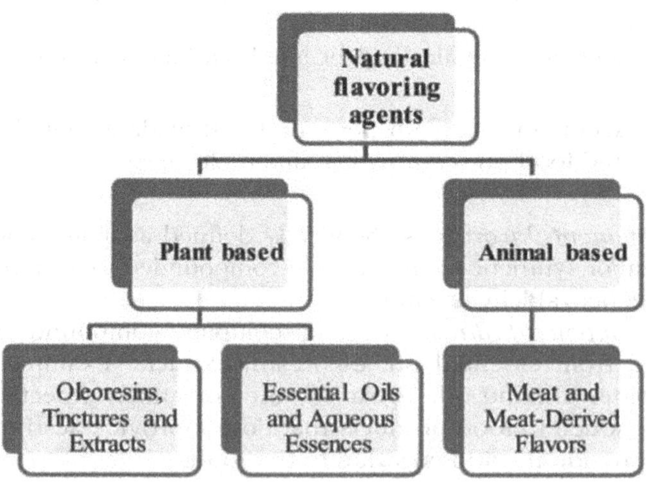

Figure 8.4 *Diagram describing sources of natural origin flavoring agents.*

Table 8.3 Plant- and Animal-Based Flavoring Agents

S. No.	Flavor Class	Subdivisions	Examples	References
1	Fruit flavor	• Citrus-type flavors (terpeny) • Berry-type flavors (non-terpeny)	Grapefruit, orange apple, raspberry, banana	Cheetham (2010)
2	Vegetable flavors	—	Lettuce, celery	Cheetham (2010)
3	Spice flavors	• Aromatic • Lachrymogenic • Hot	Cinnamon, peppermint onion, garlic pepper, ginger	Kollmannsberger et al. (2007)
4	Beverage flavors	• Unfermented flavors • Fermented flavors • Compounded flavors	Juices, milk wine, beer, tea soft drinks	Kollmannsberger et al. (2007)
5	Meat flavors	• Mammal flavors • Sea food flavors	Lean beef fish, clams	Zviely (1997)
6	Fat flavors	—	Olive oil, coconut fat, pork, butter fat	Zviely (1997)
7	Cooked flavors	• Vegetable • Broth • Fruit	Legume, potatoes beef bouillon marmalade	Cheetham (2010)
8	Processed flavors	• Smoky flavors • Broiled, fried flavors • Roasted, toasted, baked flavors	Ham processed meat product coffee, snack foods, processed cereals	Zviely (1997)
9	Stench flavors	—	Cheese	Kollmannsberger et al. (2007)

occurs naturally, and it was used for flavoring of beer until 1960 when it was found that it has hepatotoxicity property.

- Vegetarians may unknowingly be ingesting animal-derived natural flavors in processed foods.
- Compared to natural flavor, artificial flavors sometimes contain fewer chemicals because they are produced under tightly controlled laboratory conditions.

Natural-derived flavors are safer, and cost of production is high, when compared to artificial flavors. Artificial flavors are cheap, stable and devoid of the agriculture harvesting cycle (Grumezescu and Holban, 2015).

8.3.3 Naturally identical flavors

Naturally identical flavors are the substances chemically isolated from aromatic raw material or synthetically obtained. They are chemically identical to substance present in natural product. However, they do not contain any artificial flavoring substance ("Codex Committee on Food Additives," 2013). A flavoring substance is obtained by enzymatic action as well as thermal processing. **Figure 8.5** gives an illustrative presentation of derived natural products.

- *Smoke-based reaction flavors*: Natural aqueous condensate of wood smoke by controlled pyrolysis of wood components.
- *Process flavor*: Substances that are formed from natural substances upon processing, mainly heating. A common example is caramel, which is produced by heating sugars.
- *Biotechnology derived flavors*: The natural flavor produced by biotechnological techniques involves use of microorganism such as lactic acid bacterial Lactococcus, Lactobacillus, Lecuconostoc or also *Enterococcus faecium*. Recently, a carboxyl esterase from *Bacillus licheniformis* was reported to synthesize isoamyl acetate from isoamyl alcohol and p-nitro phenyl acetate in n-hexane (Berger, 2015).
- *Hydrolyzed vegetable protein (HVP)*: Produced by acid, alkaline, enzymatic hydrolysis.
- *Autolyzed yeast extracts*: Endogenous enzymes create flavor profile and generate aroma precursors.

8.3.3.1 Floral-based extracts used in flavors

Floral flavors are commonly extracted by solvent extracts, syrups and oil. In Asia the market for floral flavors like jasmine, rose, elderflower, hibiscus

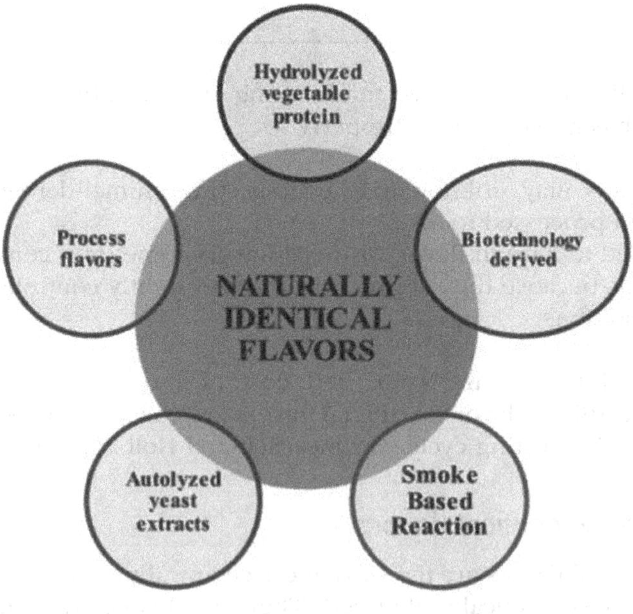

Figure 8.5 *Types of natural identical flavors.*

and chrysanthemum are used most commonly in food product and beverages, specifically rose accents are an emerging trend, gaining popularity in Europe and Asia while chamomile and hibiscus have long been known as beverage flavors (Mebakerlin and Chakravorty, 2015).

Edible flowers include:

- Lavender
- Rose
- Elderflower
- Hibiscus—dried petals of this orange tropical flower are said to have a cranberry-like taste
- Calendula
- Nasturtium petals have long been used in international cuisines and dried for use in herbal teas

The floral flavor rose is primarily seen in Middle Eastern countries as it is a basic flavoring component for a large variety of dishes. The flavor of rose can be consumed through multiple methods such as oils, waters and even directly from the plant in the form of petals or rose hips, and it has a delicate flavor that is characteristically floral, fruity, perfume like and sweet.

Possible flavor combinations with rose include:

- Rose and rhubarb—Rhubarb without the addition of any sugar is intensely tart, but once sugar is added the flavor is changed. Fruity, aromatic and candied strawberry notes develop and offset the strong aromatic, floral notes of rose.
- Rose and raspberry—Raspberry is sweet and has fruity and floral qualities that can pair with the floral rose. Overall, this combination is reminiscent of perfume.
- Rose and lychee—Tutti fruit flavor, lychees should be paired with rose as it brings out the fragrance of roses.

Applications and examples of floral flavor include:

- Violet concord grape white tea
- Elderflower lemonade protein drink
- Rose cocktail
- Orange blossom yogurt
- Honeysuckle hibiscus lemon bar
- Lavender blueberry chocolate bar

8.3.4 Artificial flavors

Artificial flavoring means chemical similarity to natural flavoring substances. The function of which is to impart flavor, which is not derived from a spice,

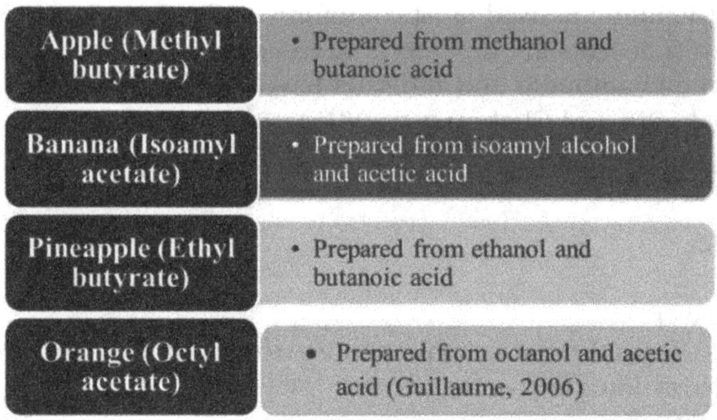

Figure 8.6 *Diagram explaining some artificial flavors with method of preparation.*

fruit or fruit juice, vegetable or vegetable juice, edible yeast, herb, bark, bud, root, leaf or similar plant material, meat, fish, poultry, eggs, dairy products or fermentation products ("CFR—Code of Federal Regulations Title 21," n.d.; Dikshit and Tallapragada, 2018).

During the era of development of organic chemistry, many chemicals had been utilized for synthesis of synthetic flavor. In 1870s, chemist cracked molecular structure of vanillin, and it was the first artificial flavor that was derived from woody pulp or petroleum. Later Dutch chemists discovered a buttery-flavor compound called diacetyl, grape flavor from methyl anthranilate. Serendipitously, peach flavor was discovered in castor oil production. By the mid of twentieth century artificial flavors were professionally started producing in mass scale by many industries. At the end of 1980s artificial flavor became less popular due to stringent safety assessment by FDA. Some examples of artificial flavors with method of preparation are given in **Figure 8.6**.

These artificial flavors are added to duplicate the natural flavor, but most of these flavor are banned in European union (EU). As this involves allergic reactions, attention deficit hyperactivity disease (ADHD) and carcinogenicities as the known and suspected health risks of artificial flavors.

8.4 Flavoring enhancer

The flavor enhancer is a substance added to the food product to intensify or modify the flavor that are already present in food products and have no taste of their own (Ashurst, 1991). The most commonly used substances in this category are monosodium L-glutamate (MSG), disodium 5′-inosinate (IMP) and disodium 5′-guanylate (GMP). Some examples of flavoring enhancer are included in **Figure 8.7**.

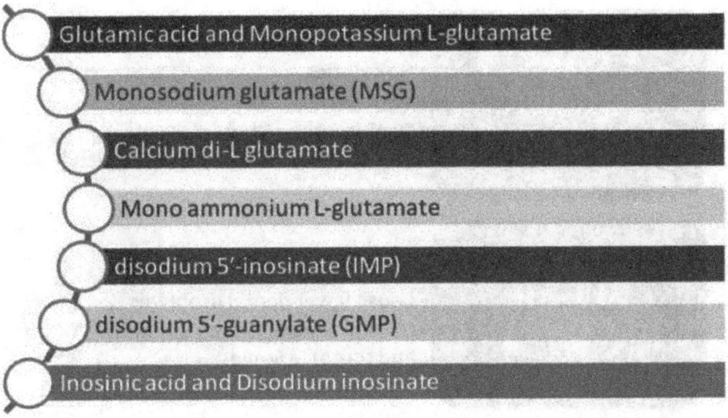

Figure 8.7 *Example of flavoring enhancer sweeteners (non-nutritive).*

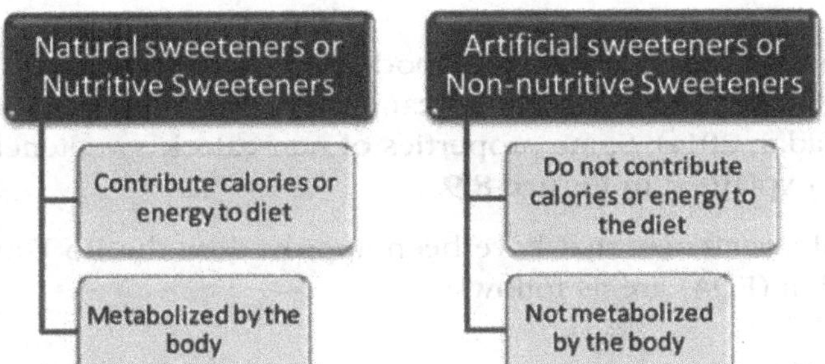

Figure 8.8 *Diagram explaining properties of natural and artificial sweeteners.*

Sweeteners are agents that provide sweetness to the food and flavorable (**Figure 8.8**). While choosing an alternative sweetener to conventional sweetener in a functional food or beverage, certain properties should be considered. It should have a similar same function and texture with different nutritional and biochemical metabolism as the conventional sweetener (Belloir et al., 2017; Madar, 2014; Raben, 2012).

Currently the non-nutritive sweeteners are more preferred over nutritive sweetener. Drawbacks of artificial sweeteners include the development of lipid dysregulation, visceral adiposity, hypertension, inflammation and clinical coronary heart disease (Chattopadhyay et al., 2014; Raben, 2012; Sardarodiyan and Hakimzadeh, 2016).

Artificial sweeteners are a synthetic- or natural-derived conventional sweetener substitute, also referred to as "intense sweetener." It needs less quantity to impart sweetness compare to conventional sweetener like sugars, syrups, molasses, sugar alcohols or polyols and honey (Weihrauch and Diehl, 2004).

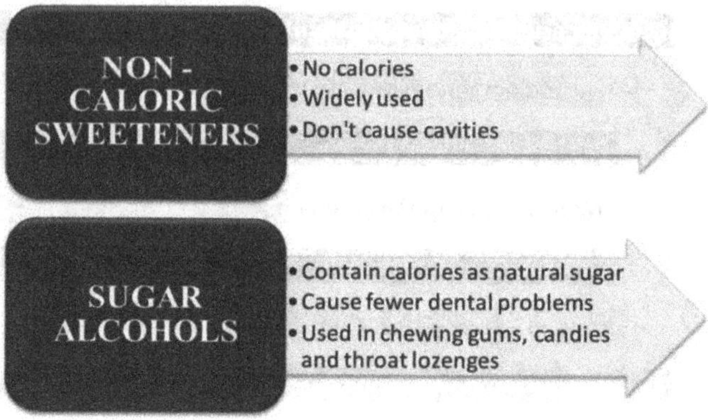

Figure 8.9 *Advantage of non-caloric and sugar-based sweeteners.*

These are mostly used in functional food like baked goods, soft drinks, powdered drink mixes, candy, puddings, canned foods, jams and jellies and dairy products (Madar, 2014). Some properties of non-caloric sweeteners and sugar alcohols are explained in **Figure 8.9**.

The artificial sweeteners that have been approved by the US Food and Drug Administration (FDA) are as follows:

- Aspartame
- Acesulfame-K
- Neotame
- Saccharin
- Sucralose

8.5 Need for flavorings in functional food

The consumption and acceptance of functional food is the flavor of that food that imparts taste to food. In adjunct, the flavor, taste and aroma of food stimulate salivary flow and consequently aid digestion and metabolism. Despite the added value of heath functionality, consumers are unlikely to buy functional foods over the long term if the foods don't taste good. Individual flavoring substances that occur naturally in food, together with the flavors generated through cooking and other sorts of preparation of food for human consumption, account for the majority of flavorings found in the daily intake of food. Even in industrialized countries, added flavoring materials represents only a minority of the flavoring materials that we consume. Flavoring substances are among the most rigidly evaluated and tested food ingredients and as such can be regarded safe under their conditions of intended use (Schrankel, 2004).

Flavors of functional food include:

1. Desirable flavor
2. Undesirable flavor (off-flavor)

Flavors are added to functional foods to counterbalance for:

1. Reductions in undesirable food ingredients such as fat, sugar and salt.
2. The loss of flavor during the processing and storage of foods (pasteurized foods).
3. Natural seasonal or geographical variations in crops.

The use of flavorings can counterbalance for supply limitations by helping to standardize the flavor of food (Hallagan and Hall, 2009).

Natural flavor and artificial flavor both are similar in term of chemical composition and effect on health, during labeling of functional food products the manufacturer don't reveal the chemical mixture of the flavor used in it, so whenever possible choose natural and whole food rather than choosing flavored food (Fahlbusch et al., 2003).

8.6 Salutary effect of natural flavor

The chemical components (volatile and non-volatile) in the natural flavors that are added to functional food also contribute some therapeutic activity. This provides great opportunities to develop novel functional foods. Although in order to get health benefits, the compound responsible for activity is necessary to withstand food processing and to be released from the food matrix on ingestion. The metabolites of compounds that must reach the target tissue for action are the challenges involved in this. Studies reveal that trepenoids, phenolics and alkaloids are the most significant part of flavor to produce novel functional food with antimicrobial activity (Kennedy and Cosnett, 2001; Ramesh and Muthuraman, 2018).

For example, chlorogenic acid which is present in coffee, mango, wine, apple juice and blueberry, gingerly in ginger, raspberry ketone in red grapes shows antimicrobial and anti-obesity. Geraniol in rose oil or citronella oil has anti-inflammatory and neurological activity. Linalool in citrus fruit shows both anti-inflammatory and antimicrobial. Vanilla beans contain more than 200 volatile compounds and the most important one is vanillin; studies showed that antioxidant and anti-mutagenic properties due to its volatile compounds are present in it (Ayseli and İpek Ayseli, 2016).

8.7 Formulated form of commercial flavors

Complex blends of aromatic chemical and essential oil available in:

- Liquid flavor—aroma chemicals dissolved or emulsified in carrier solvent like vegetable oil base, ethanol, triacetin
- Microencapsulation flavor (spray dried)
- Dry blended flavor—mixture of dried flavors, spices or herb powder
- Extracts
- Concentrates
- Pastes (Wright, 2010)

8.8 Challenges in flavoring agents

The taste and aroma is the main vision of the functional foods apart from the nutritional value to make the product demand to customer. So in this case tremendous care must be taken in considering the use of natural flavor rather than artificial flavor for flavor quality of the product. Studies also revealed that consumer will not purchase functional food in long term if the taste is not good. These are the challenges faced by functional food industry to grow significant in food sector market (Augustin and Sanguansri, 2013).

In case of functional food the challenges in flavoring includes:

- Problem of off-flavor characteristic with functional or associated ingredient.
- Variation in flavor during manufacturing functional food.
- Base interaction of natural flavor with ingredient.
- By nature, flavor chemical are more reactive.
- Spices and herbs also show bitter, astringent, grassy, and other flavor defects.

For example, functional foods like ginseng shows bitter, ginger like taste and multivitamin mineral drinks shows a meaty broth flavor and metallic taste.

So by analyzing the flavor property individually and along with functional ingredient flavorists choose appropriate flavor that adds and masks the inherent flavor of the functional food (Reineccius, 1999).

8.9 Natural flavors specifically used for dietary food supplements

Food supplements are the products that are taken orally and contain dietary ingredients for nutrition therapy and to counteract health problems (Baran, 2014).

They include:

- Vitamins
- Minerals
- Herbs
- Amino acids and so on

These supplements include various natural flavors in order to mask taste and odor for patient compliance. These includes both nutritive and non-nutritive ingredients. Essential oils such as cinnamon, peppermint, lavender, jasmine and so on are included in various food supplements that have no nutritive value but help to increase intake of nutritive food by masking pungent odor, bitter taste and more. Examples of some flavors used in various food supplements are mentioned below.

8.9.1 Eniva® Cell-ready® Multi minerals

It is a blend of natural minerals in a concentrated form. These natural flavors help for significant physiological balance of minerals to meet the patient needs and boost individual immune systems. These products contain mineral ions containing citrus flavor for a refreshing taste. It is a proprietary product that uses multi-step purified water and is believed to be readily absorbed from GIT in comparison to already marketed tablets and capsule form. **Solutomic** is clear solution of cooper ions prepared in aqueous solution with sparkling color for increased absorption.

8.9.2 Green tea

It contains various anti-oxidants and is used to decrease fatty acids in the body for the management of obesity (Koech et al., 2012).

It includes various flavors such as orange passion fruit jasmine, purple acai blueberry, white mangosteen peach, cranberry pomegranate, lemon ginseng and red goji raspberry.

8.9.3 Vitamin C

It is used as an anti-oxidant containing various natural flavors such as lemon, orange, honey and so on (Null, 2001).

8.10 Regulatory bodies of flavor

International Organization of the Flavor Industry (IOFI) is a non-government and non-profit organization of flavor association and industry and also acts as observer in WHO/FAO Codex Alimentarius Commission established in 1969.

Its association members include EFFA (Europe), FEMA (USA), JFFMA (Japan) (full form given in **Table 8.4**).

Table 8.4 Full Form of Association Members

S. No.	Organization	Full Form
1	EFFA	European Flavor and Fragrance Association
2	FEMA	Flavor and Extract Manufacturers Association
3	JFFMA	Japan Flavors and Fragrance Materials Association

Roles and responsibilities include:

- Safe use of flavorings is the flavor industry
- Sound scientific information to the industry, customers, government agencies to promote the benefits and safe use of flavors.

In 1956 international scientific expert committee established the Joint FAO/WHO Expert Committee on Food Additives (JECFA) to evaluate safety of food additive (flavor substance), contaminants, naturally occurring toxicants and residues in food. Submission process for JECFA evaluation is diagrammatically presented in **Figure 8.10**.

IOFI Code of Practice provides the basis for the IOFI Global Reference List of flavorings that are safe for the intended use, were evaluated by JECFA and its methodology of evaluation is based on:

- Exposure
- Metabolism of flavoring substances
- Evaluation in groups of substances
- Structure activity relationships
- IOFI considers as acceptable flavoring substances materials that meet one or more of the following requirements, and these materials comprise the IOFI.

Global reference list of flavorings include:

- Flavors evaluated by JECFA to represent "no safety concern under conditions of intended use."
- Flavors that were evaluated by authoritative bodies such as the European Food Safety Authority (EFSA) and the Japanese Food Safety Commission (FSC) using the same or similar methodology as used by JECFA.

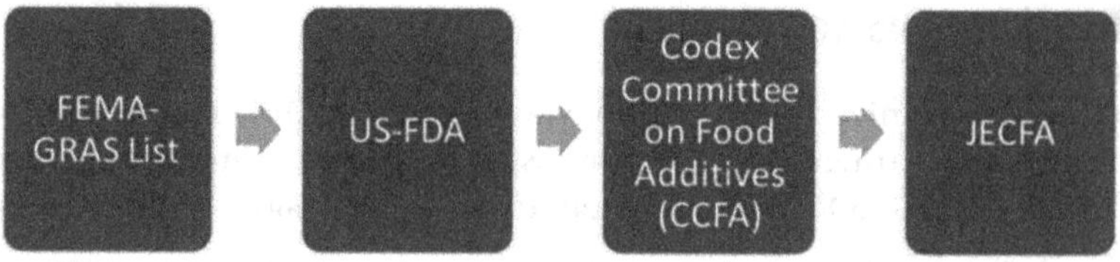

***Figure 8.10** Submission process for JECFA evaluation.*

- Flavors that are generally recognized as safe (GRAS) by the US Food and Drug Administration (US FDA) including GRAS determinations published by the expert panel of the Flavor and Extract Manufacturers Association of the United States (FEMA) ("International Organization of the Flavor Industry").

8.11 Conclusion

Functional foods are an essential part of the food chain. They are consumed daily by a large number of the population and enhance the taste of foods. Flavor is the sensory impression of functional foods, which comprises one of the most important classes of food additives. Flavors are additives used to enhance, modify and create the taste and the aroma in food products. Consumer satisfactory is the major concern for any products in the market. When the natural products are used as such, they can retain their properties such as tastes and flavors. However, the call for ready-to-eat or serve food products has been increased in the market which assigned flavors as essential additives in the food products.

Flavors are also used in medicines and especially in pediatric and geriatric formulations to provide palatability and helpful in consuming formulation. The generation of an acceptable and desire flavor is an important part that plays a significant role in the pharmaceutical industry. The selection of flavoring agent depends on the properties of the drug such as the taste and also depends on the types of excipients incorporated in the formulation. The industry is currently facing stability issues with natural flavors and some interactions. Apart from this cost of extraction, methods, labor cost and many manufacturers increase the cost of flavors. As with the increase in population, there is demand of natural origin flavors. A major hurdle is to meet the present demand for natural-based flavors. The industry is facing challenges to meet the population demand. One of the major challenges is acceptability of different tastes by different age groups and locality. Despite of all the challenges, natural flavor is still in demand due to its aroma and medicinal property as well as its non-toxicity as customer safety is the utmost concern. A list of natural origin flavors with their health benefits is shown in **Table 8.5**.

Table 8.5 List of Natural Origin Flavors with Their Health Benefits

Functional Component	Health Benefits
Soluble oat fiber	Coronary heart disease (Xu, 2012)
Soy protein	Coronary heart disease (Mojica et al., 2014)
Phytosterol/stanol esters	Coronary heart disease (Escurriol et al., 2010)
Calcium	Osteoporosis (European Foundation for Osteoporosis, 1993)
Folate-enriched foods	Neural tube defects (Laiño et al., 2015)

Acknowledgment

We are thankful to Department of Pharmaceuticals (DoP), Government of India for supporting us. NIPER-R communication number for this chapter is NIPER-R/Communication/050.

References

Arihara, K., 2014. Functional foods, in: *Encyclopedia of Meat Sciences.* doi:10.1016/B978-0-12-384731-7.00172-0.

Ashurst, P.R. (Ed.), 1991. *Food Flavorings.* Springer US, Boston, MA. doi:10.1007/978-1-4615-2161-7.

Attokaran, M., 2011. Flavors, in: *Natural Food Flavors and Colorants.* pp. 9–10. doi:10.1002/9780470959152.ch2.

Augustin, M.A., Sanguansri, L., 2013. Challenges in developing delivery systems for food additives, nutraceuticals and dietary supplements. *Encapsulation Technologies and Delivery Systems for Food Ingredients and Nutraceuticals* 19–48. doi:10.1533/9780857095909.1.19.

Ayseli, M.T., İpek Ayseli, Y., 2016. Flavors of the future: Health benefits of flavor precursors and volatile compounds in plant foods. *Trends in Food Science & Technology* 48, 69–77. doi:10.1016/J.TIFS.2015.11.005.

Baran, K.P., 2014. Dietary Supplements, in: *Encyclopedia of Toxicology: Third Edition.* pp. 128–133. doi:10.1016/B978-0-12-386454-3.00838-1.

Bech-Larsen, T., Grunert, K.G., 2003. The perceived healthiness of functional foods: A conjoint study of Danish, Finnish and American consumers' perception of functional foods. *Appetite* 40, 9–14. doi:10.1016/S0195-6663(02)00171-X.

Belloir, C., Neiers, F., Briand, L., 2017. Sweeteners and sweetness enhancers. *Current Opinion in Clinical Nutrition and Metabolic Care* doi:10.1097/MCO.0000000000000377.

Berger, R.G., 2015. Biotechnology as a source of natural volatile flavours. *Current Opinion in Food Science* 1, 38–43. doi:10.1016/j.cofs.2014.09.003.

Bloom, J., 2017. Natural and artificial flavors what's the difference? American Council on Science and Health.

Borguini, R.G., da Silva Torres, E.A.F., 2009. Tomatoes and tomato products as dietary sources of antioxidants. *Food Reviews International* doi:10.1080/87559120903155859.

Breslin, P.A.S., 2001. Human gustation and flavour. *Flavour and Fragrance Journal* 16, 439–456. doi:10.1002/ffj.1054.

CFR—Code of Federal Regulations Title 21, n.d.

Chattopadhyay, S., Raychaudhuri, U., Chakraborty, R., 2014. Artificial sweeteners—A review. *Journal of Food Science and Technology.* doi:10.1007/s13197-011-0571-1.

Chauhan S.V., and Chorawala, M.R., 2014. Probiotics, prebiotics and synbiotics. *International Journal of Pharmaceutical Sciences and Research* 3(3), 711.

Cheetham, P.S.J., 2010. Natural Sources of Flavours, in: *Food Flavour Technology: Second Edition.* pp. 127–177. doi:10.1002/9781444317770.ch5.

Codex Committee on Food Additives (WWW Document), 2013.

De Pee, S., Bloem, M.W., 2009. Current and potential role of specially formulated foods and food supplements for preventing malnutrition among 6-to 23-month-old children and for treating moderate malnutrition among 6- to 59-month-old children. *Food and Nutrition Bulletin.* doi:10.1177/15648265090303S305.

Deng, R., 2009. Food and food supplements with hypocholesterolemic effects. *Recent Patents on Food, Nutrition & Agriculture.* doi:10.2174/2212798410901010015.

Dikshit, R., Tallapragada, P., 2018. Chapter 3—Comparative Study of Natural and Artificial Flavoring Agents and Dyes A2—Grumezescu, Alexandru Mihai, in: *Handbook of Food Bioengineering.* doi:10.1016/B978-0-12-811518-3.00003-X.

Doyon, M., Labrecque, J.A., 2008. Functional foods: A conceptual definition. *British Food Journal* doi:10.1108/00070700810918036.

Escurriol, V., Cofán, M., Moreno-Iribas, C., Larrañaga, N., Martínez, C., Navarro, C., Rodríguez, L., González, C.A., Corella, D., Ros, E., 2010. Phytosterol plasma concentrations and coronary heart disease in the prospective Spanish EPIC cohort. *Journal of Lipid Research* 51, 618–624. doi:10.1194/jlr.P000471.

European Foundation for Osteoporosis, 1993. Prevention of osteoporosis: Calcium nutrition. *Osteoporosis International* 3, 163–165.

Fahlbusch, K.-G., Hammerschmidt, F.-J., Panten, J., Pickenhagen, W., Schatkowski, D., Bauer, K., Garbe, D., Surburg, H., 2003. Flavors and fragrances, in: *Ullmann's Encyclopedia of Industrial Chemistry*. doi:10.1002/14356007.a11_141.

Grajek, W., Olejnik, A., Sip, A., 2005. Probiotics, prebiotics and antioxidants as functional foods, in: *Acta Biochimica Polonica*. doi:10.1093/brain/awt108.

Grumezescu, A.M., Holban, A.M., 2015. *Natural and Artificial Flavoring Agents and Food Dyes*.

Hallagan, J.B., Hall, R.L., 2009. Under the conditions of intended use—New developments in the FEMA GRAS program and the safety assessment of flavor ingredients. *Food and Chemical Toxicology* 47, 267–278. doi:10.1016/j.fct.2008.11.011.

Holzmeister, L.A., 2009. Functional foods. Diabetes self-management. doi:10.1509/jmkr.48.SPL.S1.

Hummel, T., Seo, H.S., 2016. Orthonasal and retronasal perception, in: *Flavour: From Food to Perception*. pp. 310–318. doi:10.1002/9781118929384.ch13.

International Organization of the Flavor Industry (WWW Document), n.d.

Jain, A.P., Aggarwal, K.K., Zhang, P.Y., 2015. Omega-3 fatty acids and cardiovascular disease. *European Review for Medical and Pharmacological Sciences* doi:10.1097/MCO.0b013e3280127af0.

Kaitano, T.E., 2013. Food Additives: Flavors and Flavor Enhancers, in: *Encyclopedia of Food Safety*. pp. 466–470. doi:10.1016/B978-0-12-378612-8.00230-4.

Kennedy, J.F., Cosnett, I.R., 2001. Food Flavours Biology and Chemistry, *Carbohydrate Polymers*. doi:10.1016/S0144-8617(01)00236-3.

Koech, F.N., Wanyoko, J.K., Kerio, L.C., Khalid, R., Ngure, R.M., Bii, C. 2012. Potential antibacterial and antioxidant activity of different tea extracts. *Tea* 33, 67–77.

Kollmannsberger, H., Nitz, S., Blank, I., 2007. Flavours and Fragrances. *Flavours and Fragrances: Chemistry, Bioprocessing and Sustainability* 313–361. doi:10.1007/978-3-540-49339-6.

Kornsteiner, M., Wagner, K.H., Elmadfa, I., 2006. Tocopherols and total phenolics in 10 different nut types. *Food Chemistry*. doi:10.1016/j.foodchem.2005.07.033.

Laiño, J.E., Zelaya, H., Juárez del Valle, M., Savoy de Giori, G., LeBlanc, J.G., 2015. Milk fermented with selected strains of lactic acid bacteria is able to improve folate status of deficient rodents and also prevent folate deficiency. *Journal of Functional Foods* 17, 22–32. doi:10.1016/j.jff.2015.04.055.

Lanciotti, R., Gianotti, A., Patrignani, F., Belletti, N., Guerzoni, M., Gardini, F., 2004. Use of natural aroma compounds to improve shelf-life and safety of minimally processed fruits. *Trends in Food Science & Technology* 15, 201–208. doi:10.1016/J.TIFS.2003.10.004.

Madar, A., 2014. Artificial sweeteners versus natural sweeteners. *Bulletin of the Transilvania University of Braşov Series V. Economic Sciences* 7(1), 59–64.

Marnett, L.J., Cohen, S.M., Fukushima, S. et al. 2013. Gras: Flavoring substances 26. *Food Technology* 67(8), 38–43.

Mebakerlin, M.S., Chakravorty, S., 2015. Value addition in flowers, in: *Value Addition of Horticultural Crops: Recent Trends and Future Directions*. pp. 83–99. doi:10.1007/978-81-322-2262-0_5.

Mishellany-Dutour, A., Woda, A., Labouré, H., Bourdiol, P., Lachaze, P., Guichard, E., Feron, G., 2012. Retro-Nasal aroma release is correlated with variations in the in-mouth air cavity volume after empty deglutition. *PLoS ONE* 7, e41276. doi:10.1371/journal.pone.0041276.

Mojica, L., Dia, V.P., De Mejía, E.G., 2014. Soy Proteins, in: *Applied Food Protein Chemistry*. pp. 139–191. doi:10.1002/9781118860588.ch9.

Mouritsen, O.G., 2015. The science of taste. *Flavour* 4, 18. doi:10.1186/s13411-014-0028-3.

Nedovic, V., Kalusevic, A., Manojlovic, V., Levic, S., Bugarski, B., 2011. An overview of encapsulation technologies for food applications peer-review under responsibility of ICEF11 Executive Committee Members. *Italian Oral Surgery* 1, 1806–1815. doi:10.1016/j.profoo.2011.09.266.

Null, G., 2001. The Antioxidant Vitamin—Vitamin C. Vitamincfoundaion.org. 1–14.

Ortuño, J., Covas, M.I., Farre, M., et al. 2010. Matrix effects on the bioavailability of resveratrol in humans. *Food Chemistry*. doi:10.1016/j.foodchem.2009.11.032.

Parvez, S., Malik, K.A., Ah Kang, S., Kim, H.Y., 2006. Probiotics and their fermented food products are beneficial for health. *Journal of Applied Microbiology*. doi:10.1111/j.1365-2672.2006.02963.x.

Raben, A., 2012. Artificial sweeteners: A place in the field of functional foods? Focus on obesity and related metabolic disorders. *Current Opinion in Clinical Nutrition and Metabolic Care* 15(6), 597–604. doi:10.1097/MCO.0b013e328359678a.

Ramesh, M., Muthuraman, A., 2018. Flavoring and coloring agents: Health risks and potential problems, in: *Natural and Artificial Flavoring Agents and Food Dyes*. doi:10.1016/B978-0-12-811518-3.00001-6.

Reineccius, G., 1999. *Source Book of Flavors*. Aspen Publishers, New York.

Richardson, L., 2013. Flavour, taste and smell. *Mind and Language* 28, 322–341. doi:10.1111/mila.12020.

Sardarodiyan, M., Hakimzadeh, V., 2016. Artificial sweeteners. *International Journal of PharmTech Research*. doi:10.1007/s13749-012-0003-5.

Schieber, A., 2012. Functional foods and nutraceuticals. *Food Research International*. doi:10.1016/j.foodres.2012.02.009.

Schrankel, K.R., 2004. Safety evaluation of food flavorings, in: *Toxicology*. doi:10.1016/j.tox.2004.01.027.

Sibinovska, N., Komoni, V., Netkovska, K.A., Vranic, E., Crcarevska, M.S., Dodov, M.G., 2016. Novel approaches in treatment of Acne vulgaris: Patents related to micro/nanoparticulated carrier systems 62, 3–16.

Smith, B., 2015. The nature of flavours and flavour experiences, in: *The Oxford Handbook of Philosophy of Perception*.

Smith, R.L., Cohen, S.M., Fukushima, S., Gooderham, N.J., Hecht, S.S., Guengerich, F.P., Rietjens, I.M.C.M., Bastaki, M., Harman, C.L., McGowen, M.M., Taylor, S. V., 2018. The safety evaluation of food flavouring substances: The role of metabolic studies. *Toxicology Research* 7, 618–646. doi:10.1039/C7TX00254H.

Tosh, S.M., Miller, S.S., 2015. Oats, in: *Encyclopedia of Food and Health*. doi:10.1016/B978-0-12-384947-2.00497-9.

Tournier, C., Sulmont-Rosse, C., Guichard, E., 2007. Flavour perception: Aroma, taste and texture interactions. *Food* 1, 246–257.

Van der Schaft, P., 2015. Approaches to production of natural flavours, in: *Flavour Development, Analysis and Perception in Food and Beverages*. pp. 235–248. doi:10.1016/B978-1-78242-103-0.00011-4.

Victoria Urquiza-Martínez, M., Fenton Navarro, B., 2016. Antioxidant capacity of food. *Free Radicals and Antioxidants* 6, 1–12. doi:10.5530/fra.2016.1.1.

Vilgis, T., 2013. Texture, taste and aroma: Multi-scale materials and the gastrophysics of food. *Flavour* 2, 1–5. doi:10.1186/2044-7248-2-12.

Volpe, S.L., 2015. Functional food and health. *ACSM's Health and Fitness Journal* 19, 32–34. doi:10.1249/FIT.0000000000000087.

Weihrauch, M.R., Diehl, V., 2004. Artificial sweeteners—Do they bear a carcinogenic risk? *Annals of Oncology* doi:10.1093/annonc/mdh256.

Wright, J., 2010. Creating and formulating flavours, in: *Food Flavour Technology: Second Edition*. pp. 1–23. doi:10.1002/9781444317770.ch1.
Xu, R., 2012. Oat fibre: Overview on their main biological properties. *European Food Research and Technology*. doi:10.1007/s00217-012-1666-2.
Zviely, M., 1997. Flavours and fragrances, in: *Trends in Food Science and Technology*. pp. 277–279. doi:10.1016/S0924-2244(97)01044-3.

9

Flavors and Fragrances

Natural and Food-Grade Ingredients for Nutraceuticals Applications

Rahul Shukla, Nikhil Chauhan, Chetan Rajak, and S. J. S. Flora

Contents

9.1 Introduction

Flavor and fragrance ingredients are relatively organic compounds with distinctive pleasant odors. Therefore, they are being used as fragrance and flavoring agents in different sorts of nutraceuticals, food and beverages (Capuzzo et al., 2013). Basically, flavor and fragrances are chemical moieties whose receptors resides in the olfactory cells of the nose. Flavor is basically composed of three components, i.e. aroma (fragrance) which can be detected by sensory receptors due to its presence in olfactory cells, taste which can be perceived by taste buds available on the tongue and consistency that comprises of physical texture like mouth feel and a physical sensation that determines an individual perception. The most crucial component of flavors is fragrance (Booth, 1994). In 16th century, a Swiss physician coined the term essential oils. Fragrance and flavors (F&F) are most important components of any food product because of their positive or negative impact of the brand on the customer. Flavors are segmented based upon on the industry they serve (Bakkali et al., 2008). Commonly used flavors include fruity flavor, savory flavor, cheese and other dairy flavors.

Fruity flavors are used in fruit flavored drinks and beverages. Savory flavors are used in cooking sauces, snacks and ready to eat meals. Cheese and other flavored milk are used in yoghurt, flavored milk, cheese and other dairy product manufacturer. Fragrance of the product is described by the notes which can be sensed by use of a particular product. The notes of the fragrances are top and middle note. Top/head note is the lightest note of fragrance that can be sensed immediately, e.g. citrus fruits (orange and lemon), light fruits (grapefruit, berries) and so on. Middle/heart note is sensed when the top note evaporates. It has long lasting effect as compared to top note, e.g. rose, lemongrass, coriander and so on (Mensing and Beck, 1988).

Flavor plays an important role in promoting consumption of food products by adding to the functionality, quality and taste acceptance of food products. This prompts the flavor industry to actively participate in developing new flavor ingredients. When added in little quantities to food products, flavors have a huge impact on quality, cost and consumer satisfaction of the finished product. Flavor and fragrances are the characteristic features of fruits and vegetables that affect the consumer acceptance and sensory perception. Human perception for chemical senses like taste and odor are stimulated by consumption of food and beverages. The flavor, texture, appearance and odor of food are considered to be decisive for the selection and ingestion of a particular food (Kader, 2008). In the early 1970s, less than 1,500 flavor chemicals had been identified in food products. Currently, number of flavors has increased manifold times, surpassing around 7,000 compounds with nearly 300 compounds identified only from strawberry and over 1,000 compounds from coffee (Ehab, 2011). Global market of flavor and fragrance was approximately ~18 billion US dollar in the year 2006. Most of the flavors are of synthetic origin, and only less than 5% flavors are extracted from the natural sources. Bouquet of rose petals, lemon scent and tingle of cinnamon bark constitute some natural fragrance. Food beverages and nutraceuticals that contain artificial flavors are mostly avoided by consumers due to safety issues associated with them. Mostly flavor industry is focused mainly on switching from artificial flavors to natural-based flavors (Yaylayan, 1991).

The quality attributes of a flavor can be categorized into two ways:

1. The components which impart characteristic taste to the food
2. The essential oils which impart distinctive odor to the food and nutraceutical products.

Both the quality attributes are imparted due to metabolic processes of the organism. The metabolic enzymes present in food do not actually form a part of the flavor ingredients but act as catalytic agents accomplishing the required action for the synthesis of the flavor components. However, processing of foods has deleterious effect on the flavors present in them. The main reason behind loss in perception of flavor in processed food is due to their sensitivity to heat. Since most of the flavor components, particularly the odor/flavor components, are volatile as well as thermolabile, processing of food leads to their degradation. In contrast to odor flavor, many of the components of taste flavors are relatively non-volatile and thermosensitive, thus surviving the processing operation. In order to overcome these shortcomings, the processed food industry began to use natural and synthetic flavor fortifiers (Burnette, 1977). However, fortification with such flavors may present serious problems with respect to government regulations such as those of the Food and Drug Administration (FDA). In many instances, fortified foods are devoid of the natural flavors and qualities of the fresh material. Also fortification by natural flavor leads to higher cost due to cannibalization. In general, flavor and fragrance compounds do not form a prominent part of plant basically accounting less than 1% of entire weight of plant. Sometimes their origin might be from precursors, which constitutes a large portion of the plant. Notable examples include Maltol (a burnt sugar), vanillin, beta-ionone (berry flavors) having carbohydrate, lignin and carotenoid respectively as their precursor (Huma et al., 2007). Flavor and fragrance (F&F) chemicals are basically derived from plant's secondary metabolites that are produced typically for secondary processes like protection or propagation and in no way related to the plant's primary internal processes such as photosynthesis (Schwab et al., 2008). Quinine and caffeine produced by cinchona and coffee plant respectively are among the bitter compounds produced by the plants in order to protect them from foraging animals. Similarly, flowering plants release aromatic compounds in order to attract pollinators to the plant, ensuring their completion of reproductive cycle. Many of the released secondary metabolites reside in the plant and are chemically bound to glycosides, prior to their release as flavor or fragrance compounds. Treating plant material with enzymes designed to cleave the glycosidic bonds can release secondary metabolites which are not available, increasing the yield of flavor and fragrance materials recovered (Flores et al., 1987).

Nutraceuticals are basically food products and used as dietary supplements with certain therapeutic efficacy. The term "nutraceuticals" derives its name from nutrition and pharmaceutics. The term can be applied to dietary supplements, processed foods such as cereals, soups and beverages as well as products that are isolated from herbal extracts. They are mainly used to improve the health, delay the aging process, prevent from chronic diseases, support the structure or function of the body (Das et al., 2012). Nutraceuticals, owing to their potential nutritional as well as therapeutic efficacy, have received considerable interest. According to market research, the global nutraceuticals market has expanded exponentially and was expected to cross US$250 billion by 2018. Nutraceuticals are available in several forms including antioxidants, minerals, dietary fibers, vitamins and omega 3 poly unsaturated fatty acids (PUFA) (Chauhan and Kumar, 2013).

9.2 Flavor and fragrance

Flavor and fragrance are added to foodstuffs, beverages and nutraceuticals in order to increase the functionality as well as consumer acceptance (Booth, 1994).

9.2.1 Difference between flavors and fragrances

In most instances these two terms are used interchangeably. However, these terms could be distinguished based upon the following parameters which are discussed briefly in **Table 9.1**.

9.2.2 Physiological mechanism of flavor and fragrance

Flavor is perceived through a number of processes prior to the sensory gustation effects like taste and smell, followed by the usual sense of chemicals (Delahunty et al., 1996). The senses that elicit special features of flavor substances are due to their involvement with distinct chemical stimulant interaction from a food or beverages. It modulates the response of other

Table 9.1 Difference between Flavors and Fragrances

Additives	Description	Taste	Sensory Organ	Uses	Interrelation	References
Flavor	Flavor is composed of aroma (fragrance), taste and texture.	Distinctive taste	Tongue	To describe the food	Influenced by the fragrance	Kraft and Swift, (2005); Laing and Jinks (1996)
Fragrance	Fragrance is a component of flavor mixture.	Sweet and pleasant smell	Nose	To describe the floral scents	Not influenced by flavor	

probable senses also. **Figure 9.1** clearly depicts the physiological mechanism of flavors and fragrances through the nasal cavity. However, beginning of the perception is marked by physical appearance of food for its identification and further deciding factor whether they are qualified for consumption or not. Nasopharynx and saliva play an important role in the movement of volatile compounds. Nasopharynx helps in transportation of volatile compounds (odor and irritants) throughout the nasal cavity and its associated receptors where the saliva aids the motion of known volatile compounds to taste sensitive region of mouth cavity (Kraft and Swift, 2005). Since the senses of taste and small result directly by the action of flavor and fragrances on the gustatory and olfactory receptors in the mouth and nose respectively. These receptors get saturated at very low levels in animals. This makes the utilization of flavor and fragrances limited due to their lower effect (Laing and Jinks, 1996).

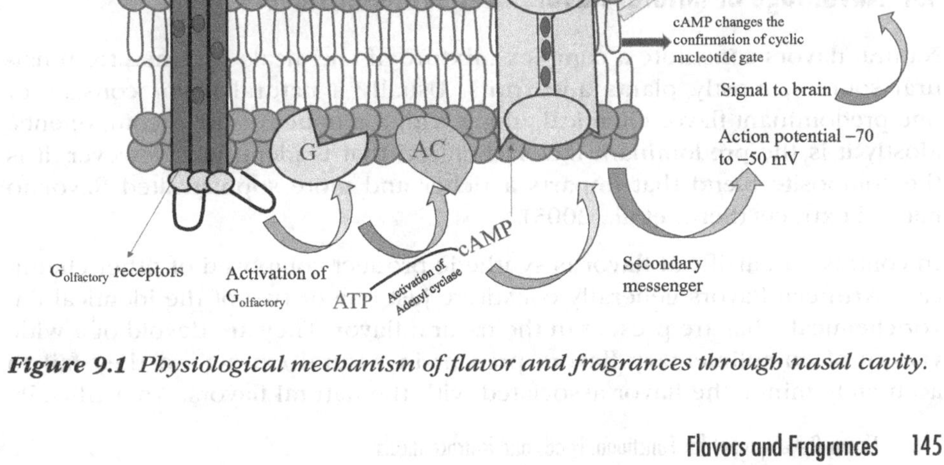

***Figure 9.1** Physiological mechanism of flavor and fragrances through nasal cavity.*

9.3 Coding of chemosensory information

9.3.1 Tastants

In general, there are five major classes of tastants have been recognized by humans including bitter, salty, sweet, sour and umami substances (Pearson, 1991). These tastants are sensed through the definite cells of the taste receptors located throughout the tongue area, especially on front, rear, side followed by the cheeks, soft palate and esophagus (Collingst, 1974). Above areas are responsible for the transmission of specific taste to the brain system thereby the signals are generated for the particular taste. The frontal region of the tongue governs the sweetness of the substances, left and right side of the tongues delivers the information about the either salty or sour tastants while the rear space examines the exposure of bitter stimuli irrespective of their sensitivity on entire location of tongue (Herness and Gilbertson, 1999).

9.3.2 Odorants

Physiological evidence suggest that exposure to a particular odorant leads to generation of three-dimensional structure in the olfactory bulb as well as different site of the brain (Wilson and Baietto, 2009). Further transmission of a specific odor is identified through the group of neurons associated to nose. Odorants act on various receptor cells to produce action potential, which further converges on certain bulbar cells and then generates a specific pattern for particular cells and activation of these cells deliver the information about the feature of odorants (Kauer and White, 2001). The bulbar cells create a map that is further transported to the next regions of the olfactory nerves, specifically in the brain, which aids in eliciting the responses to that odorant. Often, the production of participating cells varied in the nasal cavity seeing that the cells entangle in each geometry increases with the mass concentration of odorants and vice versa (Laing and Jinks, 1996).

9.4 Advantage of natural flavors over synthetic flavors

Natural flavors constitute a complex mixture of chemicals obtained from natural sources, mostly plants and fruits. Usually a natural flavor consists of one predominant flavor chemical, along with a number of other components. Mostly it is the predominant flavor chemical that is identified; however, it is the composite blend that imparts a richer and more sophisticated flavor to natural extracts (Serra et al., 2005).

In contrast, an artificial flavor is synthetic product composed of other chemicals. Artificial flavors generally constitute just one or two of the identical flavor chemicals that are present in the natural flavor. They are devoid of a wide variety of complimentary flavors present in natural extract, so they fail to accurately mimic the flavor associated with the natural flavors. An artificially flavored food would appear bland upon tasting, and the person would only be able to recognize the predominant flavor chemical (Manley, 2000); however, some are better than others, and in order to choose best out of it we need to have a better knowledge of these flavors.

9.5 Difference between natural and synthetic flavor and fragrances

Natural flavors and fragrances and synthetic flavors and fragrances could be differentiated from each other based upon some existing properties associated in terms of their sources, the way they are derived as shown in **Table 9.2**.

9.6 Some natural flavors and fragrance may be toxic

We generally have a classical notion that whatever we obtain from natural sources are beneficial for our health and are devoid of any potential side effects and toxicity. However, it is not always the case. Consider example of comfrey which was earlier used as a healing herb, later on found to contain a chemical which predisposes to liver cancer. Owing to the possibility of such potential risks in flavor compounds, their use in food or food like products is strictly regulated by a proper regulatory body for their consumption toward human population (Smith et al., 2005). In United States of America, the regulatory body so far involved is Food and Drug Administration (FDA) which monitors use of a particular flavor for the human purpose (Oser et al., 1973). All the regulatory action, are taken by them, which belong to the Food Additives Amendment, section 409 of Food and Drugs Act, 1958. An expert panel is provided by the Flavor and Extract Manufacturers Association (FEMA), which determines those flavors and the permitted levels that are granted exemptions from the section 409 and which are generally recognized as safe (GRAS) (Hallagan and Hall, 1995). Due to its potential oral toxicity and various skin problems such as potential irritation, contact sensitization, and/or photo toxicity and skin photosensitization, the International Fragrance Association (IFRA) (formed in 1973), has created several industrial guidelines

Table 9.2 Major Differences between Natural Flavor and Synthetic Flavor

Flavors	Sources	Production	Environmental Status	Safety Status	Cost	References
Natural	Plants or fruits	Extraction	Less hazardous	Partially safer	Expensive	Carocho et al. (2014)
Synthetic	Chemicals	Chemical reactions	Somewhat hazardous	Slightly toxic	Less expensive as compare to natural flavors	Hallagan and Hall (1995)

in order to restrict ingredient usage in fragrances. New fragrance ingredients, for which no adequate safety data has been reported, should be used only prior to satisfactory evaluation by competent toxicologists. The urge to eat a particular food product is determined mostly by taste and aroma. However, in certain cases it even helps us to evaluate the quality of food as well as potential contaminants. US Food and Drug Administration (21 CFR Sec. 182.10, 182.20, 482.40 and 182.50) specifies that the use of natural flavors is generally recognized as safe (Hallagan and Hall, 1995).

9.7 IFRA recommendations on fragrance ingredients

Fragrance ingredients are added to nutraceuticals, beverages and food products to enhance the palatability of products and customer acceptance. But International Fragrance Research Association (IFRA) has limited the use of some fragrance ingredients due to lack of safety data of some particular fragrance ingredients, which are mentioned in **Table 9.3**.

Table 9.3 IFRA Recommendations on Fragrance

Natural Product	Limit Status	References
Elecampane oil	Prohibited	Cadby et al. (2002); Ford et al. (2000)
Angelica root oil	Restricted; limit to 3.9%	
Berga mot oil	Restricted; limit to 2%	
Bitter orange oil	Restricted; limit to 7%	
Cassia oil	Restricted; limit to 1%	
Chinopodium oil	Prohibited	
Cinnamon bark oil	Restricted; limit to 1%	
Costus root oil	Prohibited	
Cumine oil	Restricted; limit to 2%	
Fig leaf absolute	Prohibited	
Lemon oil, cold pressed	Restricted; limit to 10%	
Lemon berbina absolute	Limit to 1%	
Lemon berbina oil	Prohibited	
Lime oil, cold pressed	Restricted; limit to 2.5%	
Merry gold oil	Restricted; limit 0.25%	
Oak moss absolute and resinoid	Restricted; limit to 3%	
Opopanax	Restricted; use only from commiphoraerythraea	
Peru balsam	Restricted; used only from myroxylon Pereirae	
Rue oil	Restricted; limit to 3.9%	
Sassafras oil	Restricted; saffrole content should not exceed 0.005%	

9.8 Classification of flavors used in food products

Mainly three types of flavors have been categorized under European Union and Australia for its utilization in foods:

1. **Natural flavoring substances**
   Substances which neither undergo any chemical modification nor change their form which is obtained from the plant materials, fruits followed by animal materials. These flavors are extracted out from the materials and are subjected to further purification process to avoid any contamination residue (Serra et al., 2005).
2. **Natural substance with identical flavoring**
   The substance produced through the little modification of natural materials or being synthesized by the process of chemicals and chemically similarity exists with flavor from natural sources (Serra et al., 2005).
3. **Artificial flavoring substances**
   The flavors that are produced through this method are likely to go for chemical procedure where they are altered by the process of chemical synthesis and do not rely on the natural materials (Maier, 1970).

9.9 Different types of flavors and fragrances

Flavor and fragrances are complex mixtures that either chemically synthesized or obtained from natural sources. Flavor and fragrances constituents are obtained from different natural sources like fruits, vegetables, spices, essential oils, etc. On the basis of natural occurrence different types of flavor and fragrances are summarized in **Table 9.4**.

Table 9.4 Different Types of Flavors and Fragrances

Flavor types	Examples	References
Fruit flavor	Grapefruit, Orange	Kader (2008)
(i) Citrus type flavors	Apple, Raspberry, Banana	
(ii) Berry type flavors		
Vegetable flavor	Lettuce, Celery	Schwab et al. (2008)
Spice flavor		Das (2015)
(i) Aromatic	Cinnamon, Peppermint	
(ii) Lachrymogenic	Onion, Garlic	
(iii) Hot	Pepper, Ginger	
Beverages flavor		
(i) Unfermented flavors	Juices, Milk	
(ii) Fermented flavors	Wine, Beer, Tea	
(iii) Compounded flavors	Soft drinks	

9.10 Production of flavors and fragrances

Basically there are four types of fruits that are of produced globally and commercialized throughout, among which citrus fruit, bananas, apples, grapes, pears, peaches, and plums are primary. However citrus fruits recently overtook grapes as the world's most widely produced fruit in the year of 1991 (Krings, 1998; Serra et al., 2005). From 1980s, China is the largest producer of fruit and vegetables, followed by India (Stringer et al., 2009). The current status of production of major fruits (globally) is shown in **Figure 9.2** and the production of flavor and fragrances (India) is shown in **Figure 9.3**. Production of flavor and fragrances obtained from natural plants is clearly depicted in **Figure 9.4**. Enzyme utilization in the production of flavor and fragrances is discussed in **Table 9.5** and **Table 9.6** depicts the production of flavor using microorganisms.

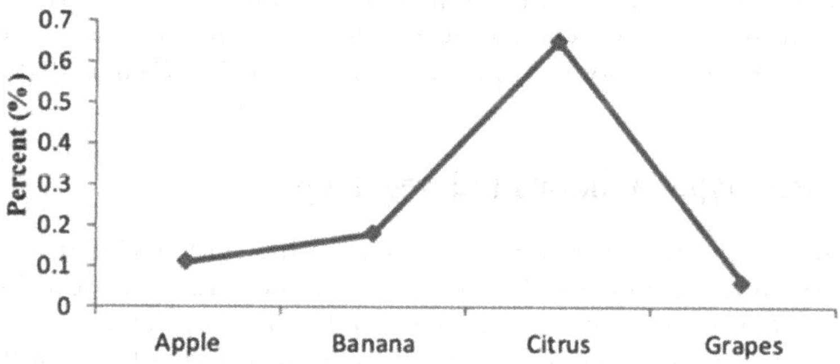

Figure 9.2 *Global production of major fruits for flavor and fragrances.*

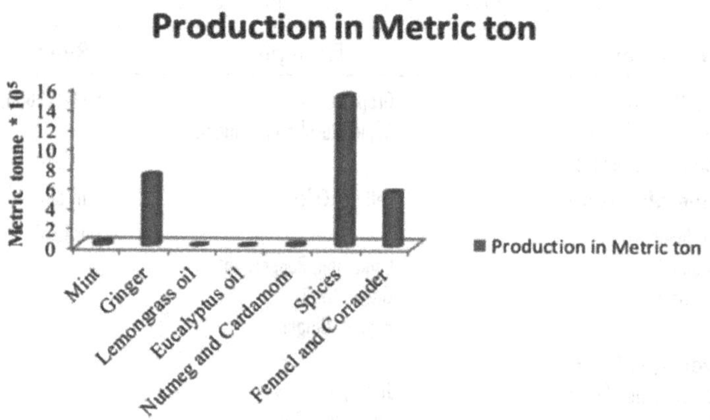

Figure 9.3 *Production of flavor and fragrance in India.*

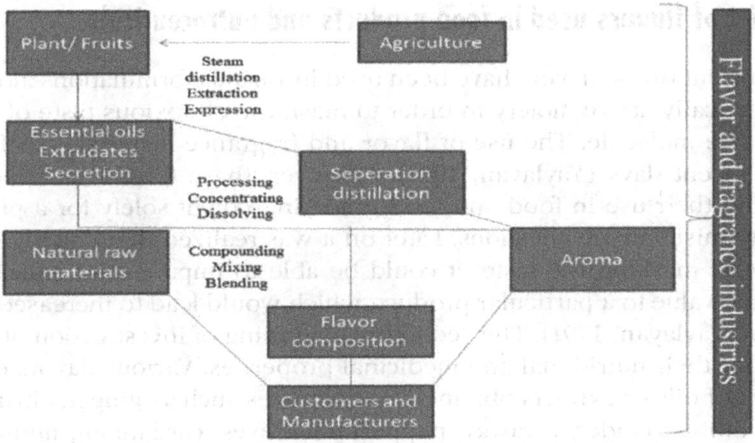

***Figure 9.4** Production of flavor and fragrances obtained from natural plants.*

Table 9.5 Enzyme Utilization in the Production of Flavor and Fragrances

Enzyme	Substrate	Flavor/Product	References
Lipase	Butyric acid with isoamyl alcohol and volatile oil	Flavor ester like isoamyl acetate (banana)	Krings (1998)
Lipase	Acetic acid/butyric acid and geraniol	Rose	
Soy lipoxygenase	Eugenol and coniferyl aldehyde	Vanillin	
Beta glucosidase, mandelonitrile	Cyanogenic glucoside amygdakin from almond and cherry	Cherry and almond tasting benzaldehyde	
Protease	Residue associated with processing of cray fish	Benzaldehyde, pyrazines	
Lipase	Butyric acid/acetic acid and citronellol	Lemon	Feron et al. (1996); Vandamme (2018)
Lipase	Ethyl valerate	Green apple flavor	
Protease	Esterification of protease	Savory flavor	
Carboxy esterase	Acetyl ester 2	Banana flavor	

Table 9.6 Production of Flavor Using Microorganisms

Substrate	Microorganism	Flavor Product	References
Eugenol	*Pseudomonas* and *Arthrobacter*	Vanillin	Feron et al. (1996); Vandamme (2018)
Apple pomance	*Tyromyceschioneus*	Modulate flavor and impart fragrance to wine	Feron et al. (1996); Vandamme (2018)
Grapes	*Saccharomyces cerevisiae*	Wine	Feron et al. (1996; Vandamme (2018)
Eugenol	*Pseudomonas putida*	Vanillic acid	Feron et al. (1996); Vandamme (2018)
Eugenol	*Rhodococcusopacus PD630*	Vanillin	Feron et al. (1996); Vandamme (2018)

9.11 List of flavors used in food products and nutraceuticals

From ancient times, flavors have been used in various formulations along with therapeutically active moiety in order to mask the obnoxious taste of biologically active molecule. The use of flavor and fragrance materials has been on rise in recent days (Yaylayan, 1991). However, these flavors lack medicinal value and their use in food and food grade ingredient solely for a particular property raised many questions. Later on it was realized that, if a given flavor apart from masking the taste, it could be able to impart some nutritious or medicinal value to a particular product, which would lead to increased market demand (Yaylayan, 1991). This led to the beginning of investigation of various flavors for their nutritional and medicinal properties. Various flavors either in the form of oils or extracts obtained from sources such as ginger, citrus fruits, fennel, dill, coriander, caraway, peppermint leaves, cardamom, nutmeg, cinnamon, etc., have been explored for their nutritional properties (Srinivasan, 2005). Liquorice extract, which is being used to mask taste of anti-nauseous medicines and wild cherry syrup *BPC*1988, though they have been used in the formulation of several preparation along with cough preparations, but are known most commonly for their flavoring characteristics. Similarly, raspberry and black currant syrups have been found to lack therapeutic value but their extract contains a little amount of ascorbic acid and thus used as a source of vitamin C; however, saffron and rose oil (used as flavoring agent for lozenges) reported to be devoid of any medicinal value (Kader, 2008).

9.12 Fruits source for flavors and fragrances

9.12.1 Vanillin

The principal flavor chemical vanillin constitutes both natural and artificially flavored vanilla. However, the bean extract constitute three other major components, vanillic acid, 4-hydroxybenzoicacidand 4-hydroxybenzaldehyde, accounting for 17 percent (by weight) of the flavor chemicals that make up vanilla (Walton et al., 2000). Although, none of these chemicals actually smell or taste like vanilla, they contribute to the flavor and scent of vanilla extract owing to the presence of flavors and scents of their own. Neither vanilla extract nor synthetic vanillin poses any health risks (Sharma and Sharma, 2007). The only difference between these two is the perceived flavor and the actual cost.

9.12.2 Grapes

Unlike vanilla, there is no single principal flavor chemical present in grapes. The flavor arises from the combined effect of many chemicals and sugars. The differences in composition of the natural and artificial flavors of grape and vanilla are profound. While the flavor of vanilla extract is solely determined by a single chemical, the flavor of grapes is the product of hundreds of naturally occurring chemical compounds. Grapes crop are on sixth position

Table 9.7 Types of Terpene Alcohol Present in Grapes and Other Natural Sources

Chemicals	Natural Occurrences	References
Linalool	Obtained from nutmeg, butter and beer	Marais (1983); Mateo and Jimenez (2000)
Citronellol	Rich in eucalyptus, coriander and apricot	Marais (1983); Mateo and Jimenez (2000)
Nerol	Carrots and rosemary	Mateo and Jimenez (2000)
Geraniol	Citrus fruits and roses	Marais (1983)
Myrcenol	Present in citrus fruits and liquorice	Mateo and Jimenez (2000)

among entire crops which are harvested in United States and have highest yield in production of fruit crops (Yang and Xiao, 2013). **Table 9.7** clearly states that types of terpene alcohol present in grapes and other natural sources.

9.12.3 Apple

Apple fruit constitutes more than 300 volatile compounds. Even though it contains hundreds of chemical constituents, the characteristic active odor and aroma of the fruit is attributed to the presence of only few of these volatile compounds. It mostly constitutes esters such as β-damascenone, butyl, isoamyl, and hexyl hexanoate, along with ethyl, propyl, and hexyl butanoates. Esters, alcohols, aldehydes, ketones and ethers, are among the most abundant volatile components, but the prominent fruity odor results mainly from the ester compounds (Dimick et al., 1983).

9.12.4 Banana

Banana fruit contains a number of volatile components. The major volatile components among these include alcohols and esters, including amyl acetate, isoamyl acetate, butyl butyrate and amyl butyrate. The volatile fraction mostly composed of esters in banana fruit (Beekwilder et al., 2004). A combination of analytic chemistry along with sensory analysis led to the identification of penten-2-one, 3-methylbutyl and 2-methylpropylesters of acetate and butyrate as the crucial constituent of banana fruit for aroma (Knee, 2002). Isopentyl acetate and isobutyl acetate have also been found to be important for aroma of banana. The ripening of banana fruit is marked by the increase in concentrations of acetates and butanoates. Olfactometric analysis further helped in detection of isoamyl alcohol, isoamyl acetate, butyl acetate and elemicine, which have been found responsible for imparting peculiar odor to banana (Golding et al., 1999).

9.12.5 Strawberry

Strawberry fruit has been reported to contain over 360 different volatile compounds. The aroma of strawberry results predominately from esters. However, alcohols, ketones, lactones and aldehydes are also found in fewer amounts.

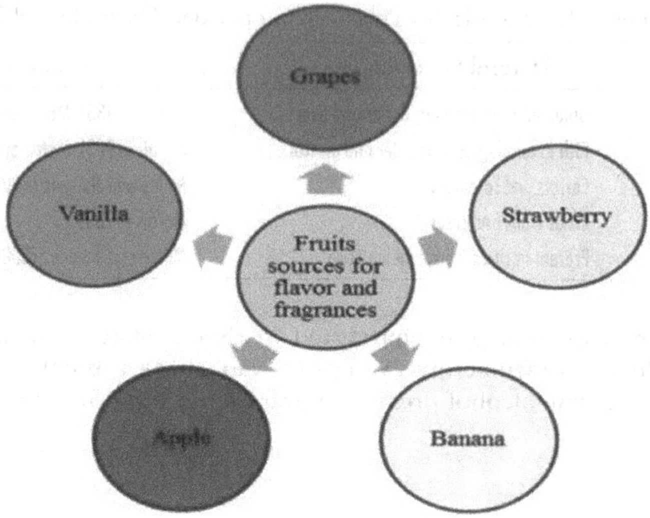

Figure 9.5 Flavor and fragrance sources in different fruits.

Strawberries contain predominantly straight esters, such as methyl and ethyl acetates, butanoates and hexanoates (Rios et al., 2014). Esters mainly impart a distinctive aroma to fruit. The fragrance of strawberry is also influenced by sulfur compounds and terpenoids. The most compounds in strawberry fruit responsible for imparting aroma as well a fruity flavor include ethyl cinnamates, methyl cinnamates, 2,5-dimethyl-4-hydroxy-3(2H)-furanone, furaneol, furaneol-beta- glucoside, dimethyl-4-methoxy—3(2H)—furanone (mesifurane), methyl and ethyl acetates, propionates and butyrates (Hathcock, 2001). Flavor of strawberry is also influenced by terpenes (Yan et al., 2018). Nutritional composition found in 100 g fresh weight strawberry are vitamins (B1, B2, C, A) and minerals like selenium, iron, calcium followed by carbohydrates (**Figure 9.5**).

9.13 List of spices and essential oils used as flavor and fragrance

Spices and herbs are being widely used from ancient times. They possess antioxidant activity due to their chemical constituents and are used for preservation of beverages, food products and nutraceuticals. Spices facilitate flavor of the beverages and influence perception of consumer. They inhibit oxidation of lipids in food products. The benefits of spices and herbs are depicted in **Figure 9.6**. The general applications of spices and herbs are shown in **Figure 9.7**. **Table 9.8** clearly states herb sources for flavor and fragrances with nutraceuticals properties.

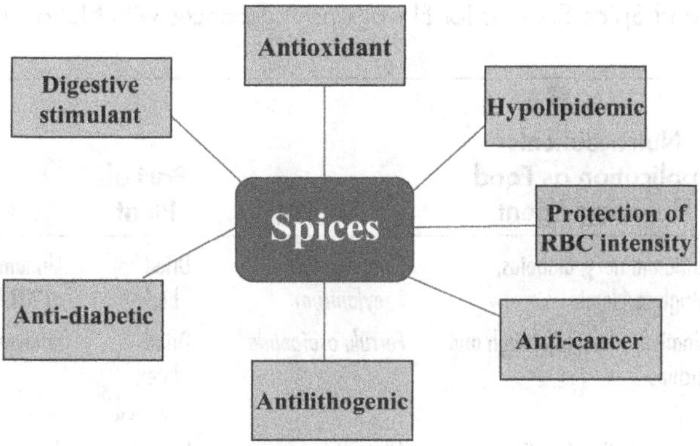

Figure 9.6 *General benefits of spices and herbs.*

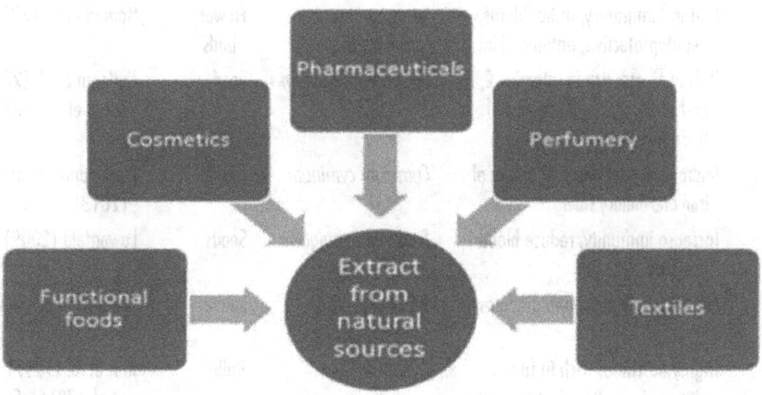

Figure 9.7 ***Application of natural extract (flavor and fragrances) in various industries.***

9.13.1 Essential oils

Essential oils are obtained from natural and are complex mixture of multi-component systems composed mainly of terpenes and some components other than terpenes (Edris, 2007). Essential oils has been found to present in several of the plant parts beginning from the root to leaves followed by seeds, fruits, berries, rhizome, flowers, bark, wood, resins and peels. Their composition might vary during the extraction for the process of getting the essential oils (Masotti et al., 2003). Most of the essential oils have the tendency to cure several disorder/ailments (Hajhashemi et al., 2003; Perry et al., 2003; Silva et al., 2003). The use of these essential oils in food industries has been widely explored in the past few decades and is expected to get the high demand in upcoming future (Bakkali et al., 2008). The application of few of these oils has been shown in **Tables 9.9** and **9.10**.

Table 9.8 Herb and Spice Sources for Flavor and Fragrances with Nutraceuticals Properties

Name of Herbs and Spices	Nutraceuticals Application as Food Flavoring Agent	Source	Part of Plant	References
Cinnamon	Anti-inflammatory, diabetes, antitriglyceridemia	*Cinnamomum-zeylanicum*	Dried bark	Muhammad and Dewettinck (2017)
Asafoetida	Carminative, whooping cough and bronchitis	*Ferrula asafoetida*	Dried latex exuded	Srinivasan (2005)
Cardamom	Diuretic properties, laxative, antidiabetic, antibacterial effect, anticancer	*Elletaria cardamomum*	Seeds	Anwar et al. (2015); Gilani et al. (2007)
Clove	Anti-inflammatory, antioxidants, hepatoprotective, antimicrobial	*Syzygium aromaticum*	Flower buds	Hamed et al. (2012)
Coriander	Rich of fibers, rich in vitamins C, K and proteins, small amount of thiamine, niacin	*Coriandrum sativum*	Seed	Nadeem et al. (2013); Sahib et al. (2013)
Cumin	Improve blood cholesterol, source of iron and dietary fiber	*Cuminum cyminum*	Seeds	Taylor and Sowbhagya (2013)
Fennels	Increase immunity, reduce blood pressure	*Foeniculam vulgare*	Seeds	Lowenfeld (2009)
Fenugreek	balance cholesterol and diabetes	*Trigonella foenum-graecum*	Leaves	Acharya et al. (2006)
Garlic	Highly nutritious, rich in fibers, vitamin C, B_6, B1, nicotinic acid, iron, antioxidant, antiviral, hypolipidemic	*Allium sativum*	Bulb	Kim et al. (1997); Ourouadi et al. (2016); Suleria et al. (2015); Touloupakis and Ghanotakis (n.d.)
Ginger	Antioxidants, antiobesity, rheumatoid arthritis and cardioprotective	*Ginger officinale*	Rhizome	Semwal et al. (2015); Srinivasan (2017)
Saffron	Antioxidants, diabetes, cancer	*Crocus sativus*	Stigma	Bathaie (2018); Mashmoul et al. (2013); Science (2014)
Turmeric	Powerful anti-inflammatory, boost the antioxidants, antifungal, anticoagulant	*Curcuma longa*	Roots	Das (2015)
Oregano	Rich in antioxidants	*Origanum vulgare*	Flowers	
Sage	Anti-inflammatory, antioxidant, vitamin B2, C, A, B1, niacin	*Salvia officinalis*	Leaves	Devansh (2012); Hamidpour et al. (2014)

Table 9.9 List of Some Essentials Oils and Their Potential Physiological Action

Essential Oils (Flavors and Fragrance)	Pharmacological Applications	References
Coriander sativum	Carminative, antioxidants	Singh et al. (2006)
Myristica fragrans	Potent hepatoprotective activity, antimicrobial activity	Lima et al. (2012); Periasamy et al. (2016)
Eucalyptus species	Analgesic and anti-inflammatory effects	
Origanum onites	Anti-angiogenic and anti-tumor activities	Biol et al. (2011); Pizzale et al. (2002)
Syzygium aromaticum	Antiseptic	Bhowmik et al. (2012)
Ginger officinale	Anti-inflammatory, antidiabetic, anticancer	Malhotra et al. (2003); Schnitzler and Koch (2007); Shukla and Singh (2007)

Table 9.10 Application of Flavor and Fragrances (Essential Oils) in Food Industries

Examples	Part of Plants	Extraction Methods	Application in Food	References
Anise	Dry ripe Seed of anise	Solvent extraction and supercritical fluid extraction	Food flavoring in fish, soups, ice-cream, cakes, sweet snacks	Kosalec et al. (2005)
Cinnamon bark	Bark	Soxhlet extraction	Widely employed in meat, confectionary and bakery foods	Bozan et al. (2003); Muthiah and Suja (2017)
Ginger officinale	Rhizome	CO_2 extraction	Sauces and beverages	Jitoe and Gara (1992)
Peppermint leaves	Leaves	Steam distillation	Chewing gum, dental products and confectionary products	Marinova and Yanishlieva (1991); Yanishlieva et al. (2006)
Rosemary	Leaves	Steam distillation under reduced pressure	Seasoning blends	Lee and Shibamoto (2002)
Jasmine	Flowers	Extracted by solvent extraction	Wines, champaign and other food products	Akram et al. (2017); Sabharwal, Swati; Sudan, Swati, Ranjan (2013); Ye et al. (2015)
Citrus fruit peels, bergamot, grapefruit, lemon, lime and orange	Peels	Cold expression	Beverages and confectionary foods	Linnemann et al. (2017)

9.14 Nutraceuticals applications as food ingredients

The concept of "nutraceutical" first came in to picture during a survey conducted in Germany, United Kingdom and France. It is observed that demand of diet is more as compared to genetic factors and workout in order to accomplish a better health benefits (Chauhan and Kumar, 2013). Nutraceuticals play an important role in combating and mitigating many disorders and illness related to lifestyle. Nutraceuticals contain certain food grade ingredients, which are used to maintain healthy life style. Nutraceuticals delay the aging process and reduces the risk of serious conditions such as cancer, heart disease, hypertension, excessive weight, high cholesterol, diabetes, osteoporosis, arthritis, insomnia, cataracts, constipation, indigestion (Ronis et al., 2018), and many other lifestyle-related disorders (Souyoul et al., 2018). **Figure 9.8** gives a broader vision of the nutraceutical market in United States billions of dollars. Nutraceuticals can be isolated and purified from plant, animal or marine sources. Some of the advantages of nutraceuticals include a longer half-life period, immediate action upon intake, ready availability, and fewer side effects (Keservani et al., 2015). There are some examples of dietary supplements which provide health benefits, summarized in **Table 9.11**.

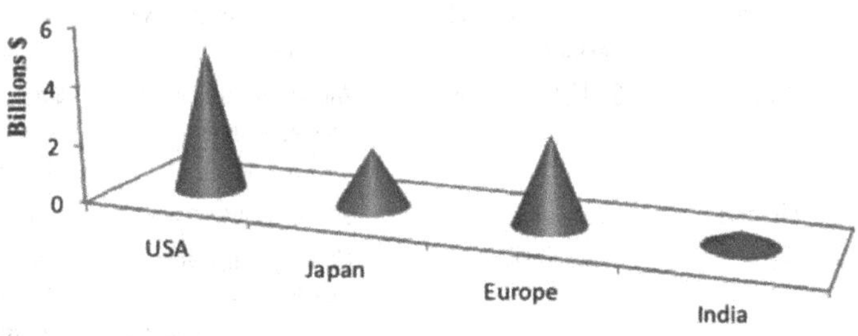

Figure 9.8 *Nutraceutical market in US billions of dollars.*

Table 9.11 Food Grade Ingredients and Their Applications in Nutraceuticals

Ingredients	Applications	References
Isoflavone	Beneficial in menopause symptoms	Caveats (2016)
Green tea extract	Free radical scavengers	Perumalla and Hettiarachchy (2011); Sarma et al. (2008)
Lutein	Macular degeneration	Mozaffarieh et al. (2003)
Carotenoids	Enhances immunity	Jaswir et al. (2011)
Saponins	Hypolipidemic	Mazza (2007); Oakenfull (1981)
Polyphenols	Free radical scavengers	Fang (2010)
β-Glucan	Anti-diabetic and hypolipidemic	Ahmad et al. (2012); Ahmad and Kaleem (2018)
Peptides	Muscle growth	Hartmann and Meisel (2007)

Table 9.12 Amount of Dietary Fibers in Different Foods and Fruits

Foods and Fruits	Level (g/100 g) Except Fructans	References
White bread	2.0	Das et al. (2012)
Brown bread	4.5	Das et al. (2012)
Cabbage	2.0	Das et al. (2012)
Baked beans	4.5	Das et al. (2012)
Boiled carrots	3.1	Das et al. (2012)
Bananas	1.9	Das et al. (2012)
Apple	2.0	Das et al. (2012)

9.14.1 Dietary fibers as nutraceuticals

Dietary fibers basically includes the nutritional supplements. Specifically, the substances obtained from various source of plants which is generally not prone to hydrolyzed by gastrointestinal enzymes. They are not hydrolyzed by the enzymes present in the digestive tract; rather, digestion occurs through gut microflora (Leclere et al., 1994). Dietary fibers are comprised of dextrin, gums, lignin, pectin and cellulose. All these components are polysaccharides but do not belong to starch. Fruits, oats, barley and beans are some of the foods rich in soluble fiber (Ronis et al., 2018). The amount of dietary fiber present in particular foods is as follows in **Table 9.12**.

9.15 Conclusion

Flavor and fragrances are essential ingredients for the industries dealing with the products intended for daily human intake. Food-based industries are investing huge amounts of money for purchasing high-quality flavors and fragrances for their products that exert health benefits which could lead to lifting their products in the market. Flavor and fragrances are crucial factors in food products because consumers sense them before consumption. In recent days, much focus has been carried out for supplying the flavor in the foods having nutrition value like antioxidants which would be impressive for the consumer. This may lead to raise market demands; hence, this could be considered as a nutraceuticals commodity. Apart from their utilization throughout the food products, their safety and tolerance limits must be taken care of as they are originated either naturally in the form of chemicals or being synthesized artificially inside the chemicals laboratory. All the existing flavors are chemicals but the assumption in regards to the safety comments that flavors belonging to natural sources are quite safer over artificially synthesized. Although the natural flavors are safer but their cost is slightly high since they are undertaken for the critical procedure which accounts for the reason of the expensiveness. Similarly artificially flavors are cheaper and used extensively but supposed to be toxic to some extent. The authority bodies dealing with flavors are caring all these aspects before they are practiced into the food products. IFRA already established the application of these flavors in a limited quantity. In future, flavor and fragrance demands may be expected to be high in each and every food products as they can deliver the small portion of nutrient to the consumer for the benefits of their health which would be aiding some medicinal value. The scope of searching for the novel flavor and fragrances are of great concerned for the researcher conducting the study on the flavor and fragrances containing nutritious components because of limited numbers of availability. Although Natural flavors and fragrances are being widely used in food products and beverages to attract the customers, but there should be more future perspectives related to the use, mechanism of action in biological medium, path for distribution, tolerance limit and scope of benefits of flavor and fragrances in nutraceuticals and food products in upcoming future.

Acknowledgment

We are thankful to Department of Pharmaceuticals (DoP), Government of India for supporting us. NIPER-R communication number for this chapter is NIPER-R/Communication/059.

References

Acharya, S.N., Thomas, J.E., Basu, S.K., 2006. Fenugreek: An "old world" crop for the "new world." *Biodiversity* 7, 27–30. doi:10.1080/14888386.2006.9712808.

Ahmad, A., Anjum, F.M., Zahoor, T., Nawaz, H., 2012. Beta glucan: A valuable functional ingredient in foods. *Critical Reviews in Food Science and Nutrition* 52(3), 201–212. doi:10.1080/10408398.2010.499806.

Ahmad, A., Kaleem, M., 2018. β-glucan as a food ingredient, in: *Biopolymers for Food Design*. Elsevier Inc. doi:10.1016/B978-0-12-811449-0.00011-6.

Akram, A., Younis, A., Akhtar, G., Ameer, K., Farooq, A., Hanif, M.A., Saeed, M., Lim, K., 2017. Comparative efficacy of various essential oil extraction techniques on oil yield and quality of *Jasminum sambac L. Science International* 5(3), 84–95. doi:10.17311/sciintl.2017.84.95.

Anwar, F., Abbas, A., Alkharfy, K.M., Gilani, A. ul H., 2015. Cardamom (*Elettaria cardamomum* Maton) oils, *Essential Oils in Food Preservation, Flavor and Safety*. Elsevier Inc. doi:10.1016/B978-0-12-416641-7.00033-X.

Bakkali, F., Averbeck, S., Averbeck, D., Idaomar, M., 2008. Biological effects of essential oils—A review. *Food and Chemical Toxicology* 46, 446–475. doi:10.1016/j.fct.2007.09.106.

Bathaie, S.Z., 2018. Saffron as a functional food and a nutraceutical using saffron and its constituents as the nutraceutics to protect against chronic diseases. *Acta Hortic* 1200, 201–204. doi:10.17660/ActaHortic.2018.1200.33.

Beekwilder, J., Alvarez-huerta, M., Neef, E., Verstappen, F.W.A., Bouwmeester, H.J., 2004. Functional characterization of enzymes forming volatile esters from strawberry and banana. *Plan Physiology* 135, 1865–1878. doi:10.1104/pp.104.042580.

Bhowmik, D., Kumar, K.P.S., Yadav, A., Srivastava, S., Paswan, S., Sankar, A., 2012. Recent trends in indian traditional herbs *Syzygium aromaticum* and its health benefits. *Journal of Pharmacognosy and Phytochemistry* 1, 13–22.
Biol, T.J., Özkan, A., Erdoğan, A., 2011. A comparative evaluation of antioxidant and anticancer activity of essential oil from *Origanum onites* (Lamiaceae) and its two major phenolic components. *Turkish Journal of Biology* 35, 735–742. doi:10.3906/biy-1011-170.
Booth, D.A., 1994. Flavour quality as cognitive psychology: The applied science of mental mechanisms relating flavour descriptions to chemical and physical stimulation patterns. *Food Quality and Preference.* doi:10.1016/0950-3293(94)90007-8.
Bozan, B., Kosar, M., Tunalier, Z., Ozturk, N., Baser, K.H., 2003. Antioxidant and free radical scavenging activities of *Rhus coriaria* and *Cinnamomum cassia* extracts. *Acta Alimentaria* 32(1), 53–61. doi:10.1556/AAlim. 32.2003.1.7.
Burnette, F.S., 1977. Peroxidase and its relationship to food flavor and quality: A review. *Journal of Food Science* 42, 1–6. doi:10.1111/j.1365-2621.1977.tb01204.x.
Cadby, P.A., Troy, W.R., Vey, M.G.H., 2002. Consumer exposure to fragrance ingredients: Providing estimates for safety evaluation. *Regulatory Toxicology and Pharmacology* 36, 246–252. doi:10.1006/rtph.2002.1581.
Capuzzo, A., Maffei, M.E., Occhipinti, A., 2013. Supercritical fluid extraction of plant flavors and fragrances. *Molecules* 18(6), 7194–7238. doi:10.3390/molecules18067194.
Carocho, M., Barreiro, M.F., Morales, P., Ferreira, I.C.F.R., 2014. Adding molecules to food, pros and cons: A review on synthetic and natural food additives. *Comprehensive Reviews in Food Science and Food Safety* 13, 377–399. doi:10.1111/1541-4337.12065.
Caveats, P., 2016. Isoflavones: Anti-Inflammatory benefit and possible caveats. *Nutrients* 8(6), 1–16. doi:10.3390/nu8060361.
Chauhan, B., Kumar, G., 2013. Current concepts and prospects of herbal nutraceutical: A review. *Journal of Advanced Pharmaceutical Technology & Research* 4(1), 4. doi:10.4103/2231-4040.107494.
Chemistry, F., 2017. Nutritional and nutraceutical quality of strawberries in relation to harvest time and crop conditions. *Journal of Agricultural and Food Chemistry* 62(25), 5749–5760. doi:10.1021/jf500769x.
Collingst, V.B., 1974. Human taste response as a function of locus of stimulation on the tongue and soft palate. *Perception & Psychophysics* 16, 169–174.
Das, K., 2015. Turmeric (*Curcuma longa*) oils, *Essential Oils in Food Preservation, Flavor and Safety.* Elsevier Inc. doi:10.1016/B978-0-12-416641-7.00095-X.
Das, L., Bhaumik, E., Raychaudhuri, U., 2012. Role of nutraceuticals in human health. *Journal of Food Science and Technology* 49, 173–183. doi:10.1007/s13197-011-0269-4.
Delahunty, C.M., Piggott, J.R., Conner, J.M., Paterson, A., 1996. Comparison of dynamic flavour release from hard cheeses and analysis of headspace volatiles from the mouth with flavour perception during consumption. *Journal of the Science of Food and Agriculture* 71(3), 273–281.
Devansh, M., 2012. *Salvia officinalis* Linn.: Relevance to modern research drive. *Planta Activa* 4, 203–207.
Dimick, P.S., Hoskin, J.C., Acree, T.E., 1983. Review of apple flavor—State of the art. *Critical Reviews in Food Science & Nutrition* 18, 37–41.
Edris, A.E., 2007. Pharmaceutical and therapeutic potentials of essential oils and their individual volatile constituents: A Review. *Phytotherapy Research: An International Journal Devoted to Pharmacological and Toxicological Evaluation of Natural Product Derivatives* 323, 308–323. doi:/10.1002/ptr.2072.
Ehab, A., 2011. *Leung's Encyclopedia of Common Natural Ingredients: Used in Food, Drugs, and Cosmetics,* John Wiley and Sons.
Fang, Z., 2010. Encapsulation of polyphenols—A review. *Trends in Food Science & Technology.* 21, 510–523. doi:10.1016/j.tifs.2010.08.003.

Feron, G., Bonnarme, P., Durand, A., 1996. Prospects for the microbial production of food flavours. *Trends in Food Science and Technology*. 71, 285–293.

Flores, H.E., Hoy, M.W., Pickard, J.J., 1987. Secondary metabolites from root cultures. *Trends Biotechnology* 5, 64–69. doi:10.1016/S0167-7799(87)80013-6.

Ford, R.A., Domeyer, B., Easterday, O., Maier, K., Middleton, J., 2000. Criteria for development of a database for safety evaluation of fragrance ingredients. *Regulatory Toxicology and Pharmacology* 31, 166–181. doi:10.1006/rtph.1999.1362.

Gilani, A.H., Jabeen, Q., Khan, A.ul, Shah, A.J., 2007. Gut modulatory, blood pressure lowering, diuretic and sedative activities of cardamom. *Journal of Ethnopharmacology* 115, 463–472. doi:10.1016/j.jep.2007.10.015.

Golding, J.B., Shearer, D., Mcglasson, W.B., Wyllie, S.G., 1999. Relationships between respiration, ethylene, and aroma production in ripening banana. *Journal of Agricultural and Food Chemistry* 47, 1646–1651. doi:10.1021/jf980906c.

Hajhashemi, V., Ghannadi, A., Sharif, B., 2003. Anti-inflammatory and analgesic properties of the leaf extracts and essential oil of *Lavandula angustifolia* Mill. *Journal of Ethnopharmacology* 89, 67–71. doi:10.1016/S0378-8741(03)00234-4.

Hallagan, J.B., Hall, R.L., 1995. FEMA GRAS—A GRAS assessment program for flavor ingredients. *Regulatory Toxicology and Pharmacology* 21(3), 422–430. doi:10.1006/rtph.1995.1057.

Hamed, S.F., Sadek, Z., Edris, A., 2012. Antioxidant and antimicrobial activities of clove bud essential oil and eugenol nanoparticles in alcohol-free microemulsion. *Journal of Oleo Science* 61, 641–648. doi:10.5650/jos.61.641.

Hamidpour, M., Hamidpour, R., Hamidpour, S., Shahlari, M., 2014. Chemistry, pharmacology, and medicinal property of sage (Salvia) to prevent and cure illnesses such as obesity, diabetes, depression, dementia, lupus, autism, heart disease, and cancer. *Journal of Traditional and Complementary Medicine* 4, 82–88. doi:10.4103/2225-4110.130373.

Hartmann, R., Meisel, H., 2007. Food-derived peptides with biological activity: From research to food applications. *Current Opinion in Biotechnology* 18, 163–169. doi:10.1016/j.copbio.2007.01.013.

Hathcock, J., 2001. Recent advances on the nutritional effects associated with the use of Garlic as a supplement dietary supplements: How they are used and regulated 1, 1114–1117.

Herness, M.S., Gilbertson, T.A., 1999. Cellular mechanisms. *Annual Review of Physiology* 61(1), 873–900.

Huma, N., Salim-Ur-Rehman, Anjum, F.M., Murtaza, M.A., Sheikh, M.A., 2007. Food fortification strategy—Preventing iron deficiency anemia: A review. *Critical Reviews in Food Science and Nutrition* 47, 259–265. 10.1080/10408390600698262.

Jaswir, I., Noviendri, D., Hasrini, R.F., Octavianti, F., 2011. Carotenoids: Sources, medicinal properties and their application in food and nutraceutical industry. *Journal of Medicinal Plants Research* 5, 7119–7131. doi:10.5897/JMPRx11.011.

Jitoe, A., Gara, W., 1992. Antioxidant activity of tropical ginger extracts and analysis of the contained curcuminoids. *Journal of Agricultural and Food Chemistry* 40, 1337–1340.

Kader, A.A., 2008. Flavor quality of fruits and vegetables 1868, 1863–1868. doi:10.1002/jsfa.

Kauer, J.S., White, J., 2001. Imaging and coding in the olfactory system. *Annual Review of Neuroscience* 24(1), 963–979.

Keservani, R.K., Gandhi, R., Vishwavidyalaya, P., 2015. Scholars research library nutraceutical and functional food as future food: A review, Der Pharmacia Lettre: Scholars Research Library, 2 (1) 106–116.

Kim, S.M., Kubota, K., Kobayashi, A., 1997. Antioxidative activity of sulfur-containing flavor compounds in garlic. *Bioscience, Biotechnology, and Biochemistry* 61, 1482–1485. doi:10.1271/bbb.61.1482.

Knee, M., 2012. *Fruit quality and its biological basis* (Vol. 9). CRC Press.

Kosalec, I., Pepeljnjak, S., Kustrak, D., 2005. Antifungal activity of fluid extract and essential oil from anise fruits (*Pimpinella anisum* L., Apiaceae). *Acta Pharmaceutica* 55, 377–85.

Kraft, P., Swift, K.A.D., (Eds.) 2005. Perspectives in flavor and perspectives in flavor and fragrance research. *Helvetica Chimica Acta*.

Krings, M.U., 1998. Biotechnological production of flavours and fragrances. *Applied Microbiology and Biotechnology* 49(1), 1–8.

Laing, D.G., Jinks, A., 1996. Flavour perception mechanisms. *Trends in Food Science & Technology* 71, 387–389.

Leclere, C.J., Champ, M., Boillot, J., Guille, G., Lecannu, G., Molis, C., Bornet, F., Krempf, M., Delort-Laval, J., Galmiche, J.P., 1994. Role of viscous guar gums in lowering the glycemic response after a solid meal. *The American Journal of Clinical Nutrition* 59, 914–921.

Lee, K.G., Shibamoto, T., 2002. Determination of antioxidant potential of volatile extracts isolated from various herbs and spices. *Journal of Agricultural and Food Chemistry* 50(17), 4947–4952.

Lima, R.K., Batista, R., Nelson, D.L., 2012. Bactericidal and antioxidant activity of essential oils from *Myristica fragrans* houtt and *Salvia microphylla* H.B.K. *The Journal of the American Oil Chemists' Society* 89(3), 523–528. doi:10.1007/s11746-011-1938-1.

Linnemann, D., Salas, Hernández, J., Vázquez, García, J., Ortiz, Aldana, I., Fernández-Vega, D.M., Río-Navarro, Be, 2017. Mexican Asthma Guidelines: GUIMA 2017 Guía Mexicana del Asma: GUIMA 2017. Rev Alerg Mex Guía Mex. del Asma Rev Alerg Mex 64, 11–128. doi:10.1078/0944-7113-00375.

Lowenfeld, A.F., 2009. Foreword. Eff. Treaties Foreign Direct Invest. Bilater. Invest. Treaties, Double Tax. Treaties, Invest. Flows 3686–3692. doi:10.1093/acprof:oso/9780195388534.002.0007.

Maier, H.G., 1970. Volatile flavoring substances in foodstuffs. *Angewandte Chemie International Edition* 9(12), (English) 9, 917–926. doi:10.1002/anie.197009171.

Malhotra, S., Singh, A.P., Zingiber, G., 2003. Medicinal properties of Ginger (*Zingiber officinale* Rosc.). *Natural Product Radiance* 2, 296–301.

Manley, C.H., 2000. Trends in industrial flavor research. *ACS Symposium Series* 756, 2–12. doi:10.1021/bk-2000-0756.

Marais, J., 1983. Terpenes in the aroma of grapes and wines: A review. *South African Journal of Enology and Viticulture* 4, 49–58.

Marinova, E.M., Yanishlieva, N. V, 1997. Antioxidative activity of extracts from selected species of the family Lamiaceae in sunflower oil. *Food Chemistry* 58, 245–248.

Mashmoul, M., Azlan, A., Khaza'ai, H., Yusof, B., Noor, S., 2013. Saffron: A natural potent antioxidant as a promising anti-obesity drug. *Antioxidants* 2, 293–308. doi:10.3390/antiox2040293.

Masotti, V., Juteau, F., Bessière, J.M., Viano, J., 2003. Seasonal and phenological variations of the essential oil from the narrow endemic species *Artemisia molinieri* and its biological activities. *Journal of Agricultural and Food Chemistry* 51, 7115–7121. doi:10.1021/jf034621y.

Mateo, J.J., Jimenez, M., 2000. Monoterpenes in grape juice and wines. *Journal of chromatography A* 881, 557–567.

Mazza, G., 2007. Saponins: Properties, applications. *Critical Reviews in Food Science and Nutrition* 47(3), 231–258 doi:10.1080/10408390600698197.

Mensing, J., Beck, C., 1988. The psychology of fragrance selection. *Perfumery* 185–204. doi:10.1007/978-94-009-1215-1.

Mozaffarieh, M., Sacu, S., Wedrich, A., 2003. The role of the carotenoids, lutein and zeaxanthin, in protecting against age-related macular degeneration: A review based on controversial evidence. *Nutrition Journal* 8, 1–8.

Muhammad, D.R.A., Dewettinck, K., 2017. Cinnamon and its derivatives as potential ingredient in functional food—A review. *International Journal of Food Properties* 20, 2237–2263. doi:10.1080/10942912.2017.1369102.

Muthiah, K., Suja, S., 2017. A study on sense, feel, think, act, relate factors of experiential marketing in retailing. *Transformations in Business & Economics* 16, 85–99. doi:10.1002/ptr.

Nadeem, M., Anjum, F.M., Khan, M.I., Tehseen, S., El-Ghorab, A., Sultan, J.I., 2013. Nutritional and medicinal aspects of coriander (*Coriandrum sativum* L.): A review. *British Food Journal* 115, 743–755. doi:10.1108/00070701311331526.

Oakenfull D. 1981. Saponins in food—a review. *Food chemistry* 7(1), 19–40.

Oser, B.L., Ford, R.A., Col, M., Horace, W., Laboratories, B., Seevers, M.H., Spencer, H.C., Colft, M., Parte, T., Woods, L.A., Commonwealth, V., Cis--ctenoate, E., 1973. Recent Progress in the Considerdtion of Flavoring Ingredients Under the Food Additives Amendment 6. GRAS Substances. *Food Technology* 27, 64–67.

Ourouadi, S., Moumene, H., Zaki, N., Boulli, A., 2016. Garlic (*Allium sativum*): A source of multiple nutraceutical and functional components (review). *Journal of Chemical, Biological and Physical Sciences* 7, 9–21.

Pearson, P., 1991. Human development and umami taste, *Physiology & Behavior* 49, 1009–1012.

Periasamy, G., Karim, A., Gibrelibanos, M., Gebremedhin, G., 2016. Nutmeg (*Myristica fragrans* Houtt.) oils, *Essential Oils in Food Preservation, Flavor and Safety*. Elsevier Inc. doi:10.1016/B978-0-12-416641-7.00069-9.

Perry, N.S.L., Bollen, C., Perry, E.K., Ballard, C., 2003. Salvia for dementia therapy: Review of pharmacological activity and pilot tolerability clinical trial. *Pharmacology Biochemistry and Behavior* 75, 651–659. doi:10.1016/S0091-3057(03)00108-4.

Perumalla, A.V.S., Hettiarachchy, N.S., 2011. Green tea and grape seed extracts—Potential applications in food safety and quality. *Food Research International* 44, 827–839. doi:10.1016/j.foodres.2011.01.022.

Pizzale, L., Bortolomeazzi, R., Vichi, S., Eva, U., 2002. Antioxidant activity of sage (*Salvia officinalis* and *S fruticosa*) and oregano (*Origanum onites* and *O indercedens*) extracts related to their phenolic compound content. *Journal of the Science of the Food and Agriculture* 82(14), 1645–1651. doi:10.1002/jsfa.1240.

Rios, V., Souza, D., Aparecida, P., Pereira, P., Lomônaco, T., Carlos, L., Lima, D.O., Pio, R., Queiroz, F., 2014. Determination of the bioactive compounds, antioxidant activity and chemical composition of Brazilian blackberry, red raspberry, strawberry, blueberry and sweet cherry fruits. *Food Chemistry* 156, 362–368. doi:10.1016/j.foodchem.2014.01.125.

Ronis, M.J.J., Pedersen, K.B., Watt, J., 2018. Adverse Effects of Nutraceuticals and Dietary Supplements. *Annual Review of Pharmacology and Toxicology* 58, 583–601. doi:10.1146/annurev-pharmtox-010617-052844.

Sabharwal, S., Sudan, S., Ranjan, V., 2013. *Jasminum sambac* Linn (Motia): A review. *International Journal of Pharmaceutical Research Bio-science* 2, 108–130.

Sahib, N.G., Anwar, F., Gilani, A.H., Hamid, A.A., Saari, N., Alkharfy, K.M., 2013. Coriander (*Coriandrum sativum* L.): A potential source of high-value components for functional foods and nutraceuticals—A review. *Physical Therapy Research* 27, 1439–1456. doi:10.1002/ptr.4897.

Sarma, D.N., Barrett, M.L., Chavez, M.L., Gardiner, P., Ko, R., Mahady, G.B., Marles, R.J., Pellicore, L.S., Giancaspro, G.I., Dog, T.L., 2008. Safety of green tea extracts : A systematic review by the US Pharmacopeia. *Drug Safety* 31, 469–484.

Schnitzler, P., Koch, C., 2007. Susceptibility of drug-resistant clinical herpes simplex virus type 1 strains to essential oils of ginger, thyme, hyssop, and sandalwood. *Antimicrobial Agents and Chemotherapy* 51, 1859–1862. doi:10.1128/AAC.00426-06.

Schwab, W., Davidovich-Rikanati, R., Lewinsohn, E., 2008. Biosynthesis of plant-derived flavor compounds. *The Plant Journal* 54, 712–732. doi:10.1111/j.1365-313X.2008.03446.x.

Science, F., 2014. Saffron: A golden condiment and a repository of nutraceutical potential. *Food Science Research Journal* 5, 59–67.

Semwal, R.B., Semwal, D.K., Combrinck, S., Viljoen, A.M., 2015. Gingerols and shogaols: Important nutraceutical principles from ginger. *Phytochemistry* 117, 554–568. 10.1016/j.phytochem.2015.07.012.

Serra, S., Fuganti, C., Brenna, E., 2005. Biocatalytic preparation of natural flavours and fragrances. *Trends Biotechnology* 23, 193–198. doi:10.1016/j.tibtech.2005.02.003.

Sharma, U.K., Sharma, N., 2007. A Comprehensive review on vanilla flavor: Extraction, isolation and quantification of vanillin and others constituents. *International Journal of Food Sciences and Nutrition* 59(4), 299–326.doi:10.1080/09687630701539350.

Shukla, Y., Singh, M., 2007. Cancer preventive properties of ginger: A brief review. *Food and Chemical Toxicology* 45(5), 683–690. doi:10.1016/j.fct.2006.11.002.

Silva, J., Abebe, W., Sousa, S.M., Duarte, V.G., Machado, M.I.L., Matos, F.J.A., 2003. Analgesic and anti-inflammatory effects of essential oils of Eucalyptus. *Journal of Ethnopharmacology* 89, 277–283. doi:10.1016/j.jep.2003.09.007.

Singh, G., Maurya, S., Lampasona, M.P. De, Catalan, C.A.N., 2006. Studies on essential oils, Part 41. Chemical composition, antifungal, antioxidant and sprout suppressant activities of coriander (*Coriandrum sativum*) essential oil and its oleoresin. *Flavour and Fragrance Journal* 21(3), 472–479. doi:10.1002/ffj.1608.

Smith, R.L., Cohen, S.M., Doull, J., Feron, V.J., Goodman, J.I., Marnett, L.J., 2005. A Procedure for the safety evaluation of natural flavor complexes used as ingredients in food: Essential oils. *Food and Chemical Toxicology* 43, 345–363.

Souyoul, S.A., Saussy, K.P., Lupo, M.P., 2018. Nutraceuticals: A review. *Dermatology and Therapy* (Heidelb). 8, 5–16. doi:10.1007/s13555-018-0221-x.

Srinivasan, K., 2005. Role of spices beyond food flavoring: nutraceuticals with multiple health effects. *Food Reviews International* 21, 167–188. doi:10.1081/FRI-200051872.

Srinivasan, K., 2017. Ginger rhizomes (*Zingiber officinale*): A spice with multiple health beneficial potentials. *PharmaNutrition* 5, 18–28. doi:10.1016/j.phanu.2017.01.001.

Stringer, R., Sang, N., Croppenstedt, A., 2009. Producers, processors, and procurement decisions: The case of vegetable supply chains in China. *World Development* 37, 1773–1780. doi:10.1016/j.worlddev.2008.08.027.

Suhaj, M., 2006. Spice antioxidants isolation and their antiradical activity: A review. *Journal of Food Composition and Analysis* 19, 531–537. doi:10.1016/j.jfca.2004.11.005.

Suleria, H.A.R., Butt, M.S., Khalid, N., Sultan, S., Raza, A., Aleem, M., Abbas, M., 2015. Garlic (*Allium sativum*): Diet based therapy of 21st century—A review. *Asian Pacific Journal of Tropical Disease* 5, 271–278. doi:10.1016/S2222-1808(14)60782-9.

Taylor, P., Sowbhagya, H.B., 2013. Nutraceutical functions of cumin (*Cuminum cyminum* L): An overview. *Critical Reviews in Food Science and Nutrition* 53(1), 37–41. doi:10.1080/10408398.2010.500223.

Touloupakis, E., Ghanotakis, D.F., n.d. Nutraceutical Use of Garlic.

Vandamme, E.J., 2018. Production of flavour by microorganisms. *Process Biochemistry* 27(4), 195–215. 9592. doi:10.1016/0032-9592(92)80020-4.

Walton, N.J., Narbad, A., Faulds, C.B., Williamson, G., 2000. Novel approaches to the biosynthesis of vanillin. *Current Opinion in Biotechnology* 11(5), 490–496.

Wilson, A.D., Baietto, M., 2009. Applications and advances in electronic-nose technologies. *Sensors* 9(7), 5099–5148. doi:10.3390/s90705099.

Yan, J. wei, Ban, Z. jun, Lu, H. yan, Li, D., Poverenov, E., Luo, Z. sheng, Li, L., 2018. The aroma volatile repertoire in strawberry fruit: A review. *Journal of the Science of Food and Agriculture* 98, 4395–4402. doi:10.1002/jsfa.9039.

Yang, J., Xiao, Y.Y., 2013. Grape phytochemicals and associated health benefits. *Critical Reviews in Food Science and Nutrition* 53, 1202–1225. doi:10.1080/10408398.2012.692408.

Yanishlieva, N. V, Marinova, E., Pokorny, J., 2006. Natural antioxidants from herbs and spices. *Lipid Science and Technology* 108(9), 776–793. doi:10.1002/ejlt.200600127.

Yaylayan, V.A., 1991. Flavor technology: Recent trends and future perspectives. *Canadian Institute of Food Science and Technology Journal* 24, 2–5. doi:10.1016/S0315-5463(91)70002-3.

Ye, Q.P., Jin, X.Y., Zhu, X.L., Lin, T.X., Hao, Z.L., Yang, Q., 2015. An Efficient Extraction Method for fragrant volatiles from *Jasminum sambac* (L.). *Journal of Oleo Science* 64, 645–652. doi:10.5650/jos.ess15014.

10

Recent Trends Used in Functional Food and Nutraceuticals Industries for Modulation of Flavors for Improving Sensory Perception

Komal Parmar and Jayvadan Patel

Contents

10.1 Introduction

Sensory perception is the capability of performing neurophysiological processing of the stimuli by the living being. This process is executed by the organs using senses like hearing, vision, taste, odor, and contact. From the sensory perspective, humans will not consume food—whether it is regular diet or dietary supplements—unless it tastes good. Understanding the need for flavor acceptance in human flavor perception, various new flavors are developed and modulated to improve the sensory perception for taste. Thus, the elementary purpose of the flavors is to impart taste to the product to be consumed orally, as they do not possess nutritional properties. Flavors may be classified as natural and synthetic. Synthetic flavors are chosen more cautiously to render diversity in taste. On the other hand, natural flavors are derived from fruits, spices and vegetables to attain a broad range of diversity in taste.

Global food flavor market is expected to witness a massive rise because of the increasing call from food and dietary supplement industry. The worldwide taste modulators market is anticipated to grow at a CAGR (compound annual growth rate) of 7.9% from 2018 to reach $1,831.3 million by 2024 (Market Research, Inc. 2018). Recent technological advancement in flavor manufacturing methodologies, and growing awareness of latest health trends have influenced the market growth of flavor. Other factors affecting the rise in flavor market can be counted as growing consumer base, increase in expenditure capacity of end consumer, and improved product quality.

The saying, "Let food be thy medicine and medicine be thy food," was first said around 2,500 years ago by Hippocrates, the father of medicine (Witkamp and van Norren 2018). The concept of functional food where in food provides therapeutic benefits is not new but is currently in trend. The idea was first described in ancient texts of India and China. However, the concept to develop functional food was first explicated in Japan in 1980s by the nation's Ministry of Health and Welfare with an aim of improving health of aging nationals (Henry 2010). Functional foods can be considered as food that provides potential positive and therapeutic benefits over basic nutrition. Another term used interchangeably for functional food is "nutraceuticals," coined by the Foundation for Innovation in 1991. They referred the term nutraceutical as any bioactive component providing health benefit (Hasler 2002). Nutraceuticals characterize the new classification that fills in the boundaries between medications and nourishment. It can act as the epitome of proactive medicine thereby preventing and lowering the risk associated with life style related diseases (Santini et al. 2017, Das et al. 2012). Global market of functional foods and nutraceutical products is growing at an enormous gait (Verbeke 2005). The worldwide revenue for functional food is projected to increase with a compound annual growth rate of about 7% during period of 2018–2022 (Technavio Research 2018).

Individuals have an innate inclination for certain kinds of dietary components and aversions for others. Perception of food flavor is achieved by sensory system via taste and smell. Recently analysts are starting to comprehend enhance discernment of flavor perception. They are making out as how current nutrition inclinations might be inconsistent with nutritive items that consolidate low nutrient value with flavors that normally happen in less-processed food products. Since the discovery of olfactory receptors responsible for the flavor perception, the understanding of the molecular mechanism of taste and smell has progressed substantially. On the other hand, consumers correlated flavor to the nutritional results linked with eating particular food (Verbeke 2006). Thus, flavor modulation seemed to be important while considering the large market of functional food and nutraceuticals, as the consumer will prefer a dietary supplement with good taste. Hence, flavor modulation enables the more palatable product with increased acceptance by the consumer. Most of the functional foods and nutraceuticals possess off-notes that are later not accepted by the consumer and, thus, need to be regulated. This regulation of off-note taste in oral dietary component can be referred as flavor modulation. Many methods can be applied to regulate the taste. Previously adding sweetener was the only solution to mask the off-notes of the nutraceuticals. Traditionally salt and sugar were considered the only flavoring agents to enhance the taste of food products. Currently, however, manufacturers employ flavor science to modulate the taste and enhance the palatability of the product.

Various natural and synthetic flavors are available in the market, which is utilized to enhance the taste. Some of the names to mention under category of synthetic flavors includes diacetyl (buttery), isoamyl acetate (banana), ethyl propionate (fruity), limonene (orange), allyl hexanoate (pineapple), methyl anthranilate (grape), ethyl vanillin (vanilla) and methyl salicylate (wintergreen). Savory flavorants include glutamic acid (monosodium glutamate, monopotassium glutamate, calcium diglutamate, magnesium diglutamate), glycine salts (sodium salt), guanylic acid salts (calcium guanylate, sodium guanylate, dipotassium guanylate), inosinic acid salts (disodium inosinate, dipotassium inosinate, calcium inosinate) and 5′ ribonucleotide salts (calcium 5′-ribonucleotides, disodium 5′-ribonucleotides) (Khodjaeva et al. 2013). Certain organic and inorganic acids can also be used to enhance the flavors, which impart slightly different sour tastes like acetic acid, ascorbic acid, citric acid, fumaric acid, lactic acid, malic acid and tartaric acid (Lugaz et al. 2005, Marsh et al. 2006, da Conceicao Neta et al. 2007a, 2007b).

The term artificial flavor or artificial flavoring means any substance, the function of which is to impart flavor, which is not derived from a spice, fruit or fruit juice, vegetable or vegetable juice, edible yeast, herb, bark, bud, root, leaf or similar plant material, meat, fish, poultry, eggs, dairy products, or fermentation products thereof (CFR-Code of Federal Regulations Title 21). Many functional foods are coming with natural flavors to provide the aesthetic taste of the supplement. For instance, concentrated omega-3, a softgel dietary supplement of Natural Nutra, contains natural orange flavor. The product supports cardiovascular, immune system, nervous system, and brain health. One softgel per day provides twice as much EPA and DHA (eicosapentaenoic acid and docosahexaenoic acid respectively) as other omega-3 products. Daily Vita Plus™ with Natural Lemon Flavor is multivitamin supplement by Vibrant Nutraceuticals.

Stress-Relax®-Suntheanine® L-Theanine by Natural Factors consist of pure L-Theanine, a unique amino acid found in green tea that is emerging as the premier natural products to support mental calmness and relaxation. The chewable tablets are flavored using natural tropical fruit.

UltraInflamX Plus 360® is a medical food by Metagenics which provides advanced support for the nutritional management of compromised function of gut in inflammatory bowel disease including ulcerative colitis and Crohn's disease. It comes with natural flavors of pineapple and banana.

Align Prebiotic plus Probiotic Gummies Supplement (containing natural fruit flavors) of Align Probiotic helps nourish good bacteria and adds more good bacteria to the human digestive system. Consumers want healthy, clean, and natural products that taste great. Thus, many such functional foods and nutraceuticals containing natural flavors are added up in market. The concept of having product with natural flavor makes consumer feel safe with enjoyable taste.

10.2 Taste inhibition/masking

The human taste sensory system can identify five major tastes such as sweet, sour, bitter, salty and umami/savory. Various receptors and transduction systems are involved in identifying each characteristic taste. When complex material is taken that contains more than one compound, a number of possible interactions take place. The one with potential taste quality will interfere with the receptors and transduction mechanism connected with another compound, thereby suppressing the taste of later and enhancing the taste of former component. Many efforts are made to develop blockers for odd-taste such as bitterness, as it would provide a means to reduce this off-note. However, 25 acknowledged bitter receptors (Behrens and Meyerhof 2006) make it more complex to develop such a universal bitter blocker. Thus, various means to mask such off-notes include addition of flavors, complex formation and encapsulation technology.

Taste-provoking chemical molecules evoke the responses at a high level usually when they are present alone. When one taste is stronger it can completely mask the other taste of different characteristic, wherein the former act as masking agent to the later one. For instance, sugar in high amounts is used in many products to mask the off-notes. Here, mixture suppression is taking place in the brain (Green et al. 2010).

The fundamental function of flavoring agents is to provide its characteristic flavor and to mask the base off-notes of the product. Flavor manufacturers are utilizing multi-functional approach to solve the flavor challenges by using masking technology. Masking agents are employed to mask the unwanted notes in a wide variety of applications. Masking agents cover up the off-notes of product without affecting the active components. Masking bitterness is possibly the priority prospect for the taste modulators. Bitterness occurs due to many naturally present bitter compounds, or through chemical reactions that may occur during storage or processing. Addition of sugar in large quantity may not help much in diet supplements considering deleterious effects on health. Flavors added can be used at certain strength because if the concentration in formulation is exceeded may result in imbalance of the taste performance. Besides instead of clogging the taste buds with high amount of flavor adding in the formulation, masking agents tricks the taste buds into not perceiving the off-notes. Using masking agents in combination with flavoring agents may provide double advantage in masking of off-notes.

Smoothenol® 2G is a research and development product of Sensient Pharmaceutical, Missouri, USA, that uses emerging science to enhance taste and mask bitterness. Smoothenol® 2G is designed to enhance the palatability of food, beverage and health products by masking undesirable off-notes commonly associated with sweeteners, caffeine, alcohol, vitamins and minerals, nutraceutical and functional elements. It works on the mechanism of modifying the sensory perception of ingredients with undesirable off-notes. It is developed using advanced knowledge of receptors and ligand design.

The Soy Delivery System (SDS) by Blue Pacific Flavors, California, USA, offers improved taste, aroma and full flavor balance and intensity of soymilk. It uses masking technology to understate or eliminate the "beany astringency" associated with soymilk containing soy protein or isolates.

Complexation is one of the technologies that can be utilized to mask the off-notes in functional food (Coupland and Hayes 2014). Cyclodextrins, a cyclic oligosaccharides derived from starch hydrolysates by enzyme biotransformation process, is widely accepted with food ingredients owing to its good water solubility and mild sweet taste (Szente and Szejtli 2004, Astray et al. 2009). Cyclodextrin molecules possess a cup shape structure that can accommodate the active nutrient thereby forming a complex. The inner surface of the cup is found to be hydrophobic while the outer surface is hydrophilic which thereby facilitates solubilization in aqueous media. **Figure 10.1** demonstrates schematic diagram of cyclodextrin complexation. From studies carried out, cyclodextrins were found to be capable of partially suppressing the bitterness of soy protein, soy protein hydrolysates and whey protein hydrolysates (Linde et al. 2009, Linde et al. 2010, Yang et al. 2012). A trial study to demonstrate effect of beta cyclodextrin on aroma release was carried out by Kant and co-investigators (2004). Results demonstrated that cyclodextrins can be used to modify the flavor delivery in food systems (Kant et al. 2004).

Microencapsulation technology is one of the approaches that can be used for taste masking in functional foods (Poncelet et al. 2011). The technology provides coating of the micronutrients, which thereby results into taste masking. There is no further requirement of adding taste masking agents (Nedovic et al. 2011). Besides taste masking, microencapsulation technology has various other benefits. It can ameliorate the bioavailability of an active nutrient.

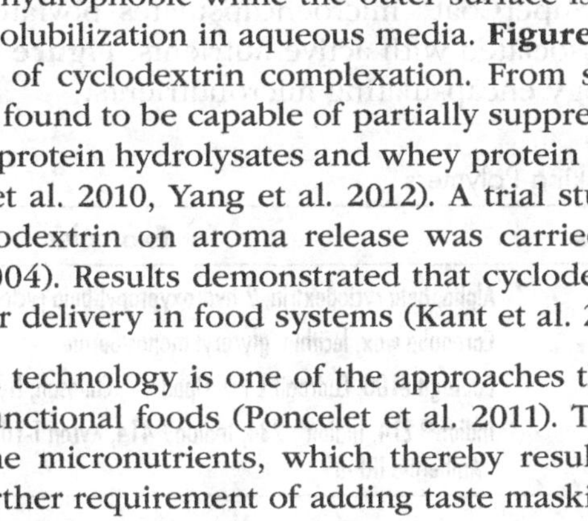

***Figure 10.1** Cyclodextrin complex with nutrient in the cavity.*

The technology involves formation of micro-constituents delivering the active nutrient at micellar level which facilitates its higher absorption. This ultimately results in better biological efficiency, thus reducing the need of higher amount of active nutrients to be taken. The microencapsulation technology apart from enhancing bioavailability also provides stability to the active nutrient protecting from the environmental and other factors affecting stability. Methodologies to conduct microencapsulation involve spray drying of emulsion and coacervation. Both methods result in the formation of a thin layer of biopolymer complex around the core material, preventing it from being tasted (Desai and Park 2007). According to a report published by Mordor Intelligence in April 2018, the global food encapsulation market is anticipated to register an estimated CAGR of 5.2%, during the forecast period (2018–2023). **Table 10.1** summarizes polymers which may be utilized for taste masking of nutraceuticals.

SuperCoat® is a proprietary microencapsulation technology of The Wright Group, Los Angeles, USA. It enhances and renders specific nutrient functions by masking unpleasant taste, odor and mouth feel, of vitamins, minerals and nutraceuticals. SuperCoat® microencapsulates obviate or greatly dilute off-notes typically associated with active nutrients. **Figure 10.2** demonstrates SuperCoat® technology encapsulating micronutrients.

Table 10.1 Taste Masking Polymers

Taste Masking Excipients	Examples
Cyclodextrin	Alpha, beta cyclodextrin, 2-hydroxypropyl-beta cyclodextrin
Lipids	Carnauba wax, lecithin, glyceryl monostearate
Polymers	Eudragit E100, Eudragit E PO, Opadry®, Chitosan, Hydroxy propyl methyl cellulose
Resins	Indion® 214, Indion® 234, Indion® 414, Kyron T-104, Kyron T 134, Tulsion 335, Amberlite IRP-69

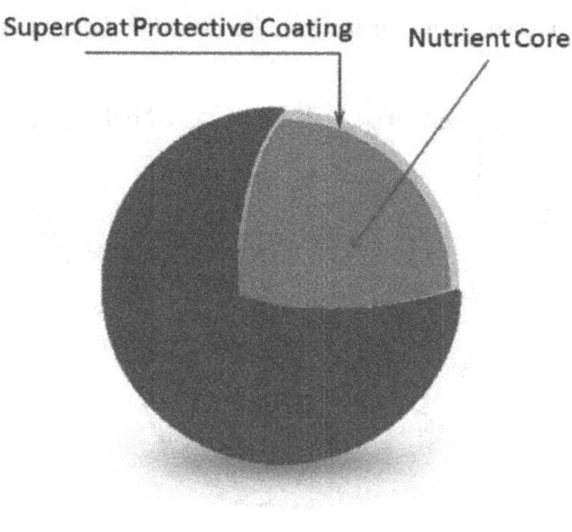

Figure 10.2 *SuperCoat® technology for taste masking of nutrients.*

10.3 Miscellaneous

In a study carried out by Uchida and co-investigators, they reported that the bitterness associated with amino acids such as L-valine, L-leucine, L-isoleucine, L-phenylalanine, L-tryptophan, L-arginine or L-lysine was disguised by addition of α,α-trehalose (carbohydrate) (Uchida et al. 2003). In another study by Doerr and his co-investigators, isomaltulose was proposed for reduction of the bitter taste of tea beverages associated to polyphenols (Doerr et al. 2007). Egg-white protein was found to inhibit off-taste of foods based on soybean milk and green vegetable juice. This was associated due to treatment of proteins with enzymes such as papain, ficin and bromelain (Kittaka et al. 2005).

10.4 Future perspective

There is no one single solution for taste modulation in functional foods and nutraceuticals. There are various approaches to flavor modulations including masking, flavor enhancement and modifiers. Combinations of different technologies can be employed such as encapsulation and flavor enhancement, selective removal of bitter molecules or biotransformation of bitter molecules, using strong tastants to produce a superior flavor in formulation. However, in the future, selective taste inhibitors for specific off-notes may be found and developed.

References

Align Prebiotic + Probiotic Gummies Supplement, Align Probiotics. https://www.aligngi.com/en-us/shop-probiotic-supplements/align-probiotic-supplement. Accessed on 30 December 2018.

Astray G, Gonzalez-Barreiro C, Mejuto JC, Rial-Otero R, Simal-Gándara J. A review on the use of cyclodextrins in foods. *Food Hydrocolloids-Journal*. 2009; 23: 1631–1640.

Behrens M, Meyerhof W. Bitter taste receptors and human bitter taste perception. *Cellular and Molecular Life Sciences*. 2006; 63(13): 1501–1509.

CFR - Code of Federal Regulations Title 21, TITLE 21--Food and Drugs, CHAPTER I--Food and Drug Administration, Department of Health and Human Services, SUBCHAPTER E--Animal Drugs, Feeds, and Related Products. https://www.accessdata.fda.gov/scripts/cdrh/cfdocs/cfcfr/cfrsearch.cfm?fr=501.22. Accessed on 28 December 2018.

Concentrated Omega-3, Softgel, Natural Nutra. https://www.natnutra.com/collections/omega-3/products/concentrated-omega-3. Accessed on 1 January 2019.

Coupland JN, Hayes JE. Physical approaches to masking bitter taste: Lessons from food and pharmaceuticals. *Pharmaceutical Research*. 2014; 31(11): 2921–2939.

da Conceicao Neta ER, Johanningsmeier SD, Drake MA, McFeeters RF. A chemical basis for sour taste perception of acid solutions and fresh-pack dill pickles. *Journal of Food Science*. 2007a; 72(6): S352–S359.

da Conceicao Neta ER, Johanningsmeier SD, McFeeters RF. The chemistry and physiology of sour taste—A review. *Journal of Food Science*. 2007b; 72(2): R33–R38.

Daily Vita Plus™, Vibrant Nutraceuticals. https://www.vibrantnutra.com/products/daily-vita-plus/. Accessed on 4 January 2019.

Das L, Bhaumik E, Raychaudhuri U, Chakraborty R. Role of nutraceuticals in human health. *Journal of Food Science and Technology*. 2012; 49(2): 173–183.

Desai KGH, Park HJ. Recent Developments in microencapsulation of food ingredients. *Journal Drying Technology*. 2005; 23(7): 1361–1394.

Doerr T, Hausmanns S, Kowalczyk J, Becker M, Killeit U. Formulation for masking bitterness, especially in tea extracts, comprising a polyphenol-containing composition and isomaltulose. *WO*. 2007; 59: 953.

Green BG, Lim J, Osterhoff F, Blacher K, Nachtigal D. Taste mixture interactions: Suppression, additivity, and the predominance of sweetness. *Physiology & Behavior*. 2010; 101(5): 731–737.

Hasler CM. Functional foods: Benefits, concerns and challenges—A position paper from the American Council on Science and Health. *The Journal of Nutrition*. 2002; 132(12): 3772–3781.

Henry CJ. Functional foods. *European Journal of Clinical Nutrition*. 2010; 64: 657–659.

Kant A, Linforth RST, Hort J, Taylor AJ. Effect of β-cyclodextrin on aroma release and flavor perception. *Journal of Agriculture and Food Chemistry*. 2004; 52(7): 2028–2035.

Khodjaeva U, Bojňanská T, Vietoris V, Sytar O, Singh R. Food additives as important part of functional food. *International Research Journal of Biological Sciences*. 2013; 2 (4): 74–86.

Kittaka R, Higashiguchi S, Yoshida Y. Taste-modifying peptides, their manufacture from egg white, and use for foods and beverages. *JP*. 2005; 336: 067.

Linde GA, Junior AL, De Faria EV, Colauto NB, De Moraes FF, Zanin GM. Taste modification of amino acids and protein hydrolysate by α-cyclodextrin. *Food Research International-Journal*. 2009; 42: 814–818.

Linde GA, Junior AL, De Faria EV, Colauto NB, De Moraes FF, Zanin GM. The use of 2D NMR to study β-cyclodextrin complexation and debittering of amino acids and peptides. *Food Research International-Journal*. 2010; 43: 187–192.

Lugaz O, Pillias AM, Boirceau-Ducept N, Faurion A. Time-intensity evaluation of acid taste in subjects with saliva high flow and low flow rates for acids of various chemical properties. *Chemical Senses*. 2005; 30(1): 89–103.

Market Research, Inc. 2018. Taste modulators market by type (Sweet Modulators, Salt Modulators, Fat Modulators), Application (Food Products, Beverages), And Geography–Global Opportunity Analysis And Industry Forecast (2018–2024). https://www.marketresearch.com/Meticulous-Research-v4061/Taste-Modulators-Type-Sweet-Salt-12070911/. Accessed on 27 December 2018.

Marsh KB, Friel EN, Gunson A, Lund C, MacRae E. Perception of flavour in standardised fruit pulps with additions of acids or sugars. *Food Quality and Preference*. 2006; 17(5): 376–386.

Mordor Intelligence Report. 2018. https://www.mordorintelligence.com/industry-reports/food-encapsulation-market. Accessed on 21 January 2019.

Nedovic V, Kalusevic A, Manojlovic V, Levic S, Bugarski B. An overview of encapsulation technologies for food applications. *Procedia Food Science*. 2011; 1: 1806–1815.

Poncelet D, Picot A, El Mafadi S. Encapsulation: an essential technology for functional food applications. *Innovations in Food Technology*. 2011; http://www.capsulae.com/media/copy_inftissuefeb2011__029884800_1734_06062011__094901400_1554_12022014.pdf. Accessed on 18 January 2019.

Santini A, Tenore GC, Novellino E. Nutraceuticals: A paradigm of proactive medicine. *European Journal of Pharmaceutical Sciences*. 2017; 96: 53–61.

Smoothenol, Sensient Pharmaceutical, Missouri, USA. https://sensientpharma.com/products/flavors/smoothenol/. Accessed on 6 January 2019.

Stress-Relax®-Suntheanine® L-Theanine, Natural Factors. https://naturalfactors.com/en-us/product/suntheanine-ltheanine-2-en-us/. Accessed on 8 January 2019.

SuperCoat® Technology, The Wright Group. https://www.thewrightgroup.net/supercoat

Szente L, Szejtli J. Cyclodextrins as food ingredients. *Trends in Food Science & Technology*. 2004; 15: 137–142.

Technavio Research. Global Functional Food Ingredients Market 2018-2022 | Growth Analysis and Forecast. Business Wire, A Berkshire Hathaway Company. https://www.businesswire.com/news/home/20181228005182/en/Global-Functional-Food-Ingredients-Market-2018-2022-Growth. Accessed on 30 December 2018.

The Soy Delivery System (SDS), Blue Pacific Flavors, California. https://www.nutraceuticalsworld.com/issues/2002-06/view_features/flavor-update/1583. Accessed on 6 January 2019.

Uchida Y, Wada S, Takeuchi K. Alpha, alpha-Trehalose containing agents and method for bitterness masking of bitter amino acids, and amino acid compositions containing the agents. *JP*. 2003; 274: 896.

UltraInflamX Plus 360®, Metagenics. https://www.metagenics.com/ultrainflamx-plus-360. Accessed on 30 December 2018.

Verbeke W. Consumer acceptance of functional foods: Socio-demographic, cognitive and attitudinal determinants. *Food Quality and Preference-Journal*. 2005; 16: 45–57.

Verbeke W. Functional foods: Consumer willingness to compromise on taste for health? *Food Quality and Preference-Journal*. 2006; 17: 126–131.

Witkamp RF, van Norren K. Let thy food be thy medicine....when possible. *European Journal of Pharmacology*. 2018; 836: 102–114.

Yang S, Mao X-Y, Li F-F, Zhang D, Leng X-J, Ren F-Z, TENG G-X. The improving effect of spray-drying encapsulation process on the bitter taste and stability of whey protein hydrolysate. *European Food Research and Technology*. 2012: 91–97.

11

The Bad Taste of Nutraceuticals

Overview of Basic Research on Bitter Taste

Bhoomika Patel, Ankita Mehta, Jayvadan Patel, and Mayur M. Patel

Contents

11.1 Introduction

The term nutraceutical was first described by Dr. Stephen De Felice in 1989 and is composed of two different terms: nutrition and pharmaceutical, a food with nutritional values that provides health benefits. According to him, nutraceuticals are intended for prevention and/or treatment of the disease(s) or disorder(s) and/or deficiency conditions (Maddi et al. 2007, Brower 1998). Health Canada defines the term nutraceuticals as "a product prepared from foods, but sold in the form of pills, or powder (potions) or in other medicinal forms, not usually associated with foods" (Wildman 2002, Esther 2000). Nutraceuticals can be isolated nutrients, herbal products, dietary supplements, processed foods and genetically engineered food (Malik et al. 2008, Dureja et al. 2003). There are two synonymous terms to nutraceuticals—dietary supplements and functional food, but these terms have very thin line differences. Difference between dietary supplements and nutraceuticals is that dietary supplements are to supplement the diet which contains one or more combination of the dietary ingredients like vitamins, minerals, amino acids, etc. (Chauhan et al. 2013). Whereas nutraceuticals are not used only as a diet supplement, but they also intend to treat or prevent the disease. Functional food is a widely used term for nutraceuticals (Chauhan et al. 2013). Functional food defines as food prepared in scientific intelligence with or without knowledge of its usefulness. When functional food benefits treatment and/or prevention of disease(s) or disorders(s), it is known as nutraceuticals (Dudeja and Gupta 2017). Health benefits from nutraceuticals include reduced side effects, increase beneficial effects and eliminate unpleasant side effects, increase health value and our diet, improvement of medical condition and easy availability with economical advantages (Zhao 2007).

Nutraceuticals are categorized based upon different factors like food and non-food sources, mechanism of action or pharmacological activity and chemical nature. Food and non-food sources are categorized as plant, animals and microbial source. Based on mechanism of action they are classified as anti-cancer, anti-diabetic, anti-inflammatory, anti-oxidant, osteogenic or bone protective, hepatoprotective, anti-obesity, anti-bacterial, anti-microbial, etc. In this chapter, we have included nutraceuticals classified based upon their chemical nature. That includes phytoconstituents like alkaloid, flavanoids, tannins, triterpenes, isothiocyanate, prebiotics, probiotics and other miscellaneous classes of nutraceuticals among which impart bitter taste followed by methods to remove bitterness from these agents.

Globally the nutraceutical market is growing exponentially, and it is expected to reach up to $578.23 in 2025. Nutraceuticals made from herbal or natural source are widely consumed and 7%–12% per year growth of the nutraceutical industries with herbal products was recorded (Shinde et al. 2014). The Indian nutraceutical market is one of the fastest growing markets in Asia-Pacific region. Asian nutraceutical largest market is in Japan, followed by China (Ajit 2012). According to a 2018 Assocham Knowledge report, the Indian nutraceuticals market is expected to grow from $4 billion in 2017 to $18 billion in 2025 and it is getting raised by 21% every year (Verma 2018). Nutraceutical products available in market are natural lycopene, green tea, garlic, glucosamine, digestive enzymes, ginko biloba, ginseng, gymnema, phycocyanine powder, etc.

11.2 Bitter taste perception and receptors

Among five basic tastes, bitter taste and its perception plays important role in defense mechanism against toxic compounds and acts as warning signal to central nervous system (CNS) (Wu et al. 2002). Bitter taste perception arises from ingestion of chemicals and their detection by receptors present in oral cavity (Glendinning 1992). For the development of bitter tasting compounds, we must consider basic science of the gustatory system. Taste buds, principal organ of the gustatory system, are made up of around 50–100 epithelial cells, and some of them are the receptor proteins which are located on the apical membranes of microvilli that extend into epithelial pore (Glendinning 1992). In tongue, taste buds are referred as foliate papillae and circumvallate papillae, distributed in posterior tongue and dorsal surface of tongue respectively. Extralingual taste buds are present in soft palate and epithelium of larynx (Hamilton and Norgren 1984, May and Hill 2006, Corson et al. 2012). These fields are innervated by specific seventh, ninth or tenth cranial nerve branch from which they extend to the rostral Nucleus Solitary Tract (NTS) in the medulla (Hamilton and Norgren 1984, May and Hill 2006, Corson et al. 2012). This pathway may differ with different mammalian species (Lundy and Norgren 2015). As like in rodents and lagomorphs, in addition to NTS, parabrachial nucleus (PBN) is also involved in oromotor and autonomic function, which terminates in the insular cortex's gustatory zone whereas in primates neurones from NTS circumvent PBN and terminate into hypothalamus (Zaidi et al. 2008, Travers and Travers 2005, Norgren 1986, Pritchard and Norgren 2003).

G-protein coupled receptors (GPCRs) and ion channels are two classes of the taste receptors. Among which GPCRs bring sweet, bitter and umami taste perceptions and ion channel receptors mediate salt and sour taste perceptions (Lehman et al. 1995, Kinnamon 2012, Niki et al. 2011, Bigiani et al. 2003, Chandrashekar et al. 2006). As the activities of these receptors and downstream signal transduction are dependent on temperature, targeting temperature of bitter tasting medicines can be used as a strategy (Ninomiya et al. 1996, Ohkuri et al. 2009, Talavera et al. 2005). In addition to this, bitter receptors are also located in sinonasal ciliated epithelium cells which get activated from bacterial stimulation and activate immune responses (Lee et al. 2012).

The bitter taste receptor family (T2Rs) consists of around 25 GPCRs, and most of them are de-orphaned, which get activated by specific bitter tasting ligands. **Figure 11.1** depicts the signal transduction mechanism behind taste perception. Ligand binding to T2Rs activates intracellular G-protein complex containing two subunits: Gαgustducin subunit and β3γ13 subunit (Chaudhari and Roper 2010, Caicedo et al. 2003, Wong et al. 1996). After activation of Gαgustducin, it gets separated from the complex and activates phosphodiesterase leading to decrease the level of cAMP. In some cells, Gαgustducin subunit also mediates sweet taste perceptions in addition to bitter taste because GPCRs transduction signals are activated by both sweet and bitter tasting ligand, which might not be useful as selective target to decrease bitterness of compound (McLaughlin et al. 1992, Wong et al. 1996). β3γ13 subunit activates phospholipase Cβ2 (PLCβ2) (Caicedo et al. 2003). This further acts on phosphatidylinositol 4,5-bisphosphate (PIP2) to generate inositol triphosphate (IP3) and diacylglycerol (DAG). IP3 activates IP3 specific receptors and leads to increase in level of Ca^{2+} from endoplasmic reticulum (Chaudhari and Roper 2010). Ca^{2+} causes membrane depolarization by opening of transient receptor potential cation channel member 5 (TRPM5). This will send signal to

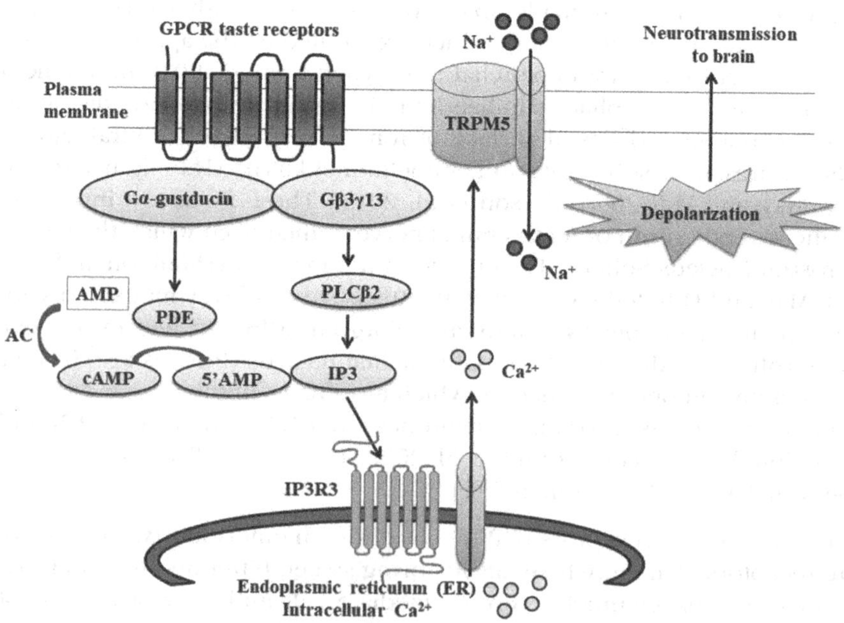

***Figure 11.1** Signal transduction mechanism of taste perception.*

the brain through neurotransmitter release (Hofmann et al. 2003). Also, from in vivo study of genetic ablation of PLCβ2 in mice has shown to lose their response to bitter stimuli which confirms role of PLCβ2 in bitter taste perception (Zhang et al. 2003). In non-gustatory tissues, T2Rs activation stimulates two opposing Ca^{2+} signaling pathway by the help of β3γ13 subunit (Zhang et al. 2013).

11.3 Bitter tasting phytoconstituents

11.3.1 Alkaloids

Alkaloids are the secondary metabolites of plant that have a nitrogen atom at any position of a molecule but not in amide or peptide linkage (Salminen et al. 2011, Mander and Liu 2010). According to W. Meisner, first person to introduce alkaloids, alkaloids are natural compounds reacting like alkalies or bases (Bruneton 1999). It is one of the largest groups of natural compounds with diverse chemical entities. Alkaloids are composed of around 12,000 natural compounds and have been widely used for various pharmacological effects for last 3,000 years while active constituents and classification of alkaloids are carried out in nineteenth century (Salminen et al. 2011, Mander and Liu 2010, Seigler 2012).

Most of the alkaloid compounds have bitter taste; point of fact many of these compounds have always been known as bitter-tasting compounds. Plant extracts and subsequent pure isolated products are bitter in taste and contribute to bitterness of natural products. Alkaloids are being used as bitter tasting standards, e.g. quinine and caffeine. Despite of bitterness of alkaloids, many food preparations include alkaloidal compounds such as tea, coffee, fruits, chocolates, alcohol, etc.

From all alkaloidal substances, caffeine is majorly found in food preparations. Caffeine is used every day, highly consumed by people in beverages like tea, coffee, soft drinks and chocolates. Major sources of caffeine are found in *Coffea Arabica* and *Coffea Robusta* and tea leaves of *Camellia sinensis*. However, caffeine is obtained from 60 different plants including coffee, tea, cocoa pods and guarana (Heckman et al. 2010). Ninety percent of caffeine consumption is from coffee and tea extracts in several populations. Also, it has been safely used from ancient times when it was added as flavor in beverages. Caffeine is moderately bitter with 0.8–1.2 mM taste threshold in water (Belitz et al. 1999, Calviño et al. 2005). It is widely used as CNS stimulant and also to improve respiration and blood circulation. It is also found in mate (Paraguayan tea) which is *Ilex paraguariensis* leaves infusion that is used to increase appetite (Calviño et al. 2005). Caffeine is present in Sterculiaceae family, genus *Cola* sp. which is used as traditional beverages. Theophylline and thobrobine are also isolated from *Cola* sp., tea leaves and *Coffea* sp., which impart the same bitter taste as that of caffeine. Bitter taste of black chocolate and cacao are due to these two alkaloids (Scharbert and Hofmann 2005). Carbonated beverages contain around 34 mg of caffeine, while in coffee and tea it is approximately 200 mg and 80 mg of caffeine content (Heckman et al. 2010). Market of caffeine beverages has been increased majorly which represent major source of caffeine consumption followed by coffee, tea and chocolates (Mitchell et al. 2014). Caffeine has CNS stimulant, mood and energy alleviating and athletic performance increasing properties. Caffeine in moderate amount is also present in many pharmaceutical forms like pain reliving medicines, weight control preparations, cold therapies, etc. Other preparations containing caffeine are jelly beans, chewing gums, pancake syrup, brewed coffee, black tea, chocolate milk, dark chocolate, dietary supplements, flavored water, etc. (Heckman et al. 2010). Caffeine also has bronchodilator, anti-oxidant, anti-aging and anti-cellulite properties (Gajewska et al. 2015). A 7.9 mg/kg body weight caffeine is regarded as safe in neonates to treat apnea (Francart et al. 2013).

Quinine and quinidine are much bitterer than caffeine. A 1 mM of quinine imparts same bitter taste as of 30–100 mM of caffeine (Frank et al. 2004). A 1:1,84,175 dilution of quinine also gives bitter taste (Bassoli et al. 2007). Quinine and quinidine are found in *Cinchona officinalis* L., Rubiaceae family as quinine hydrochloride a salt form or as an extract (Aspen et al. 1999). Quinine has antimalarial properties and there are some techniques achieved to overcome bitter taste of quinine by adding sweetening agent or by addition of bitter taste inhibitors like NaCl, tannic acid or phosphatidic acid. Quinidine has been used as an anti-arrhythmic agent. Quinine has also been used as a bittering agent in tonic drinks, bitter alcohol beverages, soft drink and bitter lemon kind of tonics and in flour confectionary (Nakamura et al. 2002, Aspen et al. 1999). Quinidine is an antiarrhythmic agent that can be used in ventricular arrhythmia.

Morphine is an opiate alkaloid and extracted from poppy straw of opium poppy of *Papaver somniferum* (Frick et al. 2005). It belongs to benzoisoquinoline class (Frick et al. 2005). It has bitter tasting properties. It has been widely used as an analgesic for severe, acute and chronic pain conditions (Liang et al. 2013). However, it has limitation of abuse, tolerance, physical dependence and persistent behavioral changes. But in safer dose, it can be used in neuropathic pain, diabetic neuropathy, restless leg syndrome, fibromyalgia and reflex sympathetic dystrophy like centralized algesia. To decrease morphine physical dependence and tolerance, it is reported to use curcumin in combination with morphine in mice (Liang et al. 2013).

α-solanine is another bitter-tasting alkaloid and its analogs like α-chachonine, β-2-chachonine and solanidine are also bitter, which are present in green potatoes of *Solanum tuberosum* and night shade plant *Solanum nigrum* (Kodamatani et al. 2005). Among these three main glycoalkaloids, α-solanine, α-chachonine and β-2-chachonine have two different organoleptic properties- bitter caffeine like taste and astringent pain sensations which at higher concentration give burning sensation lasting for 2 hr owing to the combine effect of bitter taste and pain. Bitter threshold is 0.313, 0.078, and 0.078 mg and pain stimuli at 0.625, 0.323, and 0.156 mg for α-solanine, α-chachonine and β-2-chachonine respectively. α-solanine is also found in eggplant fruits – *Solanum melongena* (Kodamatani et al. 2005).

Ephedrine or *Ma huang* is a protoalkaloid isolated from *Ephedra sinica* and *Ephedra equisetina* of Chinese species, Indian and Pakistani species *Ephedra major*, *Ephedra intermedia* and *Ephedra gerardiana* (Evans 2009). Since they don't have nitrogen in heterocyclic ring, they are not considered as typical alkaloids, but they are known as protoalkaloids (Evans 2009). Ephedrine consists of β-phenylethylamine core structure. Ephedrine is a crystalline powder and bitter in taste (Carvalho et al. 2012). Use of ephedra was begun in oriental medicines where it was used as a treatment for cough, asthma, fever, nasal congestion and headache (Foster et al. 2012, Soni et al. 2004). However, due to cardiovascular complications associated with ephedrine, United States Food and Drug Administration (US FDA) banned its products in United States (Bent et al. 2004). It is also used for weight loss and muscle gain therapy (Greenway 2001). Ephedrine is a sympathomimetic agent with similar effects to epinephrine (Dewick 2002). Its pure form is used as a bronchodilator, vasopressor and nasal decongestive agent (Dewick 2002). Other salt preparation of ephedrine such as ephedrine sulphate is used in hypotensive patients and allergic disorders like bronchial asthma. Ephedrine salt preparations are generally used for lowering blood pressure during spinal anesthesia (Robbers et al. 1996). Other applications of ephedrine are diabetic neuropathic edema, motion sickness, cough and cold, CNS stimulant, enhancement of athletic performance (Robbers et al. 1996).

Cassaine is a diterpene alkaloid, which is isolated from biological source *Erythrophleum fordii* (Leguminosae). It is a traditional Chinese drug, which has positive effects on blood circulation (Qu et al. 2006). As like other alkaloids, it also imparts bitter taste. Two main constituents of cassaine extracted from *Erythrophleum* are cassaine diterpanoid amide and amines, collectively known as cassaine alkaloids (Du et al. 2011, Qu et al. 2006). It has positive ionotropic effect on heart, which is similar to that of digitalis (Verotta et al. 1995). It inhibits Na^+K^+ATPase pump and can be useful in congestive heart failure (Tobin et al. 1975). It also has antitumor role through its cytotoxic and apoptotic action in various tumor cell lines (Hung et al. 2014, Qu et al. 2006). It inhibits VEGF induced tumor cell migration, invasion and angiogenesis process in HUVAC cell lines (Tae et al. 2017).

Capsaicin is a chemesthetic compound that is known for its pungent taste while it also gives stimulus to bitter taste which was confirmed through several studies. Capsaicin can stimulate and desensitize bitter taste which differs in amount taken for different bitter stimulus (Toh et al. 1955, Green and Hayes 2003). It is obtained from genus capsicum. It is clinically used for pain relief, weight reduction in obese patients, anti-cancer properties, neurogenic bladder, irritable bowel syndrome, dermatitis, etc. (Sharma et al. 2013).

Berberine alkaloids are used from very ancient time in Ayurvedic and Chinese medicines. Despite of bitterness of berberine alkaloids, it has been used widely. It is extracted from roots, stem bark and rhizomes of *Hydrastis Canadensis* (goldenseal), *Berberis aristata* (tree turmeric), *Coptis chinensis* (goldenthread), *Berberis aquifolium* (oregon grape), and *Berberis vulgaris* (barberry). *Berberis aristata* DC. (Berberidaceae) is used in treating numerous illnesses, mentioned in all ancient scriptures Charak, Ayurveda and Sushruta (Kirtikar and Basu 1918). *Berberin aristata* contains protoberberine and bis-isoquinoline type of alkaloids (Karimov et al. 1977). It possesses potent antimicrobial activity against bacteria, virus, fungus, protozoa, helminths and chlamydia for that it has been used in treatment of bacterial diarrhea (Joshi et al. 2011, Sharma et al. 2011). It is also reported to have hypoglycemic actions for which it can be used in diabetic patients (Pareek and Suthar 2010). After that berberine alkaloids are majorly studied in different health conditions. According to several in vitro and in vivo studies, berberine is useful in metabolic syndromes (Gupta et al. 2010), cardiovascular complications, neurological problems, hepatobiliary disorders (Rathi et al. 2013), hypolipidemic effects (Pareek and Suthar 2010, Gupta et al. 2010), urinary troubles, preventive effects on paracetamol and CCl_4 induced hepatotoxicity (Gilani et al. 1999), anti-platelet activity (Tripathi et al. 1996), ethanol induced mitochondrial damage (Sebastian and Setty 1997), ionotropic effects (Gilani et al. 1999), antioxidant properties (Singh and Kakkar 2009), anti-osteoporotic actions (Yogesh et al. 2011) and potentiation of hypnosis caused by sodium thiopentone (Dehar et al. 2012).

Pilocarpine is also a bitter alkaloid, which is a muscarinic agonist. Genus pilocarpus from Rutaceae family is only source of pilocarpine, which is extracted from leaves of pilocarpus microphyllus, also known as jaborandi (Pinheiro 1997). Pilocarpine is used as a therapy for xerostomia. Xerostomia is a condition of oral dryness caused by misfunction of salivary glands (Nusair and Rubinow 1999). Pilocarpine as a muscarinic agonist increase secretion from salivary glands which is useful in treatment of xerostomia. Pilocarpine is clinically approved and well tolerated in xerostomia patients (Nusair and Rubinow 1999). Pilocarpine is also topically used in glaucoma treatment (Heijl 1989). It causes miosis in open angle and narrow angle glaucoma. Pilocarpine is available in 1%–4% solution for treatment of glaucoma. A summary of bitter-tasting alkaloids is given in **Table 11.1**.

11.3.2 Flavanoids

Flavanoids as a dietary source was first described in 1936 by Albert Szent Gyorgyi. Flavanoids and isoflavanoids are abundantly present in plants and are secondary metabolites as like alkaloids which play protective role in plants. Flavanoids and isoflavanoids are large group of compounds classified under phenols. Isoflavanoids are mostly found in Leguminosae family (Simons 2011). They are composed of two aromatic rings attached through

Table 11.1 Summary of Different Bitter Tasting Alkaloids and Their Effects

Alkaloid	Chemical Class	Major Source	Key Actions	Reference
Caffeine	Methylxanthine alkaloid	*Coffea arabica* *Coffea robusta* *Camellia sinensis*	CNS stimulant Improve respiration	Heckman et al. (2010) Gajewska et al. (2015)
Quinine and quinidine	Quinoline alkaloid	*Cinchona officinalis L.*	Antimalarial Antiarrhythmic	Nakamura et al. (2002)
Morphine	Isoquinoline alkaloid	*Papaver somniferum*	Analgesic Use in Cough and cold	Liang et al. (2013) Nayler (1966)
α-solanine	Steroidal alkaloid	*Solanum nigrum* *Solanum tuberosum*	Antioxidant Antibacterial	Sultana et al. (1995)
Ephedrine	β-Phenylethylamine alkaloid	*Ephedra sinica and Ephedra equisetina*	Bronchodilator Vasopressor Nasal decongestant	Dewick (2002)
Cassaine	Terpenoid alkaloid	*Erythrophleum fordii*	Congestive heart failure	Tobin et al. (1975)
Capsaicin	Benzylamine alkaloid	*Capsicum*	Analgesic Antiobesity	Sharma et al. (2013)
Berberine	Isoquinoline alkaloid	*Berberis aristata DC*	Antimicrobial Antidiabetic	Pareek and Suthar (2010) Joshi et al. (2011)
Pilocarpine	Imidazole alkaloid	*Pilocarpus microphyllus*	Xerostomia Glaucoma	Nusair and Rubinow (1999) Heijl (1989)

three carbon bridge from which carbon bridge of flavanoids is in form of pyran ring (Beecher 2003). Many of the flavanoids are bitter in taste such as soybean flavanoids, which has a role in prevention of cancer, osteoporosis, cardiovascular events and menopausal complaints (Drewnowski and Gomez-Carneros 2000, Aldin et al. 2006, Okubo et al. 1992). Degree of polymerization of flavanoids determines their sensorial effects whether they are sweet, astringent or bitter in taste (Peleg et al. 1999). As like monomeric flavanoids are of bitter taste. As polymerization of flavanoids increases, bitterness decreases and astringency of the compounds increases (Peleg et al. 1999). Many of the flavanoids are reported for their use in health benefits such as anti-oxidant, anticancer, anti-diabetic, anti-inflammatory, anti-thrombotic and estrogen modulation (Ververidis et al. 2007, Soto-Vaca et al. 2012, Peleg et al. 1999). Flavanoids are reported to be consumed extremely as nutraceuticals in Asian and Western countries.

Isoflavanoids from soy products are used in nutritional food for various health effects but has limited use due to bitter and astringent taste. Soy isoflavones has wide variety of health benefits due to its anti-oxidant (Kao and Chen 2006) and phytoestrogen properties (Tikkanen and Adlercreutz 2000). It has been reported beneficial in treatment of cancer, anti-aging, osteoporosis, learning and cognitive effects in menopausal women, heart diseases, diabetes and Kawasaki disease. Soybean products contain 12 isoflavanoids isoforms and divided into four chemical forms as melonylglucoside, acetylglucosides, aglycones and unsubstituted glucosides (Messina 2010), among which few are responsible for bad taste of soy products. As an example, genistein form, daidzein form and glycitein forms are bitter in taste. Following genistein form (66%), daidzein form (24%) and glycitein form (10%), which effectively contribute to bitterness were present in soy products (Wang and Murphy 1994). In unprocessed soybeans, melonyl isoflavone glucosides (59%) and unsubstituted isoflavone glucosides (36%) are present. After processing it gets converted into acetyl isoflavone glucosides (2%) and isoflavone aglycones (3%) that are formed by dry heating and fermentation process respectively (Coward et al. 1998, Wang and Murphy 1994). Among these, literature review claims bitter taste isoflavones are melonyl glucosides (Aldin 2006), glucosides (Kudou et al. 1991, Okubo et al. 1992) and isoflavone aglycones (Drewnowski and Gomez-Carneros 2000, Kudou et al. 1991, Okubo et al. 1992). Aglycone formation occurs during soaking process by action of β-glucosidase. Fermentation process further increases isoflavone aglycone content (Murphy et al. 2002, Coward et al. 1998). Processing methods also affect contents of isoflavones in soy products. To decrease bitterness of soy products, change the processing of soybeans. Enzymatically converted isoflavones has decreased bitter taste and astringency of aglycones (Matsuda et al. 1994); another author has reported to increase in bad taste after treating with β-glucosidase (Matsuura et al. 1989). Flavanoid content in soy products depends on cultivar, weather condition and soil. As an example, soy protein from soybeans contains equal amount of genistein and daidzein, but extract from soy germs contains daidzein and glycitein in four fold amount than genistein equivalents (Nakamura et al. 2000). Genistein has anti-cancer role that acts through inhibition of tumor cell migration and proteolysis in cancer cell line studies (Farina et al. 2006). It has been reported to have protective role in postmenopausal breast cancer with HER2 mutation (Wada et al. 2013, Boucher et al. 2013, Zhang et al. 2012). Other than cancer it has other health benefits in osteoporosis (File et al. 2001, Bone et al. 2000, Wei et al. 2012), coronary heart disease (Tikkanen and Adlercreutz 2000, Beavers et al. 2012, Yamakoshi et al. 2000, Chan et al. 2008), diabetes (Gilbert and Liu 2013, Behloul and Wu 2013, Valsecchi et al. 2011) and cognitive effects (Celec et al. 2005, Kritz-Silverstein et al. 2003). Daidzein has been reported to have antitumor role in prostate cancer (Sugiyama et al. 2013) and in treatment of heart diseases (Tikkanen and Adlercreutz 2000).

Tea is also rich in bitter-tasting flavanoids (Drewnowski and Gomez-Carneros 2000). Unfermented green tea has the bitterest compounds than black tea and oolong tea. It consists of 30%–40% of phenols from which catechins content is higher followed by phenolic acids and flavanol glycosides (Graham 1992). Fermentation of green tea converts flavanoids into thearubigins, theaflavins, etc. (Drynan et al. 2010, Graham 1992). Tea flavanols are regarded as catechins which are group of compounds like catechins, epicatechin, gallocatechin, epigallocatechin, galloylated epigallocatechin gallate and epigallocatechin gallate (Peleg et al. 1999). Among these, galloylated epigallocatechins are considered as bitterest (Narukawa et al. 2010). Epigallocatechins are bitterer tasting than catechin (Kallithraka et al. 1997). Flavanoids from green tea have been widely used to treat metabolic syndromes (Yang et al. 2001), type II diabetes (Babu et al. 2006), anti-cancer (Roomi et al. 2005), cardiovascular disorders (Sueoka et al. 2001), neuroprotection (Weinreb et al. 2004), anti-viral (Weber et al. 2003), anti-bacterial (Roccaro et al. 2004), anti-arthritic (Haqqi et al. 1999), anti-angiogenic (Sartippour et al. 2002), anti-inflammatory (Donà et al. 2003), anti-oxidant (Osada et al. 2001), diet-induced obesity (Lin and Lin-Shiau 2006) and coronary disease (Tsuneki et al. 2004).

Flavanoid content of red wine is bitter in taste. The organoleptic properties of wine are dependent on aging and processing of wine. Dry red wine is the highest consuming wine which contains different phenolic compounds including flavanoids, anthocyanines, flavanols, hydroxycinnamic acid and tannins (Ribéreau-Gayon et al. 2006, Li and Sun 2017, Waterhouse 2002). Red wine contains kaempferol, quercetin, myricetin, syringetin and laricitrin flavanoids. These flavanoids are responsible for bitter taste of red wine (Price et al. 1995). Flavanol contents of red wine are (+)-catechin, (−)-epicatechin, (+)-gallocatechin and (−)-epigallocatechin (Su and Singleton 1969). Bitterness threshold of (+)-catechin and (−)-epicatechin is 290 and 232–270 mg/L respectively (Hufnagel and Hofmann 2008a; Stark et al. 2010). Red wine has major health benefits including anti-aging, heart disease and obesity (Guerrero et al. 2009). Low molecular weight catechin from red wine is found to be bitterer in taste than astringent. As molecular weight of catechin increases, astringent property increases with it (Peleg et al. 1999). Epicatechin flavan is also present in red wine which imparts more bitter taste than catechin (Thorngate and Noble 1995).

Flavanoids like quercetin, naringin and nobiletin are present in citrus fruits which imparts bitter taste. Quercetin is majorly found in our diet and has potential benefits on our health and has been extensively studied flavanoid (Hertog et al. 1992). Quercetin belongs to flavonols and present in wine (Trock et al. 1990), fresh hops (Peterson and Dwyer 1998), black tea, oolong tea (Trock et al. 1990), endive (Peterson and Dwyer 1998), several citrus fruit juices (Trock et al. 1990), capers and raw, oregano, blueberry and blackberries, cherries, elderberries and hartwort leaves (Hertog et al. 1992). It exhibits wide range of pharmacological activities including anti-oxidant (Bouayed and Bohn 2010, Kelly 2011), anti-inflammatory (Shoskes et al. 1999, Katske et al. 2001), anti-cancer (Park et al. 2005) and cardiovascular protection (Conquer et al. 1998) and neurodegenerative disease (Bellissimo et al. 2001). It is a potent anticancer agent that has reported to inhibit proliferation and angiogenesis in vitro cell line studies and in vivo models of colorectal cancer (Park et al. 2005), prostate cancer (Kim and Lee 2007), breast cancer (Conklin et al. 2007), skin cancer (Sengupta et al. 2001), ovarian cancer (Maciejczyk and Surowiak 2013), gastric cancer (Ramachandran et al. 2012) and pancreatic cancer (Angst et al. 2013). Quercetin has three ring structures with five hydroxyl groups and glycosylation process increases its hydrophilicity, bioavailability and product stability of quercetin (Williams and Grayer 2004). Naringin belongs to flavanone neohesperidoside which is very bitter in taste. Its concentration is highest in immature fruit pulp and young leaves of the plant (Del Río et al. 1998). Other than these, tengeretin (Del Río et al. 1998) and sinensetin (Sendra et al. 1988, Pupin et al. 1998) are also bitter flavones that have nutritional value.

11.3.3 Tannins

Tannins also fall under phenolic category as like flavanoids. Tannin was first delineated by Seguin as a compound that converts animal skin into leather (Mingshu et al. 2006). Tannins are present in plants as polyphenols with molecular weight between 300 and 3000 Da (Khanbabaee and van Ree 2001). They have capability to precipitate alkaloids, proteins and gelatin Da (Khanbabaee and van Ree 2001). Traditionally tannins were classified based on their hydrolyzable properties by tannases or hot water: hydrolyzable tannins and non-hydrolyzable tannins or condensed tannins (De Bruyne et al. 1999). Recent classification is based on their structural difference that includes four categories as gallotannins, ellagitannins, complex tannins and condensed tannins (Khanbabaee and van Ree 2001). Gallotannins and ellagitannins are categorized under hydrolyzable tannins which contains polyhydrolic alcohol groups and hydroxyl groups (Freudenberg 2013). While condensed tannins have more complex chemical structure and are polymerized products of flavan-3-ols and -3-diols or both. Condensed tannins are present in fruits, vegetables, plants, cocoa, red wine and several grains (Freudenberg 2013). Hydrolyzable tannins are extracted from seed pods, leaves, wood, bark and fruits of plants (Swain 1965). Tannins have wide range of health benefits as anti-carcinogenic (Kono et al. 1988), anti-oxidant, anti-mutagenic (Huang et al. 1985) and anti-microbial (Wehmer and Hausschwammstudien 1912). Anthocyanine, ellagitannin and some of the hydrolyzable tannins in pomegranate fruit has anti-cancer role (Afaq et al. 2005).

Wine contains different tannins and non-tannin compounds that impart astringent and bitter taste of wine. Condensed tannins with flavan-3-ols polymers are responsible for bitter taste perception of wine (Hufnagel and Hofmann 2008b, Noble 1994). Also, bitterness of wine is imparted through some flavanols, benzoic acid derivatives and hydroxycinnamate compounds (Sáenz-Navajas et al. 2010, Peleg and Noble 1995, Vérette et al. 1988). These agents activate bitter taste receptors TAS2Rs (Soares et al. 2013). Stereochemistry and inter-flavanoid bonding of tannins affects bitterness of wine (Peleg et al. 1999, Thorngate and Noble 1995). Other than that condensation process of tannins is also responsible for bitter taste of wine such as ethyl bridge addition to catechin, which leads to bitterness of wine (Vidal et al. 2004). Evidences suggest that galloyl content of tannins is bitterer which is contained in seed fraction of plant (Brossaud et al. 2001). This is due to hydrophobicity of galloyl group in aromatic ring. This shows difference of tannin bitterness between seed and skin fraction of plant (Brossaud et al. 2001).

Red wine prepared from grape skin and seed that contains phenolic contents of 30% and 70%, respectively (Chandrashekar et al. 2000, Noble 1998). Bitter phenolic compounds in red wine are high molecular weight tannins including mono-catechin and polymers with molecular weight less than 500 Da and catechin polymers with molecular weight more than 500 Da (Blanco et al. 1998, Brouillard et al. 1997, Singleton 1992). A study of red wine showed than epicatechin contents are bitterer in nature than catechin content of wine. Polyphenolic tannins with more than 500 Da molecular mass are more astringent than bitter in taste (Peleg et al. 1999). Red wine polyphenols are helpful in heart diseases (Sarr et al. 2006). Mechanism by which it improves disease condition is increase in level of HDL-cholesterol level, anti-oxidant properties, decrease in platelet aggregation and endothelial function restoration. Few studies have shown that daily consumption of red wine (15.5–31 g alcohol) has protective role in CVS (Sarr et al. 2006).

11.3.4 Terpenes

The term terpene was derived from turpentine, a volatile liquid from which it has been isolated (Zhao and Schlaad 2011). Terpenes are widely present as an essential oil and basic unit of terpenes is isoprene unit with five carbons (Zhang et al. 2002). They are also known as terpene hydrocarbons, which has basic formula of $(C_5H_8)n$, where n stands for number of isoprene units. Hydrocarbon terpenes are classified as: monoterpenes (n = 2), sesquiterpenes (n = 3), diterpenes (n = 4), triterpenes (n = 6) and tetraterpenes (n = 8) (Zhang et al. 2002). Terpenes are abundantly found in Labiatae, *Melissa officinalis, Agastage rugosa*, lavender and menthe spp. (Connolly and Hill 2010, Fraga 2008, Grayson 2000).

Diterpenes are 20% liquid fraction and found in green coffee which is bitter in taste as similar to caffeine. Green coffee constituents like cafestol, kahweol, 16-O-methylcafestol, cafestal, and kahweal belongs to diterpeniod class (Lee and Jeong 2007). Diterpenes are in esterified form with long chain of fatty acid present in coffee oil and they are reported to protect liver cells from oxidation stress (Lee and Jeong 2007). Cafestol and kahweal are extracted from ground coffee during brewing process and their contents are higher in espresso coffee. While, filtered, percolated and instant coffee removes major amount of these constituents (Gross et al. 1997, Urgert et al. 1995). They have wide range of pharmacological activities as nutraceuticals. These are used in obesity and weight loss therapy (Urgert et al. 1995), oxidative stress conditions (Chobotova 2009), type II diabetes (Herling et al. 1998), cardiovascular diseases like hypertension, arteriosclerosis and vascular endothelial hypofunction (Domanski et al. 2002), breast, colon, liver, rectal and skin cancers (Chobotova 2009), hepatic disorders (Gallus et al. 2002), Parkinson's and Alzheimer's disease (Cropley et al. 2012), as an immunomodulator and antibacterial agents (Lou et al. 2011).

Triterpenoids are abundantly present in plant and have wide spectrum of biological activities (Siddique and Saleem 2011). Most of the triterpenes are composed of thirty carbon atoms arranged in form of 6 isoprene units. Free triterpenes are mainly found in Leguminosae, Euphorbiaceae, Compositae, Rubiaceae, Labiatae, Meliaceae and other plant families (Meng et al. 2008). Triterpenoids are categorized as tetracyclic and pentacyclic triterpenoids. From these, pentacyclic triterpenoids have been shown to have various pharmacological effects which include hepatoprotective, anti-inflammatory, anti-arthritic, anti-cancer and immunomodulatory effects (Geetha and Varalakshmi 2001, Hu et al. 2013, Fukumitsu et al. 2016, Peláez et al. 2013). Pentacyclic triterpenoids are subcategorized as oleanane type, α-amyrin type, β-amyrin type, friedelin type, hopane type and lupine type (Lin 1977).

Limonin is a triterpene secondary metabolite present in seeds (Manners 2007), peel (Hamdan et al. 2011) and various citrus fruits like lemon, grapes, oranges, limes, pumellos and mandarins and is responsible for delayed bitter taste of the fruit (Hasegawa and Miyake 1996, Wang et al. 2016). Limonin is made by fruit tissue injury from limonin precursor. Limonin precursor is tasteless but when it gets converted to limonin it imparts bitter taste of fruit juice (Hasegawa and Miyake 1996). As like limonin, nomilin is also a triterpene which has bitter taste present in grape fruit juice, oroblanco juice and melogold juice (Hsu et al. 1998). Limonin has been reported to have in vitro and in vivo anticancer, antioxidant (Murthy et al. 2013, Vanamala et al. 2005), antiviral (Balestrieri et al. 2011), antibacterial (Vikram et al. 2012), larvicide (Bilal et al. 2012) and insecticidal (Yu et al. 2015), analgesic and anti-inflammatory activities (Kim et al. 2011), antidiabetic (Ono et al. 2011) and inhibition of osteoclastogenesis properties (Kimira et al. 2015).

11.3.5 Miscellaneous

Other than above phytoconstituents, following are used as nutraceuticals that are having off-taste. Resveratrol is a polyphenol belonging to stilbenoids class and consist of two aromatic rings bridged by ethane or ethylene residue. It is present in wine grape, which has benefits to heart and cardiovascular system (Babal et al. 2006, Chen and Pace-Asciak 1996). It works through decrease in platelet aggregation, protection of heart cells from ischemic reperfusion, decrease in TG and increase in HDL (Babal et al. 2006, Chen and Pace-Asciak 1996). It also has antioxidant, antiviral and antitumor activities (Gregory et al. 2004, Aluyen et al. 2012). Some of the organosulfur compounds are also used as nutraceuticals. Glucosinolates are natural pesticides and they have goitrogenic properties (Fenwick et al. 1983a). Sinigrin belongs to glucosinolate class which is majorly found in green mustard and cruciferous vegetables, has bitter taste (Fenwick et al. 1983b). It has anti-cancer (Bhattacharya et al. 2010, Jie et al. 2014), anti-inflammatory (Jang et al. 2015, Lee and Lee 2015), antibacterial (Zsolnai 1966), anti-fungal (Ocampo 1980), anti-oxidant (Peluffo and Radi 2007) and wound healing activities (Mazumder 2016). Other than sinigrin,

Progoitrin and Glucobrassicin are also bitter glucosinolates. Goitrin 5-vinyl-2-oxazolidine thione is a hydrolysis product of glucosinolates which tastes bitter and mainly present in cabbage and pith (Heaney and Fenwick 1980, Fenwick et al. 1983a). Brassica glucosinate, also known by mustard oil glycoside has bitter taste and it contains sinigrin, progoitrin and glucobrassicin. Glucosinate products are bitter due to these constituents and are not well accepted by consumers (Fenwick et al. 1983a).

Another group of phytoconstituents include isothiocyanates. Damage to plant tissue causes hydrolysis of glucosinolates by enzyme myrosinase. This will form isothiocyanate, hydrogen sulfate and glucose. Isothiocyanates are referred as mustard oils which have pungent taste and they are phase II enzyme activators (Fahey et al. 1998). Allyl-isothiocyanate, 3-methyl-sulfinylpropyl isothiocyanate, benzyl isothiocyanate 4-methylsulfinyl butyl isothiocyanate and phenylethyl isothiocyanate are acrid and pungent mustard oils that are mainly found in cabbage, pith and cambial cortex (Van Etten et al. 1979). Isothiocyanates are having physiological role in prostate disease, skin cancer and antimicrobial properties in food preservatives (Cheng et al. 2011, Delaquis and Mazza 1995).

11.4 Animal models for taste perception

Taste aversion paradigm is used for identification of dislikeness of taste of a particular drug (Vogel and Vogel 2013). In taste aversion method or taste aversion learning, rats are given conditioned stimuli of 0.01M of saccharine drinking solution (palatable taste) and after that unconditioned ingestion of 0.15 M LiCl solution by intraperitoneal route. In taste aversion method, ingestion of taste stimulus is combined with internal malaise, animal can remember the stimuli and after learning process it rejects ingestion (Vogel and Vogel 2013). After the test, inhibiting and facilitating (Lipinski et al. 1995, Sobel et al. 1995) as well as neuronal mechanisms (Agüero et al. 1993, Yamamoto 1993, Swank et al. 1995) are analyzed. Male Wistar rats of 220–250 g are used and given to free food and fluid access. Before 24 hr of the first session, animals are deprived of fluid. Fluid access is given only for 15 min and given conditioned taste aversion in six sessions. In first four sessions bottle contains only tap water where animals learns to drink reasonable amount of fluid. For 5th session animals are given saccharine solution which is conditioning session. Two days after 5th session, animals are given free access to tap water and no sessions are conducted. For the sixth session, before experiment animals are again deprived and then given saccharine solution and tap water in different bottles. Location bias is avoided by keeping saccharine bottles both the side of cages. Amount of fluid taken is measured which indicates drug-induced conditioned taste aversion. This can be evaluated by analyzing relative saccharine intake in drug control and vehicle control groups. Ratio of intake is calculated by saccharine solution/(saccharine solution + tap water). Fluid intake score is calculated in grams (Vogel and Vogel 2013). This method can be modified by substitution of LiCl with ethanol (Gauvin and Holloway 1992, June et al. 1992; Thiele et al. 1996, Bienkowski et al. 1997), cocaine (Van Haaren and Highes 1990, Glowa et al. 1994), morphine (Miller et al. 1990; Bardo and Valone 1994), d-amphetamine (Davies and Wellman 1990, Lin et al. 1994), naloxone (Mucha 1997), cisplatin (Mele et al. 1992), caffeine (Brockwell et al. 1991), BDZ (Neisewander et al. 1990), 9Δ-tetrahydrocannabinols(Parker and Gillies 1995), apomorphine (McAllister and Pratt 1998), cyclosporine A (Exton et al. 1998), cholecystokinin (Ervin et al. 1995, Mosher et al. 1996), nicotine (Shoaib and Stolermman 1996), quinine (Parker 1994) and dopamine D_3 agonist (Bevins et al. 1996).

Another method was to investigate effect of drugs on taste-potentiated odor aversion in rats (Willner et al. 1992). Some authors have studied taste aversion learning in ferrets (Rabin and Hunt 1992). Another study was done for individual differences in reaction to drug aversive properties which used high and low conditioned taste avoidance groups based on saccharine intake paired with morphine, lithium, fenfluramine or amphetamine (Turenne et al. 1996). Since animals have same bitter taste receptors, they react similar to humans to taste stimuli (Brasser et al. 2005). To evaluate bitter taste perceptions sensory analysis methods have been applied in animals (Peleg and Noble 1999, Vérette et al. 1988). Bitter taste receptor activation followed by Ca^{+2} releases from endoplasmic reticulum to cytoplasm is used to identify bitterness of a compound. With the help of fluorescent based calcium imaging techniques, bitterness of a compound can be analyzed by this mechanism (Kuhn et al. 2004). Newer techniques include electronic sensor technology which detects taste of food as well as pharmaceuticals by electronic tongue evaluation (Cetó et al. 2011).

11.5 Modifying bitter taste of nutraceuticals

Nutraceuticals are more beneficial in prevention as well as for treating diseases or disorders with lesser side effects but bitterness of the natural products limit their use (Gaudette and Pickering 2013). The food industries are now focusing on eliminating bitterness or decrease the level of bitterness from such products. Traditionally masking of bitter taste was made by addition of sweetening agents, salts, textures and odorants. Bitter taste inhibitory compounds can also be used to achieve better results (Gaudette and Pickering 2013). Several approaches have been proposed for modifying bitter taste of nutraceuticals (**Figure 11.2**).

11.5.1 Physical methods

11.5.1.1 Temperature treatment

Food treatment at different temperature helped reducing bitterness of several compounds. Heating, boiling, cold storage, baking and pressure cooking has been found advantageous in reducing bitter taste. Wild yam from

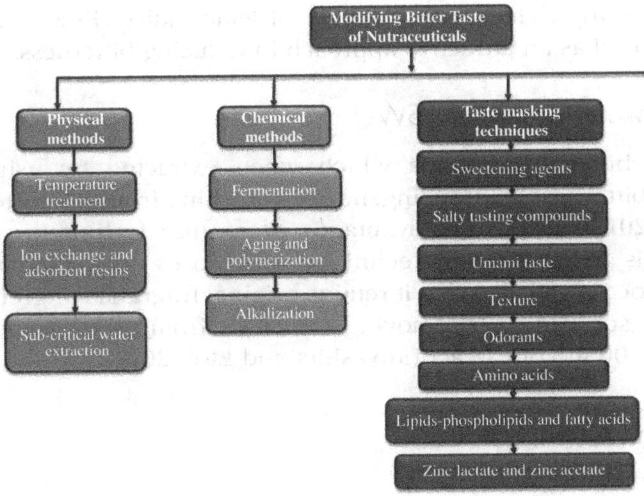

***Figure 11.2** Approaches to modify bitter taste of nutraceuticals.*

Dioscorea spp. has shown to decrease its bitterness while boiling, baking and pressure cooking (Bhandari and Kawabata 2005). Furanoid norditerpenes, disbulbins A and B are responsible for bitter taste of wild yam. Among these techniques boiling was proved to be better in reducing bitter taste of wild yam (Bhandari and Kawabata 2005). In olive oils, heating was found to be useful in decreasing bitterness due to phenolic content (García et al. 2001). Heating didn't change acidity or oxidative stability yet there was decrease in phenolic content of olive oil (García et al. 2001). In another study, effect of cold temperature was studied in reducing bitterness of olive oils. Cold storage technique was useful but it was time dependent (Yousfi et al. 2008). Depending upon type of olive oil, time duration varied from 2 to 8 weeks at 50°C temperature (Yousfi et al. 2008). Also, temperature should remain constant during the period of storage. Cold storage is more acceptable technique than heating to eliminate chances of the degradation of the useful compounds (Yousfi et al. 2008). There is still need to optimize effective temperature which varies according to product type.

11.5.1.2 Ion exchange and adsorbent resins

Bitter compounds like phenols can be adsorbed in resins to convert into non-bitter compound by trapping in polymers and precipitation following extraction with appropriate solvent (Drewnowski and Gomez-Carneros 2000). A study was carried out to remove bitter taste of limonin (terpene) in orange juice (Kola et al. 2010). Two ion exchange and resins, Amberlite XAD-16HP and Dowex Optipore L285, were used from which both were successful in reducing bitterness of orange juice but Dowex Optipore L285 altered acidity and increased pH leading to decreased solubility of soluble compounds while Amberlite

XAD-16HP did not alter any characteristics nature of lemon juice (Kola et al. 2010). Thus, it can be used as an attractive approach in reducing bitterness.

11.5.1.3 Sub-critical water extraction (SWE)

Catechins are highly bitter compounds which when extracted by using SWE lead to reduced bitterness and astringency of catechins from green tea (Miyashita and Etoh 2013). SWE has advantages over other techniques of extraction, because it is an eco-friendly technique which uses only standard water for extraction process and also that it retains original fragrance of green tea. This method was successful in removing catechins from green tea by 300 mg of catechin in 100 mL of extract (Miyashita and Etoh 2013).

11.5.2 Chemical methods

11.5.2.1 Fermentation

Researchers have proved that change in physical and chemical structure of compounds can reduce bitter taste. With the help of fermentation process, bitter taste of dark tuna meat was reduced using lactic acid fermentation with *Lactobacillus casei* subsp. Casei ATCC 393 in the presence of glucose and sodium (Bertoldi et al. 2004). Fermentation product, lactic acid effectively reduced bitterness in tuna meat. This study showed level of glucose increased fermentation process while opposite effect was seen with sodium. This study has also observed that increase in pH leads to decrease in bitterness of dark tuna meat (Bertoldi et al. 2004). However, these results didn't show similar effects in red wine study where increase in pH did not have any effect. Increase in tartaric acid slightly reduced bitter taste in red wine but on the other hand ethanol content has shown to increase in bitterness (Fontoin et al. 2008). From these diverse results of the studies, we can say that food products have their own matrix which affects their taste perception and masking properties.

11.5.2.2 Aging and polymerization

Red wine sold without aging, has high content of sugar which helps to reduce bitterness. However, aging of wine led to polymerization of phenols leading to reduce in bitter taste of wine (Noble 1998). Researchers have isolated bitter compounds, cynaropicrin and grosheimin from artichoke leaves and put them for microwave irradiation. This led to esterification of these compounds and opened lectone rings which suppressed bitter taste of these compounds (Cravotto et al. 2005).

11.5.2.3 Alkalization

Researchers have studied alkalization products of bitterants, which can reduce bitterness. In a study, cocoa alkalization led to formation of flavan-3-ol-C-glycosylation which reduced intensity of bitter taste of cocoa powder (Stark and Hofmann 2006). In another study role of C-glycosylation in reducing bitterness was confirmed using catechin (Zhang et al. 2014). They kept catechin at 2000°C for 15 min with millard models comprising of reducing sugars, D-galactose, D-glucose, D-xylose and glycine. From these reactions, six of eight products were found to be flavan-3-ol-spiro-C-glycosides and one of these spiro products suppressed bitter taste of catechin (Zhang et al. 2014).

These physical and chemical methods make an attractive approach to mask bitterness yet through these techniques, there might be loss of the nutritional compounds in the process. Thus, addition of taste masking agents to bitterants can be used as an alternative to prevent this.

11.5.3 Taste masking techniques

11.5.3.1 Sweetening agents

Sucrose and other non-nutritional sweeteners like aspartame and sucralose can be used to decrease bitterness of the product. Addition of sucrose in caffeine reduced bitter taste with increasing sucrose concentration and vice versa addition of caffeine increased bitterness (Calvino et al. 1990). Although sweet and bitter tastes are concentration dependent, low concentration of sweetening agent is not always sufficient to eliminate low concentration of bitter compound (Keast and Breslin 2002). Also with quinine, sucrose has shown to decrease its bitterness (Lawless 1982, 1986). Interestingly, before taking quinine HCL and sucrose if you take prior sucrose sample than it decreases the ability of sucrose to reduce bitterness of quinine HCL. This complex behavior of compounds can be challenging in achieving goal of reducing bitterness (Lawless 1982, 1986).

Other than sucrose, aspartame and sucralose can be used as sweeteners which are much sweeter in taste than sucrose (Wiet and Beyts 1992). Aspartame and sucralose are 200 and 500–700 times sweeter than sucrose (Wiet and Beyts 1992). These agents can reduce bitterness of pharmaceutical compounds as well as natural compounds like quinine (Suzuki et al. 2004, Nakamura et al. 2000). Higher concentration of these non-nutritive sweeteners itself can impart bitterness of a product which should be taken in consideration (Ott et al. 1991). Limitation of these sweetening agents are that sucrose has caloric values of itself which can interfere with caloric content of a natural product and non-nutritive sweeteners are found to be synthetic by consumers to use in natural products (Ott et al. 1991). These can be challenging in use of sweeteners to reduce bitter taste of nutraceuticals. Thus natural sweetening agents are more preferable which has fewer or no side effects.

11.5.3.2 Salty tasting compounds

Many studies have reported that bitter taste masking can be done by salty tasting substances (Breslin and Beauchamp 1995). Salts lack of Na^+ cation are not able to reduce bitterness. Bitter taste reduction is dependent on Na^+ cation predominantly. Other than Na^+ ion, Li^+ also has bitter taste masking properties. Several studies have reported that addition of salty compound sodium acetate in sweet-bitter mixture decreased bitterness and increased sweetness of the mixture. Na^+ containing salts work through oral peripheral physiological interaction to suppress bitterness (Breslin and Beauchamp 1995). It works through either of these mechanisms: modulation of ion channels or pumps, cell membrane stabilization and bitter taste receptor blockage, interaction with second messenger system (Keast and Breslin 2002). However, practical use of salts in nutritional food is avoided due to its negative effects on health (Sacks et al. 2001).

11.5.3.3 Umami taste

Umami taste is having savory taste quality which is imparted through glutamate containing anion salts (Löliger 2000). Among which monosodium glutamate (MSG) and NaAMP are having bitter taste masking properties (Löliger 2000). MSG has shown to decrease bitterness of quinine sulfate at suprathreshold concentration but no effect was observed at threshold concentration (Kemp and Beauchamp 1994). Other than sodium, chlorine, salicylate and gluconate anion salts did not produce the same effects (Keast et al. 2004). Other several studies have reported bitter taste masking properties of Na2AMP and MSG and inositol monophosphate combination. However, MSG has limitation of its uses as it has negative effect on health (Nakanishi et al. 2008). Studies have also confirmed safety of glutamate salts in human population (Beyreuther et al. 2007).

11.5.3.4 Texture

Modifying texture is an interesting approach to reduce bitterness. Texture modification as by increasing viscosity has been widely applied to reduce bitter taste of many compounds. Studies have reported to decrease bitterness of quinine by increasing its viscosity through addition of viscosity enhancers (Moskowitz and Arabie 1970). Cellulose gum concentrations are used to decrease intensity of bitter taste of quinine sulfate. Caffeine bitterness is reduced by 60% using gelatin matrix compared to water (Calviño et al. 1993). Other hydrocolloids like carboxymethyl cellulose (CMC), low-viscosity carboxymethyl cellulose (CMC-L), moderate-viscosity carboxymethyl cellulose (CMC-L) and xanthan, sodium alginate (Pangborn et al. 1973). Increased viscosity thought to be working through blocking of binding of bitter components to bitter taste receptor. These natural plant based viscosity enhancers also increase nutritional value of the product (Pangborn et al. 1973). However, concentration and type of gum should be considered according to type of compound. As in a study, CMC-M didn't reduce bitterness of polyphenols from grape seed (Smith et al. 1996).

11.5.3.5 Odorants

Odorant addition either suppresses or enhances taste of the product (Delwiche 2004). As an example when strawberry flavor is added to sucrose it increases sweetness of the solution but addition to NaCl doesn't affect saltiness (Frank and Byram 1988). Few studies have shown interaction of odor with bitter taste. Addition of coffee and chocolate aromas in caffeine increased its bitterness as similar with cocoa flavor increased bitter taste of cocoa beverages (Keast 2008). Further studies are needed to evaluate either odorants can mask bitterness of compounds. Since odorants have no negative effects on nutritional values of a product, they are attractive approach for taste modulation.

11.5.3.6 Amino acids

Addition of branched chained amino acids (BCAA) with fruit flavors can decrease bitterness of a product (Mukai et al. 2004). This is due to combination of sweet and sour taste and potentially because of citric acid content. Amino acids like L-isoleucine, L-valine and L-leucine are used to mask bitter taste (Mukai et al. 2004).

11.5.3.7 Lipids, phospholipids and fatty acids

Phospholipids eliminate bitterness of virgin olive oil (VOO) which has high hydrophilic polyphenol content responsible for its bitter taste (Koprivnjak et al. 2009). Granular soy lecithin significantly decreased bitterness and increase sweetness of olive oil when added in quantity of 5 gm/kg. However, phospholipid added oil is not considered as a natural product and shelf-life of a product should also be taken into consideration, during its incorporation for products (Koprivnjak et al. 2009). A study reported that Baraka cheese contains high free fatty acid level, which is responsible for bitter taste masking properties. Oleic acid also found to decrease bitterness through complexation with bitter tasting compound (Homma et al. 2012).

11.5.3.8 Zinc lactate and zinc acetate

Zinc lactate has shown to decrease bitterness in coffee and chocolate. Zinc sulfate was also effective in reducing bitter taste of tetralone responsible for bitterness in beer, quinine HCL and denatonium but was not effective in pseudoephedrine, sucrose octaacetate and dextromethorphan (Keast 2008).

Taste-masking agents should be used appropriately according to food that it should not change basic flavor of a particular food and it should not have negative effect on health (Goldberg et al. 2017).

11.5.4 Bitter taste blockers

Other than traditional techniques, "bitter blockers" or bitter taste inhibitor is a novel approach that works through complexation or encapsulation of bitter tasting compounds, by interaction with or blocking of bitter taste receptors and modulation of second messenger system or transduction pathways. Traditional bitter taste modulators had limitation of decreased healthiness of food (Gaudette and Pickering 2013). Also, bitter blockers are able to overcome bitterness of phenols where traditional modifiers failed to decrease bitter taste. Following compounds can be used as bitter taste blockers.

11.5.4.1 Cyclodextrin

Cyclodextrins (CD) are cyclic form of oligosaccharides with three subunits: α, β and γ. CD has glucose moiety attached at 6, 7 and 8 position by α-1,4-glycosidic linkage and it mimics donut-like structure (Toda et al. 1981, Szejtli and Szente 2005). Among these three subunits, β-CD has lowest water solubility because of rigid intramolecular hydrogen bonding and it falls under generally regarded as safe (GRAS) category (Astray et al. 2009). Thus it makes an attractive approach as bitter blocker. It inhibits bitter taste by inhibiting interaction of bitter compound to bitter taste receptors because it has inner hydrophobic core and hydrophilic shell that makes complex with the bitter compound. This complex gets released in stomach without interacting with bitter taste receptors (Szejtli 1988). β-CD successfully inhibited bitter taste of citrus juices (Konno et al. 1982), narginine, limonin (Konno et al. 1982), wine–derived polyphenols like *trans*-resveratrol and catechin (Renaud and de Lorgeril 1992, Frankel et al. 1993). β-CD in combination with sweetener inhibit bitterness of a product more effectively than used alone (Gaudette and Pickering 2012). It can modify sensory profile which should be taken in consideration to use it at higher concentration and also when used with other bitter blockers it may interfere with their mechanism of action (Astray et al. 2009).

11.5.4.2 Riboflavin-binding protein

Riboflavin-binding protein (RBP) is found in chick embryo which provides vitamin B12 and nutrition to the embryo (Croguennec et al. 2007). RBP is a potent inhibitor of bitter taste to number of bitter compounds like narginine, caffeine, theobromine, glycyl-L-phenylalanine, quinine HCL and denatonium. 0.2mM of RBP decreased bitterness nearly 100% of 0.125 mM quinine HCL (Maehashi et al. 2007). RBP has different mode of actions depending upon the type of bitterant. It inhibits bitterness of quinine HCL by hydrophobic interaction, and also it competes with receptor binding site (Maehashi et al. 2007). Due to its potent nature and inhibition of wide range of bitter compounds, this can be used as a novel approach for bitter blocker agent.

11.5.4.3 Flavanones

Flavanones from *Eriodictyon californicum* can decrease bitterness of wide range of bitterants (Geissman 1940, Ley et al. 2005). 100 ppm of eriodictyol and homoeriodictyol sodium salt decrease the bitter taste of 500 ppm of caffeine by around 45% and 40%, respectively. It lowers bitterness in wide range of compounds such as paracetamol, quinine, denatoinium benzoate, amarogentin, guaifenesin and salicin at around 35% (Ley et al. 2005). Lowering of bitterness might be due to inhibition of bitter taste receptors because it doesn't affect other taste perceptions (sweet or salty). However, mechanism behind its bitterness lowering property is still unclear (Gaudette and Pickering 2013). Though it has no effect on other taste perceptions, it can be used in sweet and salty tasting formulations. As it is of natural source it makes an attractive approach as bitter blocking agents.

11.5.4.4 Phosphatidic acid and β-lactoglobulin

Lipoprotein consist of phosphatidic acid (PA), a lipid and β-lactoglobulin (LG), a protein helps reducing bitter taste but as like falvanones, it doesn't affect sweet or salty taste (Katsuragi 1993, Katsuragi et al. 1995). PA is derived from soybeans and LG is from milk and eggs (Katsuragi et al. 1995). PA works through inhibition of bitter taste receptor and adsorption of bitterants (Katsuragi et al. 1996, Nakamura et al. 2002). PA decreases about 81% bitterness of quinine HCL (Nakamura et al. 2002). But PA is poorly soluble in water which can be challenging to use in aqueous preparations (Katsuragi et al. 1995). PA-LG lipoprotein also works through same mechanism as of PA. A study on frog showed to suppress bitter taste of 0.1mM quinine HCL and 1 mM papaverine by prior administration of PA-LG at concentration of 0.3%–1%. This is due to increased binding of PA-LG to bitter taste receptors (Katsuragi et al. 1996). But due to allergic properties of PA-LG, it limits its use for some of the food and beverages (Wal 1998, Sicherer et al. 2000, Savage et al. 2007).

11.5.4.5 Neodiosmin

Neodiosmin is a glycosylated flavone which is derived from a bitter flavanone neo-hesperidin but at lower concentrations it is odorless and tasteless (Del Rio et al. 1992, Guadagni et al. 1976). It is tasteless upto maximum 40 ppm concentration and it inhibits bitter taste effectively at lower concentrations (Guadagni et al. 1976). With addition of 10 ppm of neodiosmin threshold of caffeine, narginine and quinine sulphate aqueous solutions are increased by 80%, 225% and 137%, respectively (Guadagni et al. 1976, Guadagni et al. 1979). Limonin threshold is also increased by 53% and 300% in water and orange juice, respectively (Guadagni et al. 1976).

11.5.4.6 Magnesium sulfate

$MgSO_4$ is itself used as bitter stimulant and it also elicit sweet and salty tastes at the concentration of 0.78 M and 0.3 M but at lower concentration of 0.025 M it has bitter suppressing properties (Keast and Breslin 2002, Delwiche et al. 2001, Keast 2003). Also, minimum increase was observed in other taste sensation when used lower concentration of $MgSO_4$ (Lawless et al. 2003).

It suppresses bitterness of quinine HCL at around 51% without affecting basic taste perceptions (Keast 2003). Unlike zinc salts, it suppresses wide range of bitterants (Keast and Breslin 2002). Thus, using lower concentration of $MgSO_4$ can be used to suppress bitter taste of different formulations.

11.6 Conclusions

Nutraceuticals made from herbal or natural source are widely consumed due to their extended health benefits. They exhibit several advantages when taken along with conventional drugs for various disorders. Several bitter tasting nutraceuticals are available in market. With the recent advancement in food and pharmaceutical industries, various methods can be employed for taste masking in order to improve the palatability and compliance. In addition to traditional method of adding taste masking agents, newer trends are toward blocking of taste perception. This has helped lot of patients in improving access and thereby use of nutraceuticals for better health.

References

Afaq F, Saleem M, Krueger CG, Reed JD, Mukhtar H. Anthocyanin-and hydrolyzable tannin-rich pomegranate fruit extract modulates MAPK and NF-κB pathways and inhibits skin tumorigenesis in CD-1 mice. *International Journal of Cancer.* 2005;113(3):423–33.

Agüero A, Arnedo M, Gallo M, Puerto A. The functional relevance of the lateral parabrachial nucleus in lithium chloride-induced aversion learning. *Pharmacology Biochemistry and Behavior.* 1993;45(4):973–8.

Ajit S. Nutraceuticals-critical supplement for building a healthy India. Ernst and Young, FICCI task force on nutraceuticals. 2012:1–80.

Aldin E, Reitmeier HA, Murphy P. Bitterness of soy extracts containing isoflavones and saponins. *Journal of Food Science.* 2006;71(3):S211–S215.

Aluyen JK, Ton QN, Tran T, Yang AE, Gottlieb HB, Bellanger RA. Resveratrol: potential as anticancer agent. *Journal of Dietary Supplements.* 2012;9(1):45–56.

Angst E, Park JL, Moro A, Lu QY, Lu X, Li G, King J, Chen M, Reber HA, Go VL, Eibl G. The flavonoid quercetin inhibits pancreatic cancer growth in vitro and in vivo. *Pancreas.* 2013;42(2):223.

Aspen J, Gatch MB, Woods JH. Training and characterization of a quinine taste discrimination in rhesus monkeys. *Psychopharmacology.* 1999;141(3):251–257.

Astray G, Gonzalez-Barreiro C, Mejuto JC, Rial-Otero R, Simal-Gandara J. A review on the use of cyclodextrins in foods. *Food Hydrocolloids.* 2009;23(7):1631–1640.

Babal P, Kristová V, Černá A, Janega P, Pecháňová O, Danihel L, Andriantsitohaina R. Red wine polyphenols prevent endothelial damage induced by CCl4 administration. *Physiological Research.* 2006;55(3).

Babu PV, Sabitha KE, Shyamaladevi CS. Therapeutic effect of green tea extract on oxidative stress in aorta and heart of streptozotocin diabetic rats. *Chemico-Biological Interactions.* 2006;162(2):114–120.

Balestrieri E, Pizzimenti F, Ferlazzo A, Giofrè SV, Iannazzo D, Piperno A, Romeo R, Chiacchio MA, Mastino A, Macchi B. Antiviral activity of seed extract from *Citrus bergamia* towards human retroviruses. *Bioorganic & Medicinal Chemistry.* 2011;19(6):2084–2089.

Bardo MT, Valone JM. Morphine-conditioned analgesia using a taste cue: Dissociation of taste aversion and analgesia. *Psychopharmacology.* 1994;114(2):269–274.

Bassoli A, Borgonovo G, Busnelli G. *Alkaloids and the Bitter Taste.* Wiley VCH: Weinheim, Germany; 2007 October 24.

Beavers DP, Beavers KM, Miller M, Stamey J, Messina MJ. Exposure to isoflavone-containing soy products and endothelial function: A Bayesian meta-analysis of randomized controlled trials. *Nutrition, Metabolism and Cardiovascular Diseases.* 2012;22(3):182–191.

Beecher GR. Overview of dietary flavonoids: nomenclature, occurrence and intake. *The Journal of Nutrition.* 2003;133(10):3248S–3254S.

Behloul N, Wu G. Genistein: A promising therapeutic agent for obesity and diabetes treatment. *European Journal of Pharmacology.* 2013;698(1–3):31–38.

Belitz, H. D., W. Grosch, and P. Schieberle. *Food Chemistry.* ed. 378–394; 1999.

Bellissimo MI, Amado D, Abdalla DS, Ferreira EC, Cavalheiro EA, da Graça Naffah-Mazzacoratti M. Superoxide dismutase, glutathione peroxidase activities and the hydroperoxide concentration are modified in the hippocampus of epileptic rats. *Epilepsy Research.* 2001;46(2):121–128.

Bent S, Padula A, Neuhaus J. Safety and efficacy of *Citrus aurantium* for weight loss. *The American Journal of Cardiology.* 2004;94(10):1359–1361.

Bertoldi FC, Sant'Anna ES, Beirão LH. Reducing the bitterness of tuna (*Euthynnus pelamis*) dark meat with *Lactobacillus casei* subsp. casei ATCC 393. *Food Technology and Biotechnology.* 2004;42(1):41–45.

Bevins RA, Delzer TA, Bardo MT. Characterization of the conditioned taste aversion produced by 7-OH-DPAT in rats. *Pharmacology Biochemistry and Behavior.* 1996;53(3):695699.

Beyreuther K, Biesalski HK, Fernstrom JD, Grimm P, Hammes WP, Heinemann U, Kempski O, Stehle P, Steinhart H, Walker R. Consensus meeting: monosodium glutamate—an update. *European Journal of Clinical Nutrition.* 2007;61(3):304.

Bhandari MR, Kawabata J. Bitterness and toxicity in wild yam (*Dioscorea* spp.) tubers of Nepal. *Plant Foods for Human Nutrition.* 2005;60(3):129–135.

Bhattacharya A, Li Y, Wade KL, Paonessa JD, Fahey JW, Zhang Y. Allyl isothiocyanate-rich mustard seed powder inhibits bladder cancer growth and muscle invasion. *Carcinogenesis.* 2010;31(12):2105–2110.

Bienkowski P, Kuca P, Piasecki J, Kostowski W. 5-HT3 receptor antagonist, tropisetron, does not influence ethanol-induced conditioned taste aversion and conditioned place aversion. *Alcohol.* 1997;14(1):63–69.

Bigiani A, Ghiaroni V, Fieni F. Channels as taste receptors in vertebrates. *Progress in Biophysics and Molecular Biology.* 2003;83(3):193–225.

Bilal H, Akram W, Ali-Hassan S. Larvicidal activity of citrus limonoids against *Aedes albopictus* larvae. *Journal of Arthropod-Borne Diseases.* 2012;6(2):104.

Blanco VZ, Auw JM, Sims CA, O'Keefe SF. Effect of processing on phenolics of wines. In *Process-Induced Chemical Changes in Food* 1998 (pp. 327–340). Springer, Boston, MA.

Bone HG, Greenspan SL, McKeever C, Bell N, Davidson M, Downs RW, Emkey R, Meunier PJ, Miller SS, Mulloy AL, Recker RR. Alendronate and estrogen effects in postmenopausal women with low bone mineral density. *The Journal of Clinical Endocrinology & Metabolism.* 2000;85(2):720–726.

Bouayed J, Bohn T. Exogenous antioxidants—Double-edged swords in cellular redox state: Health beneficial effects at physiologic doses versus deleterious effects at high doses. *Oxidative Medicine and Cellular Longevity.* 2010;3(4):228–237.

Boucher BA, Cotterchio M, Anderson LN, Kreiger N, Kirsh VA, Thompson LU. Use of isoflavone supplements is associated with reduced postmenopausal breast cancer risk. *International Journal of Cancer.* 2013;132(6):1439–1450.

Brasser SM, Mozhui K, Smith DV. Differential covariation in taste responsiveness to bitter stimuli in rats. *Chemical Senses.* 2005;30(9):793–799.

Breslin PA, Beauchamp GK. Suppression of bitterness by sodium: Variation among bitter taste stimuli. *Chemical Senses.* 1995;20(6):609–623.

Brockwell NT, Eikelboom R, Beninger RJ. Caffeine-induced place and taste conditioning: production of dose-dependent preference and aversion. *Pharmacology Biochemistry and Behavior.* 1991;38(3):513–517.

Brossaud F, Cheynier V, Noble AC. Bitterness and astringency of grape and wine polyphenols. *Australian Journal of Grape and Wine Research.* 2001;7(1):33–39.

Brouillard R, George F, Fougerousse A. Polyphenols produced during red wine ageing. *Biofactors.* 1997;6(4):403–410.

Brower V. Nutraceuticals: Poised for a healthy slice of the healthcare market?. *Nature Biotechnology.* 1998;16(8):728.

Bruneton J. *Pharmacognosy, Phytochemistry, Medicinal Plants*, Paris, France: Lavoisier. 1999.

Caicedo A, Pereira E, Margolskee RF, Roper SD. Role of the G-protein subunit α-gustducin in taste cell responses to bitter stimuli. *Journal of Neuroscience.* 2003;23(30):9947–9952.

Calviño AM, García-Medina MR, Cometto-Muniz JE. Interactions in caffeine–sucrose and coffee–sucrose mixtures: Evidence of taste and flavor suppression. *Chemical Senses.* 1990;15(5):505–519.

Calviño AM, García-Medina MR, Cometto-Muniz JE, Rodríguez MB. Perception of sweetness and bitterness in different vehicles. *Perception & Psychophysics.* 1993;54(6):751–758.

Calviño AM, Tamasi OP, Ciappini MC. Note. Caffeine content and dynamical bitterness of Yerba Mate *Ilex paraguariensis* infusions. *Food Science and Technology International.* 2005;11(6):401–407.

Carvalho M, Carmo H, Costa VM, Capela JP, Pontes H, Remião F, Carvalho F, de Lourdes Bastos M. Toxicity of amphetamines: An update. *Archives of Toxicology.* 2012;86(8):1167–1231.

Celec P, Ostatníková D, Cagánová M, Žuchová S, Hodosy J, Putz Z, Bernadič M, Kúdela M. Endocrine and cognitive effects of short-time soybean consumption in women. *Gynecologic and Obstetric Investigation.* 2005;59(2):62–66.

Cetó X, Gutiérrez JM, Moreno-Barón L, Alegret S, Del Valle M. Voltammetric electronic tongue in the analysis of cava wines. *Electroanalysis.* 2011;23(1):72–78.

Chan YH, Lau KK, Yiu KH, Li SW, Chan HT, Fong DY, Tam S, Lau CP, Tse HF. Reduction of C-reactive protein with isoflavone supplement reverses endothelial dysfunction in patients with ischaemic stroke. *European Heart Journal.* 2008;29(22):2800–2807.

Chandrashekar J, Hoon MA, Ryba NJ, Zuker CS. The receptors and cells for mammalian taste. *Nature.* 2006;444(7117):288.

Chandrashekar J, Mueller KL, Hoon MA, Adler E, Feng L, Guo W, Zuker CS, Ryba NJ. T2Rs function as bitter taste receptors. *Cell.* 2000;100(6):703–711.

Chaudhari N, Roper SD. The cell biology of taste. *The Journal of Cell Biology.* 2010;190(3):285–296.

Chauhan B, Kumar G, Kalam N, Ansari SH. Current concepts and prospects of herbal nutraceutical: A review. *Journal of Advanced Pharmaceutical Technology & Research.* 2013;4(1):4.

Chen CK, Pace-Asciak CR. Vasorelaxing activity of resveratrol and quercetin in isolated rat aorta. *General Pharmacology: The Vascular System.* 1996;27(2):363–366.

Cheng J, Chiao J, Jin H, Zhong C, inventors; Wuxi JC. Pharmaceutical Tech Co Ltd, assignee. Use of isothiocyanates compounds in treating prostatic diseases and skin cancer. United States patent US 8,039,511. 2011 October 18.

Chobotova K. Aging and cancer: Converging routes to disease prevention. *Integrative Cancer Therapies.* 2009;8(2):115–122.

Conklin CM, Bechberger JF, MacFabe D, Guthrie N, Kurowska EM, Naus CC. Genistein and quercetin increase connexin43 and suppress growth of breast cancer cells. *Carcinogenesis.* 2007;28(1):93–100.

Connolly JD, Hill RA. Triterpenoids. *Natural Product Reports*. 2010;27(1):79–132.
Conquer JA, Maiani G, Azzini E, Raguzzini A, Holub BJ. Supplementation with quercetin markedly increases plasma quercetin concentration without effect on selected risk factors for heart disease in healthy subjects. *The Journal of Nutrition*. 1998;128(3):593–597.
Corson J, Aldridge A, Wilmoth K, Erisir A. A survey of oral cavity afferents to the rat nucleus tractus solitarii. *Journal of Comparative Neurology*. 2012;520(3):495–527.
Coward L, Smith M, Kirk M, Barnes S. Chemical modification of isoflavones in soyfoods during cooking and processing. *The American Journal of Clinical Nutrition*. 1998;68(6):1486S–1491S.
Cravotto G, Nano GM, Binello A, Spagliardi P, Seu G. Chemical and biological modification of cynaropicrin and grosheimin: A structure–bitterness relationship study. *Journal of the Science of Food and Agriculture*. 2005;85(10):1757–1764.
Croguennec T, Guérin-Dubiard C, Nau F. Riboflavin-binding protein (flavoprotein). In *Bioactive Egg Compounds* 2007 (pp. 69–74). Springer, Berlin, Germany.
Cropley V, Croft R, Silber B, Neale C, Scholey A, Stough C, Schmitt J. Does coffee enriched with chlorogenic acids improve mood and cognition after acute administration in healthy elderly? A pilot study. *Psychopharmacology*. 2012;219(3):737–749.
Davies BT, Wellman PJ. Conditioned taste reactivity in rats after phenylpropanolamine, d-amphetamine or lithium chloride. *Pharmacology Biochemistry and Behavior*. 1990;36(4):973–977.
De Bruyne T, Pieters L, Deelstra H, Vlietinck A. Condensed vegetable tannins: Biodiversity in structure and biological activities. *Biochemical Systematics and Ecology*. 1999;27(4):445–459.
Dehar NA, Walia RA, Ratol SU. Potentiation of thiopentone sodium induced hypnosis by *Berberis aristata* in rodents. *Asian Journal of Pharmaceutical and Clinical Research*. 2012;5(1):131–133.
Del Río JA, Arcas MC, Benavente O, Sabater F, Ortuño A. Changes of polymethoxylated flavones levels during development of *Citrus aurantium* (cv. Sevillano) fruits. *Planta Medica*. 1998;64(06):575–576.
Del Rio JA, Benavente O, Castillo J, Borrego F. Neodiosmin, a flavone glycoside of *Citrus aurantium*. *Phytochemistry*. 1992;31(2):723–724.
Delaquis PJ, Mazza G. Antimicrobial properties of isothiocyanates in food preservation. *Food Technology*. 1995;49(11):73–84.
Delwiche J. The impact of perceptual interactions on perceived flavor. InFood Quality Preference, 2004.
Delwiche JF, Buletic Z, Breslin PA. Covariation in individuals' sensitivities to bitter compounds: Evidence supporting multiple receptor/transduction mechanisms. *Perception & Psychophysics*. 2001;63(5):761–776.
Dewick PM. *Medicinal Natural Products: A Biosynthetic Approach*. John Wiley & Sons; 2002 January 3.
Domanski M, Mitchell G, Pfeffer M, Neaton JD, Norman J, Svendsen K, Grimm R, Cohen J, Stamler J, MRFIT Research Group. Pulse pressure and cardiovascular disease–related mortality: follow-up study of the Multiple Risk Factor Intervention Trial (MRFIT). *Jama*. 2002;287(20):2677–2683.
Donà M, Dell'Aica I, Calabrese F, Benelli R, Morini M, Albini A, Garbisa S. Neutrophil restraint by green tea: Inhibition of inflammation, associated angiogenesis, and pulmonary fibrosis. *The Journal of Immunology*. 2003;170(8):4335–4341.
Drewnowski A, Gomez-Carneros C. Bitter taste, phytonutrients, and the consumer: A review. *The American Journal of Clinical Nutrition*. 2000;72(6):1424–1435.
Drynan JW, Clifford MN, Obuchowicz J, Kuhnert N. The chemistry of low molecular weight black tea polyphenols. *Natural Product Reports*. 2010;27(3):417–462.
Du D, Fang L, Qu J, Yu S, Ma S, Lv H, Liu J, Liu Y, Wang J, Wang X. Oleanane-type triterpene saponins and cassaine-type diterpenoids from *Erythrophleum fordii*. *Planta Medica*. 2011;77(14):1631–1638.

Dudeja P, Gupta RK. Nutraceuticals. InFood Safety in the 21st Century 2017 January 1 (pp. 491–496). Academic Press Cambridge, MA.

Dureja H, Kaushik D, Kumar V. Developments in nutraceuticals. *Indian Journal of Pharmacology*. 2003;35(6):363–372.

Ervin GN, Birkemo LS, Johnson MF, Conger LK, Mosher JT, Menius JA. The effects of anorectic and aversive agents on deprivation-induced feeding and taste aversion conditioning in rats. *Journal of Pharmacology and Experimental Therapeutics*. 1995;273(3):1203–1210.

Esther B. What is nutraceutical. *The Pharmaceutical Journal*. 2000;265:57–58.

Evans WC. Trease and Evans' Pharmacognosy E-Book. *Elsevier Health Sciences*; 2009 May 27.

Exton MS, von Hörsten S, Vöge J, Westermann J, Schult M, Nagel E, Schedlowski M. Conditioned taste aversion produced by cyclosporine A: concomitant reduction in lymphoid organ weight and splenocyte proliferation. *Physiology & Behavior*. 1998 January 1;63(2):241–247.

Fahey JW, Stephenson KK, Talalay P. Glucosinolates, myrosinase, and isothiocyanates: Three reasons for eating brassica vegetables. In ACS Symposium Series 1998 (Vol. 701, pp. 16–22). American Chemical Society.

Farina HG, Pomies M, Alonso DF, Gomez DE. Antitumor and antiangiogenic activity of soy isoflavone genistein in mouse models of melanoma and breast cancer. *Oncology Reports*. 2006;16(4):885–891.

Fenwick GR, Griffiths NM, Heaney RK. Bitterness in Brussels sprouts (*Brassica oleracea* L. var. gemmifera): The role of glucosinolates and their breakdown products. *Journal of the Science of Food and Agriculture*. 1983b;34(1):73–80.

Fenwick GR, Heaney RK, Mullin WJ, VanEtten CH. Glucosinolates and their breakdown products in food and food plants. *CRC Critical Reviews in Food Science and Nutrition*. 1983a;18(2):123–201.

File SE, Jarrett N, Fluck E, Duffy R, Casey K, Wiseman H. Eating soya improves human memory. *Psychopharmacology*. 2001;157(4):430–436.

Fontoin H, Saucier C, Teissedre PL, Glories Y. Effect of pH, ethanol and acidity on astringency and bitterness of grape seed tannin oligomers in model wine solution. *Food Quality and Preference*. 2008;19(3):286–291.

Foster S. Tyler's honest herbal: A sensible guide to the use of herbs and related remedies. Routledge, Abingdon, UK; 2012 November 12.

Fraga BM. Natural sesquiterpenoids. *Natural Product Reports*. 2008;25(6):1180–1209.

Francart SJ, Allen MK, Stegall-Zanation J. Apnea of prematurity: caffeine dose optimization. *The Journal of Pediatric Pharmacology and Therapeutics*. 2013;18(1):45–52.

Frank ME, Bouverat BP, MacKinnon BI, Hettinger TP. The distinctiveness of ionic and nonionic bitter stimuli. *Physiology & Behavior*. 2004;80(4):421–431.

Frank RA, Byram J. Taste–smell interactions are tastant and odorant dependent. *Chemical Senses*. 1988;13(3):445–455.

Frankel EN, German JB, Kinsella JE, Parks E, Kanner J. Inhibition of oxidation of human low-density lipoprotein by phenolic substances in red wine. *The Lancet*. 1993;341(8843):454–457.

Freudenberg KJ. *Die Chemie der natürlichen Gerbstoffe*. Springer-Verlag Berlin, Germany; 2013 July 29.

Frick S, Kramell R, Schmidt J, Fist AJ, Kutchan TM. Comparative qualitative and quantitative determination of alkaloids in narcotic and condiment *Papaver somniferum* cultivars. *Journal of Natural Products*. 2005;68(5):666–673.

Fukumitsu S, Villareal MO, Fujitsuka T, Aida K, Isoda H. Anti-inflammatory and anti-arthritic effects of pentacyclic triterpenoids maslinic acid through NF-κB inactivation. *Molecular Nutrition & Food Research*. 2016;60(2):399–409.

Gajewska M, Paini A, Benito JS, Burton J, Worth A, Urani C, Briesen H, Schramm KW. In vitro-to-in vivo correlation of the skin penetration, liver clearance and hepatotoxicity of caffeine. *Food and Chemical Toxicology*. 2015;75:39–49.

Gallus S, Tavani A, Negri E, La Vecchia C. Does coffee protect against liver cirrhosis? *Annals of Epidemiology*. 2002;12(3):202–205.

García JM, Yousfi K, Mateos R, Olmo M, Cert A. Reduction of oil bitterness by heating of olive (*Olea europaea*) fruits. *Journal of Agricultural and Food Chemistry*. 2001;49(9):4231–4235.

Gaudette NJ, Pickering GJ. Optimizing the orosensory properties of model functional beverages: The influence of novel sweeteners, odorants, bitter blockers, and their mixtures on (+)-catechin. *Journal of Food Science*. 2012;77(6):S226–S232.

Gaudette NJ, Pickering GJ. Modifying bitterness in functional food systems. *Critical Reviews in Food Science and Nutrition*. 2013;53(5):464–481.

Gauvin DV, Holloway FA. Ethanol tolerance developed during intoxicated operant performance in rats prevents subsequent ethanol-induced conditioned taste aversion. *Alcohol*. 1992;9(2):167–170.

Geetha T, Varalakshmi P. Anti-inflammatory activity of lupeol and lupeol linoleate in rats. *Journal of Ethnopharmacology*. 2001;76(1):77–80.

Geissman TA. The isolation of eriodictyol and homoeriodictyol. An improved procedure. *Journal of the American Chemical Society*. 1940;62(11):3258–3259.

Gilani AH, Janbaz KH, Aziz N, Herzig MJ, Kazmi MM, Choudhary MI, Herzig JW. Possible mechanism of selective inotropic activity of the n-butanolic fraction from *Berberis aristata* fruit. *General Pharmacology: The Vascular System*. 1999;33(5): 407–414.

Gilbert ER, Liu D. Anti-diabetic functions of soy isoflavone genistein: Mechanisms underlying its effects on pancreatic β-cell function. *Food & Function*. 2013;4(2):200–212.

Glendinning JI. Effect of salivary proline-rich proteins on ingestive responses to tannic acid in mice. *Chemical Senses*. 1992;17(1):1–2.

Glowa JR, Shaw AE, Riley AL. Cocaine-induced conditioned taste aversions: Comparisons between effects in LEW/N and F344/N rat strains. *Psychopharmacology*. 1994;114(2): 229–232.

Goldberg E, Grant J, Aliani M, Eskin MN. 10 Methods for Removing Bitterness in Functional Foods and Nutraceuticals. *Bitterness: Perception, Chemistry and Food Processing*. John Wiley & Sons, Inc., USA; 2017 April 17.

Graham HN. Green tea composition, consumption, and polyphenol chemistry. *Preventive Medicine*. 1992;21(3):334–350.

Grayson DH. Monoterpenoids. *Natural Product Reports*. 2000;17(4):385–419.

Green BG, Hayes JE. Capsaicin as a probe of the relationship between bitter taste and chemesthesis. *Physiology & Behavior*. 2003;79(4–5):811–821.

Greenway FL. The safety and efficacy of pharmaceutical and herbal caffeine and ephedrine use as a weight loss agent. *Obesity Reviews*. 2001;2(3):199–211.

Gregory D, Hargett D, Holmes D, Money E, Bachenheimer SL. Efficient replication by herpes simplex virus type 1 involves activation of the IκB kinase-IκB-p65 pathway. *Journal of Virology*. 2004;78(24):13582–13590.

Gross G, Jaccaud E, Huggett AC. Analysis of the content of the diterpenes cafestol and kahweol in coffee brews. *Food and Chemical Toxicology*. 1997;35(6):547–554.

Guadagni DG, Horowitz RM, Gentili B, Maier VP, inventors; US Department of Agriculture, assignee. Method of reducing bitterness and off-after-taste. United States patent US 4,154,862. 1979 May 15.

Guadagni DG, Maier VP, Turnbaugh JG. Effect of neodiosmin on threshold and bitterness of limonin in water and orange juice. *Journal of Food Science*. 1976;41(3): 681–684.

Guerrero RF, Garcia-Parrilla MC, Puertas B, Cantos-Villar E. Wine, resveratrol and health: A review. *Natural Product Communications*. 2009;4(5):635–658.

Gupta JK, Mishra P, Rani A, Mazumder PM. Blood glucose lowering potential of stem bark of Berberis aristata Dc in alloxan-induced diabetic rats. *Iranian Journal of Pharmacology and Therapeutics*. 2010;9(1):21–30.

Hamdan D, El-Readi MZ, Tahrani A, Herrmann F, Kaufmann D, Farrag N, El-Shazly A, Wink M. Secondary metabolites of ponderosa lemon (*Citrus pyriformis*) and their antioxidant, anti-inflammatory, and cytotoxic activities. *Zeitschrift für Naturforschung C*. 2011;66(7–8):385–393.

Hamilton RB, Norgren R. Central projections of gustatory nerves in the rat. *Journal of Comparative Neurology*. 1984;222(4):560–577.

Haqqi TM, Anthony DD, Gupta S, Ahmad N, Lee MS, Kumar GK, Mukhtar H. Prevention of collagen-induced arthritis in mice by a polyphenolic fraction from green tea. *Proceedings of the National Academy of Sciences*. 1999;96(8):4524–4529.

Hasegawa S, Miyake M. Biochemistry and biological functions of citrus limonoids. *Food Reviews International*. 1996;12(4):413–435.

Heaney RK, Fenwick GR. Glucosinolates in Brassica vegetables. Analysis of 22 varieties of Brussels sprout (*Brassica oleracea* var. gemmifera). *Journal of the Science of Food and Agriculture*. 1980;31(8):785–793.

Heckman MA, Weil J, De Mejia EG. Caffeine (1, 3, 7-trimethylxanthine) in foods: A comprehensive review on consumption, functionality, safety, and regulatory matters. *Journal of Food Science*. 2010;75(3):R77–R87.

Heijl A. Visual field changes in early glaucoma and how to recognize them (summary). *Survey of Ophthalmology*. 1989;33:403–404.

Herling AW, Burger HJ, Schwab D, Hemmerle H, Below P, Schubert G. Pharmacodynamic profile of a novel inhibitor of the hepatic glucose-6-phosphatase system. *American Journal of Physiology-Gastrointestinal and Liver Physiology*. 1998;274(6):G1087–G1093.

Hertog MG, Hollman PC, Katan MB. Content of potentially anticarcinogenic flavonoids of 28 vegetables and 9 fruits commonly consumed in the Netherlands. *Journal of Agri-Cultural and Food Chemistry*. 1992;40(12):2379–2383.

Hofmann T, Chubanov V, Gudermann T, Montell C. TRPM5 is a voltage-modulated and Ca^{2+}-activated monovalent selective cation channel. *Current Biology*. 2003;13(13):1153–1158.

Homma R, Yamashita H, Funaki J, Ueda R, Sakurai T, Ishimaru Y, Abe K, Asakura T. Identification of bitterness-masking compounds from cheese. *Journal of Agricultural and Food Chemistry*. 2012;60(18):4492–4499.

Hsu WJ, Berhow M, Robertson GH, Hasegawa S. Limonoids and flavonoids in juices of Oroblanco and Melogold grapefruit hybrids. *Journal of Food Science*. 1998;63(1):57–60.

https://www.drugs.com/dosage/pilocarpine-ophthalmic.html (accessed July 2, 2019).

Hu J, Zhao J, Chen W, Lin S, Zhang J, Hong Z. Hepatoprotection of 1 β-hydroxyeuscaphic acid–the major constituent from *Rubus aleaefolius* against CCl4-induced injury in hepatocytes cells. *Pharmaceutical Biology*. 2013;51(6):686–690.

Huang MT, Chang RL, Wood AW, Newmark HL, Sayer JM, Yagi H, Jerina DM, Conney AH. Inhibition of the mutagenicity of bay-region diol-epoxides of polycyclic aromatic hydrocarbons by tannic acid, hydroxylated anthraquinones and hydroxylated cinnamic acid derivatives. *Carcinogenesis*. 1985;6(2):237–242.

Hufnagel JC, Hofmann T. Orosensory-directed identification of astringent mouthfeel and bitter-tasting compounds in red wine. *Journal of Agricultural and Food Chemistry*. 2008a;56(4):1376–1386.

Hufnagel JC, Hofmann T. Quantitative reconstruction of the nonvolatile sensometabolome of a red wine. *Journal of Agricultural and Food Chemistry*. 2008b;56(19):9190–9199.

Hung TM, Cuong TD, Kim JA, Lee JH, Woo MH, Min BS. In vitro apoptotic effect of cassaine-type diterpene amides from Erythrophleum fordii on PC-3 prostate cancer cells. *Bioorganic & Medicinal Chemistry Letters*. 2014;24(21):4989–4994.

Hung TM, Cuong TD, Kim JA, Tae N, Lee JH, Min BS. Cassaine diterpene alkaloids from *Erythrophleum fordii* and their anti-angiogenic effect. *Bioorganic & Medicinal Chemistry Letters*. 2014;24(1):168–172.

Jang YJ, Pyo S. Anti-atherosclerotic effect of sinigrin in ApoE-deficient mice. *The FASEB Journal*. 2015;29(1_supplement):609–611.

Jie M, Cheung WM, Yu V, Zhou Y, Tong PH, Ho JW. Anti-proliferative activities of sinigrin on carcinogen-induced hepatotoxicity in rats. *PloS one.* 2014;9(10):e110145.

Joshi PV, Shirkhedkar AA, Prakash K, Maheshwari VL. Antidiarrheal activity, chemical and toxicity profile of *Berberis aristata. Pharmaceutical Biology.* 2011;49(1):94–100.

June HL, June PL, Domangue KR, Hicks LH, Lummis GH, Lewis MJ. Failure of Ro15-4513 to alter an ethanol-induced taste aversion. *Pharmacology Biochemistry and Behavior.* 1992;41(2):455–460.

Kallithraka S, Bakker J, Clifford MN. Evaluation of bitterness and astringency of (+)-Catechin and (-)-Epicatechin in red wine and in model solution. *Journal of Sensory Studies.* 1997;12(1):25–37.

Kao TH, Chen BH. Functional components in soybean cake and their effects on antioxidant activity. *Journal of Agricultural and Food Chemistry.* 2006;54(20):7544–7555.

Karimov A, Telezhenetskaya MV, Lutfullin KL, Yunusov SY. Berberis alkaloids. The new alkaloid oblongamine. *Chemistry of Natural Compounds.* 1977;13(1):68–70.

Katske F, Shoskes DA, Sender M, Poliakin R, Gagliano K, Rajfer J. Treatment of interstitial cystitis with a quercetin supplement. *Techniques in Urology.* 2001;7(1):44–46.

Katsuragi Y. Specific inhibitor for bitter taste. *Nature.* 1993;365:213–214.

Katsuragi Y, Sugiura Y, Lee C, Otsuji K, Kurihara K. Selective inhibition of bitter taste of various drugs by lipoprotein. *Pharmaceutical Research.* 1995;12(5):658–662.

Katsuragi Y, Yasumasu T, Kurihara K. Lipoprotein that selectively inhibits taste nerve responses to bitter substances. *Brain Research.* 1996;713(1–2):240–245.

Keast RS. The effect of zinc on human taste perception. *Journal of Food Science.* 2003;68(5):1871–1877.

Keast RS. Modification of the bitterness of caffeine. *Food Quality and Preference.* 2008;19(5):465–472.

Keast RS, Breslin PA. Cross-adaptation and bitterness inhibition of L-tryptophan, L-phenylalanine and urea: further support for shared peripheral physiology. *Chemical Senses.* 2002;27(2):123–131.

Keast RS, Breslin PA. An overview of binary taste–taste interactions. *Food Quality and Preference.* 2003;14(2):111–124.

Keast RS, Canty TM, Breslin PA. The influence of sodium salts on binary mixtures of bitter-tasting compounds. *Chemical Senses.* 2004;29(5):431–439.

Kelly GS. Quercetin. *Alternative Medicine Review.* 2011;16(2):172–194.

Kemp SE, Beauchamp GK. Flavor modification by sodium chloride and monosodium glutamate. *Journal of Food Science.* 1994;59(3):682–686.

Khanbabaee K, van Ree T. Tannins: Classification and definition. *Natural Product Reports.* 2001;18(6):641–649.

Kim J, Jayaprakasha GK, Muthuchamy M, Patil BS. Structure–function relationships of citrus limonoids on p38 MAP kinase activity in human aortic smooth muscle cells. *European Journal of Pharmacology.* 2011;670(1):44–49.

Kim YH, Lee YJ. TRAIL apoptosis is enhanced by quercetin through Akt dephosphorylation. *Journal of Cellular Biochemistry.* 2007;100(4):998–1009.

Kimira Y, Taniuchi Y, Nakatani S, Sekiguchi Y, Kim HJ, Shimizu J, Ebata M, Wada M, Matsumoto A, Mano H. Citrus limonoid nomilin inhibits osteoclastogenesis in vitro by suppression of NFATc1 and MAPK signaling pathways. *Phytomedicine.* 2015;22(12):1120–1124.

Kinnamon SC. Taste receptor signalling—from tongues to lungs. *Acta physiologica.* 2012;204(2):158–168.

Kirtikar KR, Basu BD. Indian medicinal plants. *Indian Medicinal Plants.* Lalit Mohan Basu Publisher, Allahabad, India; 1918.

Kodamatani H, Saito K, Niina N, Yamazaki S, Tanaka Y. Simple and sensitive method for determination of glycoalkaloids in potato tubers by high-performance liquid chromatography with chemiluminescence detection. *Journal of Chromatography A.* 2005;1100(1):26–31.

Kola O, Kaya C, Duran H, Altan A. Removal of limonin bitterness by treatment of ion exchange and adsorbent resins. *Food Science and Biotechnology*. 2010;19(2):411–416.

Konno A, Misaki M, Toda J, Wada T, Yasumatsu K. Bitterness reduction of naringin and limonin by β-cyclodextrin. *Agricultural and Biological Chemistry*. 1982;46(9):2203–2208.

Kono S, Ikeda M, Tokudome S, Kuratsune M. A case-control study of gastric cancer and diet in northern Kyushu, Japan. *Japanese Journal of Cancer Research*. 1988;79(10):1067–1074.

Koprivnjak O, Škevin D, Petričević S, Bubola KB, Mokrovčak Ž. Bitterness, odor properties and volatile compounds of virgin olive oil with phospholipids addition. *LWT-Food Science and Technology*. 2009;42(1):50–55.

Kritz-Silverstein D, Von Mühlen D, Barrett-Connor E, Bressel MA. Isoflavones and cognitive function in older women: The SOy and Postmenopausal Health In Aging (SOPHIA) Study. *Menopause*. 2003;10(3):196–202.

Kudou S, Fleury Y, Welti D, Magnolato D, Uchida T, Kitamura K, Okubo K. Malonyl isoflavone glycosides in soybean seeds (*Glycine max* Merrill). *Agricultural and Biological Chemistry*. 1991;55(9):2227–2233.

Kuhn C, Bufe B, Winnig M, Hofmann T, Frank O, Behrens M, Lewtschenko T, Slack JP, Ward CD, Meyerhof W. Bitter taste receptors for saccharin and acesulfame K. *Journal of Neuroscience*. 2004;24(45):10260–10265.

Lawless H. Paradoxical adaptation to taste mixtures. *Physiology & Behavior*. 1982;29(1):149–52.

Lawless HT. Sensory interactions in mixtures. *Journal of Sensory Studies*. 1986;1(3–4):259–274.

Lawless HT, Rapacki F, Horne J, Hayes A. The taste of calcium and magnesium salts and anionic modifications. *Food Quality and Preference*. 2003;14(4):319–325.

Lee HW, Lee KR. Effect of sinigrin on vascular cell adhesion molecule-1 expression in TNF-α-stimulated mouse vascular smooth muscle cells via downregulation of NF-κB signaling pathways. *The FASEB Journal*. 2015;29(1_supplement):593–615.

Lee KJ, Jeong HG. Protective effects of kahweol and cafestol against hydrogen peroxide-induced oxidative stress and DNA damage. *Toxicology Letters*. 2007;173(2):80–87.

Lee RJ, Xiong G, Kofonow JM, Chen B, Lysenko A, Jiang P, Abraham V, Doghramji L, Adappa ND, Palmer JN, Kennedy DW. T2R38 taste receptor polymorphisms underlie susceptibility to upper respiratory infection. *The Journal of clinical investigation*. 2012;122(11):4145–4159.

Lehman CD, Bartoshuk LM, Catalanotto FC, Kveton JF, Lowlicht RA. Effect of anesthesia of the chorda tympani nerve on taste perception in humans. *Physiology & Behavior*. 1995;57(5):943–951.

Ley JP, Krammer G, Reinders G, Gatfield IL, Bertram HJ. Evaluation of bitter masking flavanones from Herba Santa (*Eriodictyon californicum* (H. & A.) Torr., Hydrophyllaceae). *Journal of Agricultural and Food Chemistry*. 2005;53(15):6061–6066.

Li L, Sun B. Grape and wine polymeric polyphenols: Their importance in enology. *Critical Reviews in Food Science and Nutrition*. 2017;1–7.

Liang DY, Li X, Clark JD. Epigenetic regulation of opioid-induced hyperalgesia, dependence, and tolerance in mice. *The Journal of Pain*. 2013;14(1):36–47.

Lin HQ, McGregor IS, Atrens DM, Christie MJ, Jackson DM. Contrasting effects of dopaminergic blockade on MDMA and d-amphetamine conditioned taste aversions. *Pharmacology Biochemistry and Behavior*. 1994;47(2):369–374.

Lin JK, Lin-Shiau SY. Mechanisms of hypolipidemic and anti-obesity effects of tea and tea polyphenols. *Molecular Nutrition & Food Research*. 2006;50(2):211–217.

Lin QS. Chinese herbal ingredients chemical. Sciences Press, Beijing; 1977:308–309.

Lipinski WJ, Rusiniak KW, Hilliard M, Davis RE. Nerve growth factor facilitates conditioned taste aversion learning in normal rats. *Brain Research*. 1995;692(1–2):143–153.

Löliger J. Function and importance of glutamate for savory foods. *The Journal of nutrition.* 2000;130(4):915S–920S.
Lou Z, Wang H, Zhu S, Ma C, Wang Z. Antibacterial activity and mechanism of action of chlorogenic acid. *Journal of Food Science.* 2011;76(6):M398–M403.
Lundy RF, Norgren R. Gustatory system. In *The Rat Nervous System* (Fourth Edition). Elsevier Inc., Amsterdam, Netherlands; 2015 (pp. 733–760).
Maciejczyk A, Surowiak P. Quercetin inhibits proliferation and increases sensitivity of ovarian cancer cells to cisplatin and paclitaxel. *Ginekologia Polska.* 2013;84(7).
Maddi VS, Aragade PD, Digge VG, Nitalikar MN. Importance of nutraceuticals in health management. *Pharmacognosy Reviews.* 2007;1(2):377.
Maehashi K, Matano M, Nonaka M, Udaka S, Yamamoto Y. Riboflavin-binding protein is a novel bitter inhibitor. *Chemical Senses.* 2007;33(1):57–63.
Malik A, Kumar P, Kaushik N, Singh A. The potentials of nutraceuticals. Pharmainfo. net. 2008;6:151–183.
Mander L, Liu HW. *Comprehensive natural products II: Chemistry and Biology.* Elsevier, Amsterdam, the Netherlands; 2010 March 5.
Manners GD. Citrus limonoids: analysis, bioactivity, and biomedical prospects. *Journal of Agricultural and Food Chemistry.* 2007;55(21):8285–8294.
Matsuda S, Norimoto F, Matsumoto Y, Ohba R, Teramoto Y, Ohta N, Ueda S. Solubilization of a novel isoflavone glycoside-hydrolyzing β-glucosidase from *Lactobacillus casei* subsp. rhamnosus. *Journal of Fermentation and Bioengineering.* 1994;77(4):439–441.
Matsuura M, Obata A, Fukushima D. Objectionable flavor of soy milk developed during the soaking of soybeans and its control. *Journal of Food Science.* 1989;54(3):602–605.
May OL, Hill DL. Gustatory terminal field organization and developmental plasticity in the nucleus of the solitary tract revealed through triple-fluorescence labeling. *Journal of Comparative Neurology.* 2006;497(4):658–669.
Mazumder A, Dwivedi A, du Preez JL, du Plessis J. In vitro wound healing and cytotoxic effects of sinigrin–phytosome complex. *International Journal of Pharmaceutics.* 2016;498(1–2):283–293.
McAllister KH, Pratt JA. GR205171 blocks apomorphine and amphetamine-induced conditioned taste aversions. *European Journal of Pharmacology.* 1998;353(2–3):141–148.
McLaughlin SK, McKinnon PJ, Margolskee RF. Gustducin is a taste-cell-specific G protein closely related to the transducins. *Nature.* 1992;357(6379):563.
Mele PC, McDonough JR, McLean DB, O'Halloran KP. Cisplatin-induced conditioned taste aversion: Attenuation by dexamethasone but not zacopride or GR38032F. *European Journal of Pharmacology.* 1992;218(2–3):229–236.
Meng LL, Huang CS, Liu HX. Advances in research on natural triterpenoids with bioactivities [J]. *Guihaia.* 2008;28:856–860.
Messina M. A brief historical overview of the past two decades of soy and isoflavone research, 2. *The Journal of Nutrition.* 2010;140(7):1350S–1354S.
Miller JS, Kelly KS, Neisewander JL, McCoy DF, Bardo MT. Conditioning of morphine-induced taste aversion and analgesia. *Psychopharmacology.* 1990;101(4):472–480.
Mingshu L, Kai Y, Qiang H, Dongying J. Biodegradation of gallotannins and ellagitannins. *Journal of Basic Microbiology.* 2006;46(1):68–84.
Mitchell DC, Knight CA, Hockenberry J, Teplansky R, Hartman TJ. Beverage caffeine intakes in the US. *Food and Chemical Toxicology.* 2014;63:136–142.
Miyashita T, Etoh H. Improvement of the bitterness and astringency of green tea by subcritical water extraction. *Food Science and Technology Research.* 2013;19(3):471–478.
Mosher JT, Johnson MF, Birkemo LS, Ervin GN. Several roles of CCKA and CCKB receptor subtypes in CCK-8-induced and LiCl-induced taste aversion conditioning. *Peptides.* 1996;17(3):483–488.
Moskowitz HR, Arabie P. Taste intensity as a function of stimulus concentration and solvent viscosity. *Journal of Texture Studies.* 1970;1(4):502–510.

Mucha RF. Preferences for tastes paired with a nicotine antagonist in rats chronically treated with nicotine. *Pharmacology Biochemistry and Behavior*. 1997;56(2):175–179.

Mukai J, Miyanaga Y, Ishizaka T, Asaka K, Nakai Y, Tsuji E, Uchida T. Quantitative taste evaluation of total enteral nutrients. *Chemical and Pharmaceutical Bulletin*. 2004;52(12):1416–1421.

Murphy PA, Barua K, Hauck CC. Solvent extraction selection in the determination of isoflavones in soy foods. *Journal of Chromatography B*. 2002;777(1–2):129–138.

Murthy KN, Jayaprakasha GK, Patil BS. Citrus limonoids and curcumin additively inhibit human colon cancer cells. *Food & function*. 2013;4(5):803–810.

Nakamura T, Tanigake A, Miyanaga Y, Ogawa T, Akiyoshi T, Matsuyama K, Uchida T. The effect of various substances on the suppression of the bitterness of quinine–human gustatory sensation, binding, and taste sensor studies. *Chemical and Pharmaceutical Bulletin*. 2002;50(12):1589–1593.

Nakamura Y, Tsuji S, Tonogai Y. Determination of the levels of isoflavonoids in soybeans and soy-derived foods and estimation of isoflavonoids in the Japanese daily intake. *Journal of AOAC International*. 2000;83(3):635–644.

Nakanishi Y, Tsuneyama K, Fujimoto M, Salunga TL, Nomoto K, An JL, Takano Y, Iizuka S, Nagata M, Suzuki W, Shimada T. Monosodium glutamate (MSG): A villain and promoter of liver inflammation and dysplasia. *Journal of Autoimmunity*. 2008;30(1–2):42–50.

Narukawa M, Kimata H, Noga C, Watanabe T. Taste characterisation of green tea catechins. *International Journal of Food Science & Technology*. 2010;45(8):1579–1585.

Neisewander JL, McDougall SA, Bowling SL, Bardo MT. Conditioned taste aversion and place preference with buspirone and gepirone. *Psychopharmacology*. 1990;100(4):485–490.

Niki M, Jyotaki M, Yoshida R, Ninomiya Y. Reciprocal modulation of sweet taste by leptin and endocannabinoids. In *Sensory and Metabolic Control of Energy Balance* (pp. 101–114). Springer, Berlin, Germany.

Ninomiya Y, Fukami Y, Yamazaki K, Beauchamp GK. Amiloride inhibition of chorda tympani responses to NaCI and its temperature dependency in mice. Brain research. 1996;708(1–2):153–158.

Noble AC. Bitterness in wine. *Physiology & Behavior*. 1994;56(6):1251–1255.

Noble AC. 12-Why Do Wines Taste Bitter and Feel Astringent?. InACS symposium series 1998 (Vol. 714, pp. 156–165). Washington, DC: American Chemical Society.

Norgren R. Taste and the autonomic nervous system. *In Interaction of the Chemical Senses with Nutrition*, 1986 (pp. 323–346).

Nusair S, Rubinow A. The use of oral pilocarpine in xerostomia and Sjögren's syndrome. In Seminars in arthritis and rheumatism 1999 Jun 1 (Vol. 28, No. 6, pp. 360–367). WB Saunders.

Ocampo JA, Martin J, Hayman DS. Influence of plant interactions on vesicular-arbuscular mycorrhizal infections. I. Host and non-host plants grown together. *New Phytologist*. 1980;84(1):27–35.

Ohkuri T, Yasumatsu K, Horio N, Jyotaki M, Margolskee RF, Ninomiya Y. Multiple sweet receptors and transduction pathways revealed in knockout mice by temperature dependence and gurmarin sensitivity. *American Journal of Physiology-Regulatory, Integrative and Comparative Physiology*. 2009;296(4):R960–R971.

Okubo K, Iijima M, Kobayashi Y, Yoshikoshi M, Uchida T, Kudou S. Components responsible for the undesirable taste of soybean seeds. *Bioscience, Biotechnology, and Biochemistry*. 1992;56(1):99–103.

Ono E, Inoue J, Hashidume T, Shimizu M, Sato R. Anti-obesity and anti-hyperglycemic effects of the dietary citrus limonoid nomilin in mice fed a high-fat diet. *Biochemical and Biophysical Research Communications*. 2011;410(3):677–681.

Osada K, Takahashi M, Hoshina S, Nakamura M, Nakamura S, Sugano M. Tea catechins inhibit cholesterol oxidation accompanying oxidation of low density lipoprotein in vitro. *Comparative Biochemistry and Physiology Part C: Toxicology & Pharmacology*. 2001;128(2):153–164.

Ott DB, Edwards CL, Palmer SJ. Perceived taste intensity and duration of nutritive and non-nutritive sweeteners in water using time-intensity (T-I) evaluations. *Journal of Food Science*. 1991;56(2):535–542.

Pangborn RM, Trabue IM, Szczesniak AS. Effect of hydrocolloids on oral viscosity and basic taste intensities. *Journal of Texture Studies*. 1973;4(2):224–241.

Pareek A, Suthar M. Antidiabetic activity of extract of *Berberis aristata* root in streptozotocin induced diabetic rats. *Pharmacology online*. 2010;2:179–185.

Park CH, Chang JY, Hahm ER, Park S, Kim HK, Yang CH. Quercetin, a potent inhibitor against β-catenin/Tcf signaling in SW480 colon cancer cells. *Biochemical and Biophysical Research Communications*. 2005;328(1):227–234.

Parker LA (1994) Aversive taste reactivity: Reactivity to quinine predicts aversive reactivity to lithium-paired sucrose solution. *Pharmacology Biochemistry and Behavior*. 47:73–75.

Parker LA, Gillies T. THC-induced place and taste aversions in Lewis and Sprague-Dawley rats. *Behavioral Neuroscience*. 1995;109(1):71.

Peláez GL, Sierra JA, Alzate F, Holzgrabe U, Ramirez-Pineda JR. Pentacyclic triterpenes from *Cecropia telenitida* with immunomodulatory activity on dendritic cells. *Revista Brasileira de Farmacognosia*. 2013;23(5):754–761.

Peleg H, Gacon K, Schlich P, Noble AC. Bitterness and astringency of flavan-3-ol monomers, dimers and trimers. *Journal of the Science of Food and Agriculture*. 1999;79(8):1123–1128.

Peleg H, Noble AC. Perceptual properties of benzoic acid derivatives. *Chemical Senses*. 1995;20(4):393–400.

Peleg H, Noble AC. Effect of viscosity, temperature and pH on astringency in cranberry juice. *Food Quality and Preference*. 1999;10(4–5):343–347.

Peluffo G, Radi R. Biochemistry of protein tyrosine nitration in cardiovascular pathology. *Cardiovascular Research*. 2007;75(2):291–302.

Peterson J, Dwyer J. Flavonoids: dietary occurrence and biochemical activity. *Nutrition Research*. 1998;18(12):1995–2018.

Pinheiro CU. Jaborandi (*Pilocarpus* sp., Rutaceae): A wild species. *Economic Botany*. 1997;51(1):49–58.

Price SF, Breen PJ, Valladao M, Watson BT. Cluster sun exposure and quercetin in Pinot noir grapes and wine. *American Journal of Enology and Viticulture*. 1995;46(2):187–194.

Pritchard TC, Norgren R. Gustatory system. In *The Human Nervous System*, Second Edition 2003 December 1 (pp. 1171–1196). Elsevier, Amsterdam, the Netherlands.

Pupin AM, Dennis MJ, Toledo MC. Polymethoxylated flavones in Brazilian orange juice. *Food Chemistry*. 1998;63(4):513–518.

Qu J, Hu YC, Yu SS, Chen XG, Li Y. New cassaine diterpenoid amides with cytotoxic activities from the bark of *Erythrophleum fordii*. *Planta Medica*. 2006;72(05):442–449.

Rabin BM, Hunt WA. Relationship between vomiting and taste aversion learning in the ferret: Studies with ionizing radiation, lithium chloride, and amphetamine. *Behavioral and Neural Biology*. 1992;58(2):83–93.

Ramachandran L, Manu KA, Shanmugam MK, Li F, Siveen KS, Vali S, Kapoor S, Abbasi T, Surana R, Smoot DT, Ashktorab H. Isorhamnetin inhibits proliferation, invasion, and induces apoptosis through the modulation of peroxisome proliferator-activated receptor-gamma activation pathway in gastric cancer. *Journal of Biological Chemistry*. 2012:jbc-M112.

Rathi B, Sahu J, Koul S, Kosha RL. Detailed pharmacognostical studies on *Berberis aristata* DC plant. *Ancient Science of Life*. 2013;32(4):234.

Renaud SD, de Lorgeril M. Wine, alcohol, platelets, and the French paradox for coronary heart disease. *The Lancet*. 1992;339(8808):1523–1526.

Ribéreau-Gayon P, Dubourdieu D, Donèche B, Lonvaud A, editors. *Handbook of Enology, Volume 1: The Microbiology of Wine and Vinifications*. John Wiley & Sons, Hoboken, NJ; 2006 May 1.

Robbers, J.E., Speedie, M.K. and Tyler, V.E., *Pharmacognosy and pharmacobiotechnology*. 1996.

Roccaro AS, Blanco AR, Giuliano F, Rusciano D, Enea V. Epigallocatechin-gallate enhances the activity of tetracycline in staphylococci by inhibiting its efflux from bacterial cells. *Antimicrobial Agents and Chemotherapy*. 2004;48(6):1968–1973.

Roomi MW, Ivanov V, Kalinovsky T, Niedzwiecki A, Rath M. In vitro and in vivo antitumorigenic activity of a mixture of lysine, proline, ascorbic acid, and green tea extract on human breast cancer lines MDA-MB-231 and MCF-7. *Medical Oncology*. 2005;22(2):129–138.

Sacks FM, Svetkey LP, Vollmer WM, Appel LJ, Bray GA, Harsha D, Obarzanek E, Conlin PR, Miller ER, Simons-Morton DG, Karanja N. Effects on blood pressure of reduced dietary sodium and the Dietary Approaches to Stop Hypertension (DASH) diet. *New England Journal of Medicine*. 2001;344(1):3–10.

Sáenz-Navajas MP, Ferreira V, Dizy M, Fernández-Zurbano P. Characterization of taste-active fractions in red wine combining HPLC fractionation, sensory analysis and ultra performance liquid chromatography coupled with mass spectrometry detection. *Analytica Chimica Acta*. 2010;673(2):151–159.

Salminen KA, Meyer A, Jerabkova L, Korhonen LE, Rahnasto M, Juvonen RO, Imming P, Raunio H. Inhibition of human drug metabolizing cytochrome P450 enzymes by plant isoquinoline alkaloids. *Phytomedicine*. 2011;18(6):533–538.

Sarr M, Chataigneau M, Martins S, Schott C, El Bedoui J, Oak MH, Muller B, Chataigneau T, Schini-Kerth VB. Red wine polyphenols prevent angiotensin II-induced hypertension and endothelial dysfunction in rats: role of NADPH oxidase. Cardiovascular *Research*. 2006;71(4):794–802.

Sartippour MR, Shao ZM, Heber D, Beatty P, Zhang L, Liu C, Ellis L, Liu W, Go VL, Brooks MN. Green tea inhibits vascular endothelial growth factor (VEGF) induction in human breast cancer cells. *The Journal of Nutrition*. 2002;132(8):2307–2311.

Savage JH, Matsui EC, Skripak JM, Wood RA. The natural history of egg allergy. *Journal of Allergy and Clinical Immunology*. 2007;120(6):1413–1417.

Scharbert S, Hofmann T. Molecular definition of black tea taste by means of quantitative studies, taste reconstitution, and omission experiments. *Journal of Agricultural and Food Chemistry*. 2005;53(13):5377–5384.

Sebastian T, Setty OH. Protective effect of *Berberis aristata* against ethanol-induced mitochondrial damage in rats. *Journal of Clinical Biochemistry and Nutrition*. 1997;23(1):1-3.

Seigler DS. *Plant secondary metabolism*. Springer Science & Business Media Berlin, Germany; 2012 December 6.

Sendra JM, Navarro JL, Izquierdo L. C18 solid-phase isolation and high-performance liquid chromatography/ultraviolet diode array determination of fully methoxylated flavones in citrus juices. *Journal of Chromatographic Science*. 1988;26(9):443–448.

Sengupta A, Ghosh S, Das S. Modulation of DMBA induced genotoxicity in bone marrow by quercetin during skin carcinogenesis. *Journal of Experimental & Clinical Cancer Research: CR*. 2001;20(1):131–134.

Sharma C, Aneja KR, Kasera R. Screening of *Berberis aristata* DC. for antimicrobial potential against the pathogens causing ear infection. *International Journal of Pharmacology*. 2011;7(4):536–541.

Sharma SK, Vij AS, Sharma M. Mechanisms and clinical uses of capsaicin. *European Journal of Pharmacology*. 2013;720(1–3):55–62.

Shinde N, Bangar B, Deshmukh S, Kumbhar P. Nutraceuticals: A Review on current status. *Research Journal of Pharmacy and Technology*. 2014;7(1):110–113.

Shoaib M, Stolerman IP. The NMDA antagonist dizocilpine (M K801) attenuates tolerance to nicotine in rats. *Journal of Psychopharmacology*. 1996;10(3):214–218.

Shoskes DA, Zeitlin SI, Shahed A, Rajfer J. Quercetin in men with category III chronic prostatitis: A preliminary prospective, double-blind, placebo-controlled trial. *Urology*. 1999;54(6):960–963.

Sicherer SH, Sampson HA, Burks AW. Peanut and soy allergy: A clinical and therapeutic dilemma. *Allergy.* 2000;55(6):515–521.
Siddique HR, Saleem M. Beneficial health effects of lupeol triterpene: A review of preclinical studies. *Life Sciences.* 2011;88(7–8):285–293.
Simons R. Prenylated isoflavonoids from soya and licorice: Analysis, induction and in vitro estrogenicity. 2011.
Singh J, Kakkar P. Antihyperglycemic and antioxidant effect of *Berberis aristata* root extract and its role in regulating carbohydrate metabolism in diabetic rats. *Journal of Ethnopharmacology.* 2009;123(1):22–26.
Singleton VL. Tannins and the qualities of wines. In *Plant Polyphenols* 1992 (pp. 859–880). Springer, Boston, MA.
Smith AK, June H, Noble AC. Effects of viscosity on the bitterness and astringency of grape seed tannin. *Food Quality and Preference.* 1996;7(3–4):161–166.
Soares S, Kohl S, Thalmann S, Mateus N, Meyerhof W, De Freitas V. Different phenolic compounds activate distinct human bitter taste receptors. *Journal of Agricultural and Food Chemistry.* 2013;61(7):1525–1533.
Sobel BF, Wetherington CL, Riley AL. The contribution of within-session averaging of drug-and vehicle-appropriate responding to the graded dose-response function in drug discrimination learning. *Behavioural Pharmacology.* 1995.
Soni MG, Carabin IG, Griffiths JC, Burdock GA. Safety of ephedra: Lessons learned. *Toxicology Letters.* 2004;150(1):97–110.
Soto-Vaca A, Gutierrez A, Losso JN, Xu Z, Finley JW. Evolution of phenolic compounds from color and flavor problems to health benefits. *Journal of Agricultural and Food Chemistry.* 2012;60(27):6658–6677.
Stark T, Hofmann T. Application of a molecular sensory science approach to alkalized cocoa (*Theobroma cacao*): structure determination and sensory activity of nonenzymatically C-glycosylated flavan-3-ols. *Journal of Agricultural and Food Chemistry.* 2006;54(25):9510–9521.
Stark T, Wollmann N, Wenker K, Losch S, Glabasnia A, Hofmann T. Matrix-calibrated LC-MS/MS quantitation and sensory evaluation of oak ellagitannins and their transformation products in red wines. *Journal of Agricultural and Food Chemistry.* 2010;58(10):6360–6369.
Su CT, Singleton VL. Identification of three flavan-3-ols from grapes. *Phytochemistry.* 1969;8(8):1553–1558.
Sueoka N, Suganuma M, Sueoka E, Okabe S, Matsuyama S, Imai K, Nakachi K, Fujiki H. A new function of green tea: prevention of lifestyle-related diseases. *Annals of the New York Academy of Sciences.* 2001;928(1):274–280.
Sugiyama Y, Masumori N, Fukuta F, Yoneta A, Hida T, Yamashita T, Minatoya M, Nagata Y, Mori M, Tsuji H, Akaza H. Influence of isoflavone intake and equol-producing intestinal flora on prostate cancer risk. *Asian Pacific Journal of Cancer Prevention.* 2013;14(1):1–4.
Suzuki H, Onishi H, Hisamatsu S, Masuda K, Takahashi Y, Iwata M, Machida Y. Acetaminophen-containing chewable tablets with suppressed bitterness and improved oral feeling. *International Journal of Pharmaceutics.* 2004;278(1):51–61.
Swain T. Plant biochemistry. by J. Bonner and J. Varner, Academic Press, New York and London, UK. 1965:552.
Swank MW, Schafe GE, Bernstein IL. c-Fos induction in response to taste stimuli previously paired with amphetamine or LiCl during taste aversion learning. *Brain Research.* 1995;673(2):251–261.
Szejtli J. Cyclodextrin Technology. Kluwer Academic Publishers Dordrecht. The Netherlands Google Scholar. 1988.
Szejtli J, Szente L. Elimination of bitter, disgusting tastes of drugs and foods by cyclodextrins. *European Journal of Pharmaceutics and Biopharmaceutics.* 2005;61(3):115–125.

Tae N, Hung TM, Kim O, Kim N, Lee S, Na S, Min BS, Lee JH. A cassaine diterpene alkaloid, 3β-acetyl-nor-erythrophlamide, suppresses VEGF-induced angiogenesis and tumor growth via inhibiting eNOS activation. *Oncotarget.* 2017;8(54):92346.

Talavera K, Yasumatsu K, Voets T, Droogmans G, Shigemura N, Ninomiya Y, Margolskee RF, Nilius B. Heat activation of TRPM5 underlies thermal sensitivity of sweet taste. *Nature.* 2005;438(7070):1022.

Thiele TE, Roitman MF, Bernstein LL. c-Fos induction in rat brainstem in response to ethanol-and lithium chloride-induced conditioned taste aversions. *Alcoholism: Clinical and Experimental Research.* 1996;20(6):1023–1028.

Thorngate III JH, Noble AC. Sensory evaluation of bitterness and astringency of 3R (−)-epicatechin and 3S (+)-catechin. *Journal of the Science of Food and Agriculture.* 1995;67(4):531–535.

Tikkanen MJ, Adlercreutz H. Dietary soy-derived isoflavone phytoestrogens: could they have a role in coronary heart disease prevention?. *Biochemical Pharmacology.* 2000;60(1):1–5.

Tobin T, Akera T, Brody SL, Ku D, Brody TM. Cassaine: mechanism of inhibition of Na^{++} K+-ATPase and relationship of this inhibition to cardiotonic actions. *European Journal of Pharmacology.* 1975;32(2):133–145.

Toda J, Misaki M, Konno A, Wada T, Yasumatsu K. Interaction of cyclodextrins with taste substances [Schardinger dextrins]. 1981.

Toh CC, Lee TS, Kiang AK. The pharmacological actions of capsaicin and analogues. *British Journal of Pharmacology and Chemotherapy.* 1955;10(2):175–182.

Travers SP, Travers JB. Reflex topography in the nucleus of the solitary tract. *Chemical Senses.* 2005;30(suppl_1):i180-i181.

Tripathi YB, Shukla SD. Berberis aristata inhibits PAF induced aggregation of rabbit platelets. *Phytotherapy Research.* 1996;10(7):628–630.

Trock B, Lanza E, Greenwald P. Dietary fiber, vegetables, and colon cancer: critical review and meta-analyses of the epidemiologic evidence. *JNCI: Journal of the National Cancer Institute.* 1990;82(8):650–661.

Tsuneki H, Ishizuka M, Terasawa M, Wu JB, Sasaoka T, Kimura I. Effect of green tea on blood glucose levels and serum proteomic patterns in diabetic (db/db) mice and on glucose metabolism in healthy humans. *BMC Pharmacology.* 2004;4(1):18.

Turenne SD, Miles C, Parker LA, Siegel S. Individual differences in reactivity to the rewarding/aversive properties of drugs: assessment by taste and place conditioning. *Pharmacology Biochemistry and Behavior.* 1996;53(3):511–516.

Urgert R, van der Weg G, Kosmeijer-Schuil TG, van de Bovenkamp P, Hovenier R, Katan MB. Levels of the cholesterol-elevating diterpenes cafestol and kahweol in various coffee brews. *Journal of Agricultural and Food Chemistry.* 1995;43(8):2167–2172.

Valsecchi AE, Franchi S, Panerai AE, Rossi A, Sacerdote P, Colleoni M. The soy isoflavone genistein reverses oxidative and inflammatory state, neuropathic pain, neurotrophic and vasculature deficits in diabetes mouse model. *European journal of pharmacology.* 2011;650(2):694–702.

Van Haaren F, Hughes CE. Cocaine-induced conditioned taste aversions in male and female Wistar rats. *Pharmacology Biochemistry and Behavior.* 1990;37(4):693–696.

Vanamala J, Leonardi T, Patil BS, Taddeo SS, Murphy ME, Pike LM, Chapkin RS, Lupton JR, Turner ND. Suppression of colon carcinogenesis by bioactive compounds in grapefruit. *Carcinogenesis.* 2005;27(6):1257–1265.

Van Etten CH, Daxenbichler ME, Kwolek WF, Williams PH. Distribution of glucosinolates in the pith, cambial-cortex, and leaves of the head in cabbage, *Brassica oleracea* L. *Journal of Agricultural and Food Chemistry.* 1979;27(3):648–650.

Vérette E, Noble AC, Somers TC. Hydroxycinnamates of *Vitis vinifera*: Sensory assessment in relation to bitterness in white wines. *Journal of the Science of Food and Agriculture.* 1988;45(3):267–272.

Verma S. Nutraceutical in India: How big is the market? Are they really beneficial? 2018.

Verotta L, Aburjai T, Rogers CB, Dorigo P, Maragno I, Fraccarollo D, Santostasi G, Gaion RM, Floreani M, Carpenedo F. Chemical and pharmacological characterization of *Erythrophleum lasianthum* alkaloids. *Planta Medica*. 1995;61(03):271–274.

Ververidis F, Trantas E, Douglas C, Vollmer G, Kretzschmar G, Panopoulos N. Biotechnology of flavonoids and other phenylpropanoid-derived natural products. Part I: Chemical diversity, impacts on plant biology and human health. *Biotechnology Journal: Healthcare Nutrition Technology*. 2007;2(10):1214–1234.

Vidal S, Francis L, Noble A, Kwiatkowski M, Cheynier V, Waters E. Taste and mouth-feel properties of different types of tannin-like polyphenolic compounds and anthocyanins in wine. *Analytica Chimica Acta*. 2004;513(1):57–65.

Vikram A, Jayaprakasha GK, Jesudhasan PR, Pillai SD, Patil BS. Obacunone represses Salmonella Pathogenicity Island 1 and 2 in envZ dependent fashion. *Applied and Environmental Microbiology*. 2012:AEM-01326.

Vogel HG, Vogel WH, editors. *Drug Discovery and Evaluation: Pharmacological Assays*. Springer Science & Business Media, Berlin, Germany; 2013 April 17.

Wada K, Nakamura K, Tamai Y, Tsuji M, Kawachi T, Hori A, Takeyama N, Tanabashi S, Matsushita S, Tokimitsu N, Nagata C. Soy isoflavone intake and breast cancer risk in Japan: From the Takayama study. *International Journal of Cancer*. 2013;133(4):952–960.

Wal JM. Cow's milk allergens. *Allergy*. 1998;53(11):1013–1022.

Wang HJ, Murphy PA. Isoflavone content in commercial soybean foods. *Journal of Agricultural and Food Chemistry*. 1994;42(8):1666–1673.

Wang S, Tu H, Wan J, Chen W, Liu X, Luo J, Xu J, Zhang H. Spatio-temporal distribution and natural variation of metabolites in citrus fruits. *Food Chemistry*. 2016;199:8–17.

Waterhouse AL. Wine phenolics. *Annals of the New York Academy of Sciences*. 2002;957(1):21–36.

Weber JM, Ruzindana-Umunyana A, Imbeault L, Sircar S. Inhibition of adenovirus infection and adenain by green tea catechins. *Antiviral Research*. 2003;58(2):167–173.

Wehmer C, Hausschwammstudien II. Der wachstumshemmende Einfluss von Gerbsauren auf Merulius lachrymans in seiner Beziehung zur Resistenz des Eichenholzes gegen Hausschwamm. *Mycologisches Centralbl. I*. 1912:138–148.

Wei P, Liu M, Chen Y, Chen DC. Systematic review of soy isoflavone supplements on osteoporosis in women. *Asian Pacific Journal of Tropical Medicine*. 2012;5(3):243–248.

Weinreb O, Mandel S, Amit T, Youdim MB. Neurological mechanisms of green tea polyphenols in Alzheimer's and Parkinson's diseases. *The Journal of Nutritional Biochemistry*. 2004;15(9):506–516.

Wiet SG, Beyts PK. Sensory characteristics of sucralose and other high intensity sweeteners. *Journal of Food Science*. 1992;57(4):1014–1019.

Wildman RE. *Handbook of Nutraceuticals and Functional Foods*. CRC Press Boca Raton, FL; 2002 January 1.

Williams CA, Grayer RJ. Anthocyanins and other flavonoids. *Natural Product Reports*. 2004;21(4):539–573.

Willner J, Gallagher M, Graham PW, Crooks GB. N-methyl-D-aspartate antagonist D-APV selectively disrupts taste-potentiated odor aversion learning. *Behavioral Neuroscience*. 1992;106(2):315.

Wong GT, Gannon KS, Margolskee RF. Transduction of bitter and sweet taste by gustducin. *Nature*. 1996;381(6585):796.

Wu SV, Rozengurt N, Yang M, Young SH, Sinnett-Smith J, Rozengurt E. Expression of bitter taste receptors of the T2R family in the gastrointestinal tract and enteroendocrine STC-1 cells. *Proceedings of the National Academy of Sciences*. 2002;99(4):2392–2397.

Yamakoshi J, Piskula MK, Izumi T, Tobe K, Saito M, Kataoka S, Obata A, Kikuchi M. Isoflavone aglycone–rich extract without soy protein attenuates atherosclerosis development in cholesterol-fed rabbits. *The Journal of Nutrition*. 2000;130(8):1887–1893.

Yamamoto T. Neural mechanisms of taste aversion learning. *Neuroscience Research*. 1993;16(3):181–185.

Yang MH, Wang CH, Chen HL. Green, oolong and black tea extracts modulate lipid metabolism in hyperlipidemia rats fed high-sucrose diet. *The Journal of Nutritional Biochemistry*. 2001;12(1):14–20.

Yogesh HS, Chandrashekhar VM, Katti HR, Ganapaty S, Raghavendra HL, Gowda GK, Goplakhrishna B. Anti-osteoporotic activity of aqueous-methanol extract of *Berberis aristata* in ovariectomized rats. *Journal of Ethnopharmacology*. 2011;134(2):334–338.

Yousfi K, Cayuela JA, García JM. Reduction of virgin olive oil bitterness by fruit cold storage. *Journal of Agricultural and Food Chemistry*. 2008;56(21):10085–10091.

Yu X, Ding G, Zhi X, Xu H. Insight into reduction of obacunone, and their ester derivatives as insecticidal agents against *Mythimna separata* Walker. *Bioorganic & Medicinal Chemistry Letters*. 2015;25(1):25–29.

Zaidi FN, Todd K, Enquist L, Whitehead MC. Types of taste circuits synaptically linked to a few geniculate ganglion neurons. *Journal of Comparative Neurology*. 2008;511(6):753–772.

Zhang CH, Lifshitz LM, Uy KF, Ikebe M, Fogarty KE, ZhuGe R. The cellular and molecular basis of bitter tastant-induced bronchodilation. *PLoS Biology*. 2013;11(3):e1001501.

Zhang H, Qiu M, Chen Y, Chen J, Sun Y, Wang C, Fong HH. Plant terpenes. Phytochemistry and pharmacognosy. *Encyclopedia of Lyfe Support Systems*. 2002.

Zhang L, Xia Y, Peterson DG. Identification of bitter modulating Maillard-catechin reaction products. *Journal of agricultural and food chemistry*. 2014;62(33):8470–8477.

Zhang Y, Hoon MA, Chandrashekar J, Mueller KL, Cook B, Wu D, Zuker CS, Ryba NJ. Coding of sweet, bitter, and umami tastes: different receptor cells sharing similar signaling pathways. *Cell*. 2003;112(3):293–301.

Zhang YF, Kang HB, Li BL, Zhang RM. Positive effects of soy isoflavone food on survival of breast cancer patients in China. *Asian Pacific Journal of Cancer Prevention*. 2012;13(2):479–482.

Zhao J. Nutraceuticals, nutritional therapy, phytonutrients, and phytotherapy for improvement of human health: A perspective on plant biotechnology application. *Recent Patents on Biotechnology*. 2007;1(1):75–97.

Zhao J, Schlaad H. Synthesis of terpene-based polymers. In *Bio-synthetic Polymer Conjugates* 2011 (pp. 151–190). Springer Berlin, Germany.

Zsolnai T. The antimicrobial activity of thiocyanates and isothiocyantes. 1. *Arzneimittel-Forschung*. 1966;16(7):870–876.

12

Taste Assessment Trials for Sensory Analysis of Oral Nutraceutical Products and Functional Foods

Madan Mohan Gupta

Contents

12.1 Introduction

Taste is basically a sensation, and it is felt when a drug/food product reacts chemically with taste receptor in the mouth especially with the cells located on taste buds those are found on the tongue in the oral cavity so the flavor is basically combination of both taste and smell. The taste is realization of sensation when a substance such as food, beverages, nutraceutical or medicine is placed in the oral cavity and it is the ability to take action to dissolve molecules and ions "gatekeeper to the body." The molecule of substances interacts with taste receptor located on the tongue and give different taste like bitter, sweet or other taste sensation, when they dissolve in saliva (Dalton et al., 2000; Schutz & Cardello, 2001).

This sensation of taste of product is due to signal transduction from the receptor organs for taste, which is taste buds, and these taste buds contains very sensitive nerve endings those produce and transmit electrical impulses via the seventh, ninth and tenth cranial nerves to those areas of the brain that are for the perception of taste. Biologically the taste (gustation) is defined a chemical reaction that is derived from sensory responses from the four main taste perceptions: salt, sour, bitter, and sweet (Amipara & Gupta, 2013; Amerine et al., 1965).

Two other perceptions (umami and trigeminal) should be included when considering taste of the substance. Umami or savory taste is one of the all five basic tastes (together with sweet, sour, bitter and salty). The umami taste is experienced by the people through taste receptors that typically respond to glutamate that is widely available in meat broths and fermented products. This is commonly added to some foods in the form monosodium glutamate. The taste umami having its own receptor and nowadays umami taste is considered as a distinct taste. The perception trigeminal is a burning sensation derived from spices and peppers types foods (Lawless & Heymann, 2010; Roy, 1992).

12.2 Nutraceuticals

The meaning of the term nutraceutical is a combination of nutrition and pharmaceutical so the term nutraceutical is applied to products that are isolated from herbs, diet supplements and from processed foods like cereals, soups and beverages. The nutraceuticals are generally used as oral products and derived from different food sources. The nutraceuticals claim to improve health by preventing different chronic disease as well as retard aging and increase life expectancy so the nutraceuticals are products, used as nutrition value product as well as medicine, and it has physiological benefit or provides protection against chronic disease. Currently in global market nutraceuticals have received significant interest because of nutritional value, safety and therapeutic effects of these (Kalra, 2003; Hardy, 2000).

A nutraceutical may be a medicinally active food or naturally rich, such as turmeric powder, ginger powder or soybeans. Nutraceutical may also be a specific component of a food, such as the omega-3 which may be derived from salmon and other cold-water fish and this is claim in controlling the cholesterol level.

12.3 Functional foods

The functional food may be given for health promotion or disease prevention. The functional food may be derived by adding new ingredient (s) to make a more effective or by adding more amount of existing ingredients. In appearance the functional food looks like a conventional food and designed to have specific physiological benefits as well as it may reduce the risk of some chronic disease if they are consumed as part of a regular diet.

Functional foods have a potentially positive effect on human health beyond basic nutrition. Some people opinion is about that functional foods, which can promote optimal health and reduce the risk of disease.

Functional foods provide additional or increased benefits as compared to their basic nutritional value. Some of the functional foods contains specific functional ingredient like probiotics, or plant stanols and sterols (Li & Zhang, 2001; Gupta et al., 2011, Schifferstein, 2012; Nahata, 1999).

12.4 Types of taste

Four Primary Tastes

1. Sour
2. Salty
3. Bitter
4. Sweetness

For taste sensation, the threshold concentration varies as per the taste and tongue area. For sweet taste the threshold concentration requirement is 0.5% and the tongue area is tip of the tongue. For salt, sour and bitter taste the threshold concentration requirement is 0.25%, 0.007%, 0.00005% respectively. The tongue area for taste sensation used for salt, sour and bitter is edge and upper front, alongside back, back respectively (Amerine et al., 1965; Kemp et al., 2009; Nakamura et al., 2003; Ogawa et al., 2004).

12.5 Methods to test taste buds

For sensory analysis one of important part is to test the taste buds of healthy volunteers and for this food color is required. To perform this test food color required to put on the tip of tongue by the help of cotton and a reinforcement rig should be out over the tongue. After this by the help of a magnifying glass pink dots that are inside the rings should be counted and noted.

These pink dots are basically fungi form papillae and due to their specific property, these pink dots do not take up the food color. On the tongue these papillae looks like and this provide the information about tasters, a person called excellent testers or sensitive for taste if he is having more than 30 papillae on tongue while a person is called as a very less sensitive or bad tasters or non-tester if having less than 15 papillae (Lim, 2011; Kroll, 1990; Stone & Sidel, 2004).

Factors Affecting the Perception of Bitterness

- Taste interactions
- Medium of presentation
- Viscosity and temperature
- Oral chemical irritants
- Mode of presentation and taste modifiers
- Salivary status and age

12.6 General taste masking approaches in nutraceuticals and functional food

Taste masking is one of an important characteristic for a good nutraceutical and functional foods. In recent years a significant work has been done for taste masking and their pharmaceutical formulations (Lalji & Gupta, 2013). The palatability of the substances has been improved due to reduction in bitterness.

An effective taste masking process should have the following properties

1. Use of minimum equipment and processing steps
2. Require minimum number of excipients for taste masking
3. No effect on bioavailability of nutraceuticals
4. High margin of safety, economic and readily available excipients
5. Essay and fast process to reduce manufacturing cost
6. Capable to work at room temperature

Different techniques are available for masking the bitter substance including physical and chemical methods those act as barrier between substance and taste buds and to prevent the interaction with taste buds. Some of the simple methods used for masking the bitter taste are use of sweeteners, flavors and amino acids, but these are not capable to mask the bitter taste of all foods, so other complex techniques, like resin based complex, i.e., ion exchange resin process, cyclodextrin complexation, polymer coating, etc., are used (Avari & Bhalekar, 2004; Gupta et al., 2014; Barthoshuk, 2000; Borodkin & Sundberg, 1971; Nunn & Williams, 2005).

12.7 Sensory analysis in the food industry

For the new product development, modification in the recipe as well as for evaluation of food products sensory analysis is required and this analysis also important for quality control as well as marketing of the product by the company. Different types of sensory analysis tests have been devised to fulfill specific objectives for the food. These tests are grouped into three categories (Fergonezi-Nery et al., 2002; Srivasatva, 2012, Gupta & Patel 2013).

12.7.1 Sensory analysis tests

1. Preference tests
2. Difference tests
3. Descriptive tests

12.7.1.1 Preference tests

Preference tests provide the information about whether testers like or dislike a product. Preference tests are used in the food industry to find out the opinion of consumer about the product (Bagger & Bondesson, 1989).

Following things should be checked during the preference test

- The consumer likes a product or not and how much they like it
- The product is acceptable or not
- Comparison between product whether it is as good as like other product or not
- Consumer will prefer this or not
- What consumer like most in the product and what attribute they prefer

Preference tests are often referred to as “acceptance” or “consumer” tests.

12.7.1.2 Difference tests

In the difference tests various questions are asked and the purpose of these is to find out whether or not a difference is existing between two or more products. The food industry uses this test to find out the answer of the following questions:

- Is this product same as other existing product or different?
- If different, how great is the difference?
- What difference is noticed by people?
- How tasters describe this difference?
- How many testers would detect this difference?

Difference tests are sometimes called "discrimination" tests (Yixun et al., 2015; Barra, 1999).

12.7.1.3 Descriptive tests

This test describes the product quality based on perceived sensory characteristics provided by volunteers (Dandagi et al., 2005; Green et al 1996; Lim & Fujimaru, 2010). This test is used in food industry and it will provide the following information about the product:

- The taste of this product
- Perceived sensory characteristics/attributes of the product
- How a change in process/formulation/packaging/storage condition affects the sensory quality of this product
- Difference between products regarding in sensory quality
- Which sensory attributes have the most-marked differences

12.8 Taste assessment trials

The trials for taste assessment for food and drug involve an estimation of gustatory sensation response by healthy volunteers who are involved in control process and protocol of taste assessment, and these trials are also known as psychophysical evaluation, physiological evaluation, sensory test and sensation tests. The taste assessment trials are designed by using the statistical tools to minimize or reduce the bias in responses within and between human volunteers. These can be used for assessing nutraceuticals and ingredient impact as well as intensity so that full profiles can be produced. Taste assessment trials are important for oral formulation development that are for pediatric patients (Anand et al., 2008; Garg & Gupta, 2013; Kemp et al., 2009; Smith & John, 1999).

12.8.1 Sensory evaluation

The taste is a subjective observation and it depends on individuals, the apparent taste may vary according to different degrees. To measure the taste threshold accurately and reproducibly it is required to set up a well-controlled experiment. To quantitatively evaluate taste sensation, following methods have been reported in literature to evaluate taste sensation quantitatively (Amerine et al., 1965; Gupta et al., 2014; Saito et al., 1999; Zwislocki & Goodman, 1980; Chatterjee et al., 2018).

- Human taste panel studies
- Measurement of frog taste nerve responses
- Electronic tongue/Multichannel taste sensor
- Spectrophotometric evaluation/D30s value

12.8.2 Human taste panel studies

For psychophysical rating of the gustatory stimuli, the human taste panel studies are carried out and in this study the response of gustatory sensation of food product or others are estimated in testers (human volunteers) so this study is also called as physiological evaluation, gustatory sensation testes or taste trials. This study is designed in such a way that sensitivity of product is measured with minimum chances of biasness within and between testers. This methodology for sensory analysis is divided into five different types: discrimination taste, scaling taste, experts testers, affective taste and descriptive test. The trained human volunteers assess the taste quality and intensity of product on different adjective scales as well as different properties of test product like intensity, sweet, bitter, metallic, cooling, hot etc. Each adjective can be rated on an intensity scale ranging from zero to four or perhaps even up to nine points on score sheets.

In this study, 5–10 trained human volunteers are required for taste evaluation and these testers are trained for by tasteless to very bitter taste by reference solution and instruction are generally given to them that after each test rinse the mouth with fresh water. The value from 0 to 5 is assigned them for their level of bitterness of the test product and reference product (Jones et al., 1955; Gupta & Saini, 2009; Anand et al., 2008).

Evaluation of taste masking broadly categorized into subjective and objective methods like for preference test (subjective method), difference test is used as objective method and for hedonic scale, single attribute test, duo trio test are used as objective method.

12.8.3 Measurement of frog taste nerve responses

For taste assessment of the product frog taste nerve response method is also available and in this method an adult bull frog is selected for test and anaesthetized intraperitoneally. After anaesthesia glossopharyngeal nerve is located and cut proximally. To amplify an AC-amplifier are used while for to integrate the nerve impulses, an electronic integrator are used. The peak height of the integrated response is recorded as the magnitude of response.

12.8.4 Electronic tongue

Currently this method is alternative way for sensory analysis for pharmaceuticals, food products as well as for functional foods and it is not required any human volunteers for testing. In this method magnitude of product bitterness is automatically detected by an electronic device that is called an electronic tongue (e-tongue). This technique for taste sensation provides qualitative information of product bitterness.

This electronic tongue has three components – probe membrane, transducer and computer for analysis. The probe membrane acts as taste buds, transducer acts as neural transmission and computer acts a cognition in the thalamus so as to compare to natural process. The electronic tongue has all in-vitro processes so it is like an *in vivo* and *in-vitro* correlation for taste analysis.

Generally an e- tongue system has few to dozens of sensors of a single type and out of all the most common is potentiometric and voltammetry. In the electronic tongue each sensor responds to a given product or analyte but with a different degree. If you use a second analyte, the response pattern will be different and based on the response patterns, you will be able to distinguish between analytes (**Table 12.1**). Electronic tongue refers for liquid analysis; if you use the group of sensors for gas phase sensing, then it is known as electronic nose (Breslin & Beauchamp, 1995; Gupta et al., 2014; Green et al., 1996).

If the test product is in liquid condition it can be directly analyzed without any preparation, but for solid it is required to perform dissolution make it solution then it can be used for e-tongue. The process of e-tongue method is as follows:

- Dipping of sensor and reference electrode in test solution beaker
- Voltage applied between sensor and a reference electrode to complete the circuit
- Oxidization in the solution takes place that is measured as current response
- Response is amplified by catalytic surface treatments
- E-tongue's software measured and recorded the response of this process
- Result obtained by mathematical treatment of data

Table 12.1 Relationship between Human Tongue and E-Tongue

Level	Human Tongue	E-Tongue
Receptor	Taste buds	Probe membrane
Circuit	Neural transmission	Transducer
Perceptual	Cognition in the thalamus	Computer and statistical analysis

Key benefit of e-tongue evaluation

- When limited information about the product taste is available, method is useful to quantify bitterness of food product
- Helps in development of suitable bitter matched placebos for blinded clinical testing
- Conduct comparator studies (Benchmark analysis)
- Developing optimized tastemasked formulations
- For the food product and other flavored product used as quality control function

12.8.5 Spectrophotometric method

This method is based on the principle that if calculated concentration of active moiety is less then threshold concentration, the bitter taste of product would be masked in vivo. In this method the concentration of active moiety is measured with the help of a-spectrometer. To measure the concentration, a fixed weight amount of taste-masked product is put in the 10 ml of syringe with 10 ml of distilled water and shaken properly for mixing. This solution is then filtered through a membrane filter and used for estimating the concentration of taste masked product. Usually in this method taste is evaluated by objective or analytical method or subjective or hedonic method (Green et al., 1993; Ho, 1984).

12.9 Nine-point hedonic scale

A popular scale nine-point hedonic used for measuring food acceptability, which was developed by David Peryam and his colleagues at the Quartermaster Food and Container Institute of the US Armed Forces. This scale has various scale parameters, so it was adopted and accepted very fast by the food industries.

The affective test for the hedonic nine-point scale highest order for extremely likely while the lowest order for the dislike extremely a product. The hedonic scale is only used to get an idea about product, and it is a relative liking. On the basis of scale, the most liked product can be selected for further development.

Due to categorical nature and limited choice nine-point hedonic point scale is accepted widely by food industries and it is easy to use by the participants (testers) as well as for researchers.

The handling of the nine-point hedonic scale data is easier when compare to other methods, which require measuring lines or recording magnitude estimates that may include fractions (Peryam & Girardot, 1952; Peryam & Pilgrim, 1957; Rosas et al., 2005; Mojet et al., 2003; Salazar et al., 2000; Cardello & Schutz, 2004).

Hedonic nine-point scale:

1. Like extremely
2. Like very much
3. Like moderately
4. Like slightly
5. Neither like nor dislike
6. Dislike slightly
7. Dislike moderately
8. Dislike very much
9. Dislike extremely

12.9.1 Product preference test

Example: Product: Testing the competitor product

Instruction to participants: Dear participants, as a representative of consuming population you will get a number of product samples for evaluation containing different competitor products. Kindly read the instructions carefully and then give answer. During the test please do not talk to other participants and give your personal opinion as well as if any comments for the question.

After test rinse you month with fresh water and be ready for next product testing.

You will try three brands of product A.

Preference Ranking Test (Table 12.2)
Requirement for Sensory trial in General
Approval from authority

To perform the sensory trial it is required to prepare a standard protocol and submitted to concern regulatory authority of respective country (if any). Trial protocol is first prepared and submitted to committee for sensory trials; such as in the case of drug trials, it is required to submit to an institutional review board (IRB) or ethical committee. Human volunteers consent form also should be submitted to the required authority for approval, and this form should have all the information about the volunteers such as age, sex and so on (Katsuragi et al., 1995; Jones et al., 1995).

Table 12.2 Ranking of Different Sample in Taste Analysis

Ranking (1–3)	Sample No	Comments
1		
2		
3		

12.9.2 Selection of human volunteers

For taste analysis the human volunteers are identified and screened. The volunteers are selected for study based on the inclusion and exclusion criteria. Before the study the volunteers are informed about the study, study procedure as well as about the risk factors (if any) involved in the study (Mojet et al., 2003; Jain et al., 2016).

12.9.3 Training of human volunteers

Training of participating human volunteers is required to conduct taste assessment in proper way without considering tasters personal preferences. The training sessions include different activities related to taste assessment, like sample application method, tasting of sample, intensity rating, use of appropriate scale and expectoration of samples. Volunteers assess the taste equality, intensity and in some cases temporal profile of samples. It is required to properly explain and trained to the human volunteers about the correct use of scale (Green & Hayes, 2004; Anand et al., 2008; Wichchukit & O'Mahony, 2014).

Human volunteers shall pass basic sensory training tests to become a member of the product panel team.

12.9.4 Standard stimuli

To prepare a standard solution, highly pure nutraceutical or functional food and deionized water are used for stimulus application. Freshly prepared solutions are used for taste assessment, but if storage of standard solution is needed they should be stored at refrigeration conditions but brought to room temperature at the time of use (Lugaz et al., 2005; Keast et al., 2003).

12.10 Benefits of sensory analysis for food industry

In the present global market because of high completion and more opportunities, the food industries business is accelerating. The demand of new product with good quality, better acceptance by all age consumers, extended shelf life and economic cost is increasing day by day. To prepare the new food product with all qualities, it is required that industries must know the consumer attitude and perception. This can be measured by sensory analysis so sensory analysis is an important part of food industries (de Saavedra et al., 2000; Ishizaka et al., 2004; Hutto & Bratton, 1999; Ho, 1984).

The sensory analysis has many aims and it can be used for:

- New food product development
- Information about the competitor product which is in the market

- Quality assurance and quality control of the final product and provide information about set specification
- Information consumer attitude and perception about the product
- Information about taint identification
- The recipe as well as process
- Shelf life evaluation
- Competitor benchmarking

References

Amerine, MA, Pangborn, RM and Roessler EB. 1965. *Principles of Sensory Evaluation of Food*. Academic Press, New York.

Amipara, LV, Gupta, MM. 2013. Oral disintegrating tablet of antihypertensive drug. *J Drug Deliv Ther* 3: 85–92.

Anand, V, Kharab, V, Kataria, M, Kukkar, V, Choudhary, PK. 2008. Taste assessment trails for sensory analysis of oral pharmaceutical products, *Pak J Pharm Sci* 21: 438–450.

Avari, J, Bhalekar, M. 2004. Cation exchange resins for taste masking and rapid dissolution of sparfloxacin. *Indian Drugs* 41: 19–23.

Bagger, SD, Bondesson, G. 1989. Taste Evaluation and compliance of two pediatric formulations of phenoxymethylpenicillin in children. *Scan. J Prim Health Care* 7: 87–92.

Barra, J, Lescure, F, Doelker, E. 1999. Taste masking as a consequence of the organisation of powder mixes. *Pharm Acta Helv* 74: 37–42.

Barthoshuk, L. 2000. Comparing sensory experiences across individuals: Recent psychophysical advances illuminate genetic variation in taste perception. *Chem Senses* 25: 447–460.

Borodkin, S, Sundberg, D. 1971. Polycarboxylic acid ion exchange resin adsorbates for taste coverage in chewable tablets. *J Pharm Sci* 60: 609–623.

Breslin, P, Beauchamp, G. 1995. Suppression of bitterness by sodium: Variation among bitter taste stimuli. *Chem Senses* 20: 609–623.

Cardello, AV, Schutz HG. 2004. Numerical scale-point location for constructing the LAM (labeled affective magnitude) scale. *J Sens Stud* 19: 341–346.

Chatterjee, A, Gupta, MM, Srivastava, B. 2018. Crystallo-co-agglomeration of Valsartan for Improved Solubility and Powder Flowability. *Asian J Pharmaceutics* 12: 182–195.

Dalton, P, Doolittle, N, Nagata, H, Breslin, P A S. 2000. The merging of the senses: Integration of subthreshold taste and smell. *Nat Neurosci* 3: 431–432.

Dandagi, P, Sreenivas, S, Manvi F, Patil, M, Mastiolmath, V, Gadad A. 2005. Taste masked of loxacin mouth disintegrating tablets. *Indian Drugs* 42: 52–55.

de Saavedra, MS, Cuadra, S. 2000. Application of a sensorial response model to the design of an oral liquid pharmaceutical dosage form. *Drug Dev Ind Pharm* 26: 55–60.

Fergonezi-Nery, M, Prudencio-Ferreira, S, Kedor, E, Baracat, M, Brinholi F. 2002. Sensory evaluation of albendazole supensions. *Brazilian Arch Biol Tech* 46: 457–463.

Garg, A, Gupta, MM. 2013. Taste masking and formulation development & evaluation of mouth dissolving tablets of levocetrizine hydrochloride. *J Drug DelivTher* 3:123–130.

Green, B, Dalton, P, Cowart, B, Rankin, K, Higgins, J. 1996. Evaluating the labeled magnitude scale for measuring sensation of taste and smell. *Chem Senses* 21: 323–334.

Green, B, Hayes, J. 2004. Individual differences in perception of bitterness from capsaicin, piperine and zingerone. *Chem Senses* 29: 53–60.

Green, B, Shaffer, G, Gilmore, M. 1993. A semantically labelled magnitude scale of oral sensation with apparent ratio properties. *Chem Senses* 18: 683–702.

Gupta, MM, Gupta, N, Chauhan, BS, Pandey, S. 2014. Fast disintegrating combination tablet of taste masked levocentrizinedihydrochloride and monetilukast sodium: Formulation design development and evaluation. *J Pharmaceutics* Article ID 568320.

Gupta, MM, Pandey S, Chauhan, BS, Gupta, N. 2014. Design, development and evaluation of rosuvastatin calcium and diltiazem hydrochloride bilayer tablet using combination concept of sustained layer with conventional layer. *Turkish J Pharm Sci* 11: 269–284.

Gupta, MM, Patel, M, Kedawat, M. 2011. Enhancement of dissolution rate of rapidly dissolving oral film of meclizine hydrochloride by complexation of meclizine hydrochloride with β-cyclodextrine. *J Appl Pharm Sci* 1: 150–153.

Gupta, MM, Patel, V. 2013. Formulation and evaluation of oral dispersible tablet of cinnarizine. *J Drug DelivTher* 3: 12–17.

Gupta, MM, Saini, TR. 2009. Preformulation parameters characterization to design, development and formulation of vancomycin hydrochloride tablets for pseudomembranous colitis. *Int J Pharm Res Dev* 1: 1–7.

Hardy G. 2000. Nutraceuticals and functional foods: Introduction and meaning. *Nutrition* 16: 688–689.

Ho, N. 1984. Physical model approach to taste studies of drugs and pharmaceutical formulations. *J Pharm Sci* 73: 600–603.

Hutto, C, Bratton, T. 1999. Palatability and cost comparison of five liquid corticosteroid formulations. *J Pediatr Oncol Nurs* 6: 74–77.

Ishizaka, T, Miyanaga, Y, Mukai, J, Asaka, K, Nakai, Y, Tsuji, E, Uchida, T. 2004. Bitterness evaluation of medicines for pediatric use by a taste sensor. *Chem Pharm Bull* 52: 943–948.

Jain, M, Pareek, A, Bagdi G, Aahmed, D, Ahmd A. 2016. Taste masking methods for bitter drug-A review. *Int J Pharm Life Sci* 1: 336–339.

Jones, LV, Peryam DR, Thurstone, LL. 1955. Development of a scale for measuring soldiers' food preferences. *Food Res* 20: 512–520.

Kalra EK. 2003. Nutraceutical—Definition and introduction. *AAPS Pharm Sci* 5: E25.

Katsuragi, Y, Sugiura, Y, Lee, C, Otsuji, K, Kurihara, K. 1995. Selective inhibition of bitter taste of various drugs by lipoprotein. *Pharm Res* 12: 658–662.

Keast, R, Bournazel, M, Breslin, P. 2003. A psychophysical investigation of binary bitter compound Interactions. *Chem Senses* 28: 301–313.

Kemp, SE, Hollowood T and Hort J. 2009. *Sensory Evaluation: A Practical Handbook*. Wiley–Blackwell, Chichester, UK.

Kroll, BJ. 1990. Evaluating rating scales for sensory testing with children. *Food Technol* 44: 78–80, 82, 84, 86.

Lalji, V, Gupta, MM. 2013. Oral disintegrating tablet of antihypertensive drug. *J Drug Deliv Ther* 3: 85–92.

Lawless, HT and Heymann H. 2010. *Sensory Evaluation of Food: Principles and Practices*. 2nd edition. Springer, New York.

Li SQ, Zhang QH. 2001. Advances in the development of functional foods from buckwheat. *Crit Rev Food Sci Nutr* 41: 451–64.

Lim, J. 2011. Hedonic scaling: A review of methods and theory. *Food Qual Pref* 22: 733–747.

Lim, J, Fujimaru T. 2010. Evaluation of the labelled hedonic scale under different experimental conditions. *Food Qual Pref* 21: 521–530.

Lugaz, O, Pillias, A-M, Boireau-Ducept, N, Faurion, A. 2005. Time–intensity evaluation of acid taste in subjects with saliva high flow and low flow rates for acids of various chemical properties. *Chem Senses* 30: 89–103.

Mojet, J, Heidema, J, Christ-Hazelhof, E. 2003. Taste perception with age: Generic or specific losses in suprathreshold intensities of five taste qualities? *Chem Senses* 28: 397–413.

Nahata, M. 1999. Lack of pediatric drug formulations. *Pediatrics* 104: 607–609.

Nakamura, T, Akiyoshi, T, Tanaka, N, Shinozuka, K, Matzno, S, Nakabayashi, T, Matsuyama, K, Kashiwayanagi, M, and Uchida T 2003. Effect of quinine solutions on intracellular Ca2+ levels in neuro-2a cells-conventional physiological method for the evaluation of bitterness. *Biol Pharm Bull* 26: 1637–1640.

Nunn, T, Williams, J. 2005. Formulation of medicines for children. *Br J Clin Pharmacol* 59: 674–676.

Ogawa, T, Nakamura, T, Tsuji, E, Miyanaga, Y, Nakagawa, H, Hirabayashi, H, Uchida, T. 2004. The combination effect of L-arginine and NaCl on bitterness suppression of amino acid solutions. *Chem Pharm Bull* 52: 172–177.

Peryam, DR, Girardot, NF. 1952. Advanced taste test method. *Food Eng* 24: 58–61.

Peryam DR, Pilgrim, FJ. 1957. Hedonic scale method of measuring food preferences. *Food Technol* 11: 9–14.

Rosas, NM, Angulo O, O'Mahony M. 2005. How well does the 9-point hedonic scale predict purchase frequency? *J Sens Stud* 20: 313–331.

Roy, G. 1992. Bitterness: Reduction and inhibition. *Trends Food Sci Nutr* 3: 85–91.

Saito, M, Hoshi, M, Igarshi, A, Ogata, H, Edo, K. 1999. The marked inhibition of the bitter taste of polymyxin B sulfate and trimethoprim x sulfamethoxazole by flavored BMI-60 in pediatric patients. *Biol Pharm Bull* 22: 997–998.

Salazar, d, Saavedra, M, Saavedra, C, 2000. Application of a sensorial response model to the design of an oral liquid pharmaceutical dosage form. *Drug Dev Ind Pharm.* 26: 55–60.

Schifferstein, HNJ. 2012. Labelled magnitude scales: A critical review. *Food Qual Pref* 26: 151–158.

Schutz, HG, Cardello AV. 2001. A labelled affective magnitude (LAM) scale for assessing food liking/disliking. *J Sens Stud* 16: 117–159.

Smith, DV, John, SJ. 1999. Neural coding of gustatory information. *Curr Opin Neurol* 9: 427–435.

Srivasatva, A. 2012. Recent Technologies for the taste masking of the bitter drugs. *Int J Pharma Prof Res* 3: 578–86.

Stone, II and Sidel, JL. 2004. *Sensory Evaluation Practices*, 3rd edition. Elsevier Academic Press, London, UK.

Wichchukit, S, O'Mahony, M. 2014. The 9-point hedonic scale and hedonic ranking in food science: Some reappraisals and alternatives. *J Sci Food Agric* doi:10.1002/jsfa.6993.

Yixun, X, Fang, Z, Michael, O'M. 2015. Is the discrepancy between numbers derived from verbal and numerical protocols for 9-point hedonic scales an artifact of product choice? *J Sens Stud* 30: 269–279.

Zwislocki, JJ, Goodman, DA. 1980. Absolute scaling of sensory magnitudes: A validation. *Percept Psychophys* 28: 28–38.

13

Quality Control in Flavor Industry

Komal Parmar and Jayvadan Patel

Contents

13.1 Introduction

Quality control in general means controlling of quality of the material before final packaging; it also includes control of the processes used in manufacture, storage requirements and transport facilities. Quality control in the flavor industry includes sensory and analytical testing, which might be facilitated in the same premises of the company or may be located in two separate laboratories. Both tests are conducted on the raw materials as well as finished flavors before used in any formulations designed to ensure physical, chemical, organoleptic and microbiological stability. All raw materials must meet recognized industry quality standards approved by US Food and Drug Administration.

Flavors add a sensory pleasure to human beings during the feeding process. Flavor has been defined as a mingled but unitary experience that includes the sensations of taste, smell, and pressure, and often cutaneous sensations such as warmth, color, or mild pain (Barrett et al. 2010). Flavor is further described by aroma and taste in which aroma is a volatile component sensed by receptors located in nose while taste is experienced with the help of taste receptors when substance is chewed. While color and appearance make an initial quality attribute for the acceptance of product to be taken orally, of all the three, flavor might have the largest impact on the acceptability of the product.

13.2 Sensory evaluations

Advanced sensory evaluation tools have been developed as a highly practicable tool for the flavor researchers. Sensory evaluation is the sensory science that is related with presentation of the stimulus (in this case a flavor compound or flavor) to the human subject and then evaluation of the human subject's response toward the stimulus. The response is encoded into numerical value and then statistically analyzed further. The sensory investigator then works together with the product development team to interpret the results and reach to a conclusion. Affective sensory evaluation is an important test method as it demonstrates the human perception toward the flavor and its acceptability. Test is carried out on large number of subjects (>50) to minimize the error in order to ameliorate the statistical robustness of the test method. Sensory science is a unique scientific discipline in which human subjects are required to carry out evaluations that pose certain challenges. Here training of the sensory investigator in number of fields is pre-requisite for successful sensory evaluations.

Analytical techniques for sensory evaluation include engagement of trained panellists. The techniques include discriminatory tests comprising of difference test like triangle test (O'Mahony and Goldstein 1986, de la Luz Reyes Vega et al. 1995, Radovich et al. 2004, Saavedra-Garcia et al. 2015) and threshold test like sub-threshold or additive effects (Harris 1960, Keith and Powers 1968, Meilgaard 1993, Wise and Breslin 2013, Zhang et al. 2016) and descriptive analysis (Lee and Chambers 2007).

Forss (1981) demonstrated the relationship between sensory evaluation and flavor chemical science, and further provided sensory evaluations of flavor ingredients. Moskowitz (1983) presented a study of psychophysics as a tool for sensory evaluation that was qualitative in nature. Delwiche (2009) also mentioned psychological considerations in sensory analysis. Psychophysics deals with the psychological response of the subject toward the physical stimulus (Snyder et al. 2006, Lahne 2016).

In this way of evaluation, flavor samples are brought to the laboratory where sensory evaluation has to be performed. Record of the flavor samples including name, amount of flavor, manufacturing date, manufacturer name and so on are kept and informed to the analyst. A holding sample is kept to work as a standard for the comparison of the new manufactured batch of flavor. Manufactured batch of flavor and standard are both evaluated for color, odor and flavor. Tasting medium utilized for the evaluation of flavors includes water, sugar syrup or sugar syrup plus citric acid solution. Water is used as tasting media for flavors like meat, onion, garlic, butter, coffee and other spice components. Sugar syrup is used as tasting media for caramel, banana, wintergreen, mint and vanilla flavors. Tasting media containing citric acid is used for tasting fruit flavors. Other tasting media are also utilized depending on the type of utilization. For example, vanilla flavor for dairy applications are evaluated in presence of milk. White sauce might be used to ascertain meat flavors. Vegetable oils might be used for tasting butter flavors. Thus, many other tasting media can be utilized to evaluate various types of flavors. Water utilized in the evaluation must be free of any other taste or odor. Rationale for using different types of tasting media lies in the cause that flavor should be tasted in the media which is analogous to the final finished product. Scores are allocated to variable tastes of flavor while evaluation and accordingly the flavor is categorized based on the score obtained (Clapperton 1974, Guenther et al. 2001, Barrett et al. 2010, Mamede et al. 2015). While evaluating the flavor, flavor character and flavor strength, which is equally important to be assessed.

13.2.1 Scaling methodology

A new method pertaining to scaling methodology includes magnitude estimation which refers to a class of psychophysical scaling method (Moskowitz 1977, Burseg et al. 2012). It is the simple method in which the panellists without any restriction designate a number to the response of the stimulus intensity, they experience while tasting a flavor sample. The ratios of the numbers denoted to the stimuli demonstrates ratio of sensory representation of humans. Before the use of the magnitude estimation method, investigators used an arbitrary scaling in order to quantify the stimuli response. In this the scale length varies and is denoted by qualitatively using words like low, medium, moderate, high, extreme, or by some other qualitative word. The acceptance testing of the flavor was evaluated on the basis of a 9-point hedonic scale for a set of attributes. These nine points include: 1, not present;

2, threshold; 3, slight; 4, slight to moderate; 5, moderate; 6, moderate to strong; 7, strong; 8, strong to extreme; and 9, extreme. Li et al. (2014) demonstrated an experiment to optimize a coffee-flavored dairy beverage using a nine-point hedonic scale method. However, by this later method there was chance of non-linearity in the results obtained from various panellists. Thus, in the subsequent time of research the scale has been adapted with more points corresponding to more distinction in the intensity of the stimulus experienced. Liggett et al. (2008) studied the impact of flavor properties using a 15-point scale for evaluation of liking swiss cheese product. Flavor attributes in 15-point scale was widely differentiated suggesting specificity of intensity of stimulus to be noted.

Triangular test method consists of three samples in which two are identical and one is different (**Figure 13.1**). The panellist needs to identify the different sample. The objective of the test method is to identify the odd sample and commit a preference. Comments on all the samples are recorded in the triangle test form, a sample form is shown in **Figure 13.2**.

Keith and Powers (1968) carried out threshold trials for determination of flavor thresholds based on concentration. This suggests an approach in which the panellists provide a score that represents the percentage of the concentration of flavor that was detectable, and very next low level of concentration was non-detectable.

Wise and Breslin (2013) demonstrated recognition thresholds for citric acid and sodium chloride.

Zhang et al. (2016) carried out threshold recognition trials for the sensory perception of Szechuan pepper. Results suggested that the response of pungency among assessors was dependent on concentration of pungent extract solution.

Gillette (1984) demonstrated applications of descriptive analysis for sensory evaluation. Descriptive analysis is a valuable approach which provides important data regarding aroma and is used efficiently for product and process development.

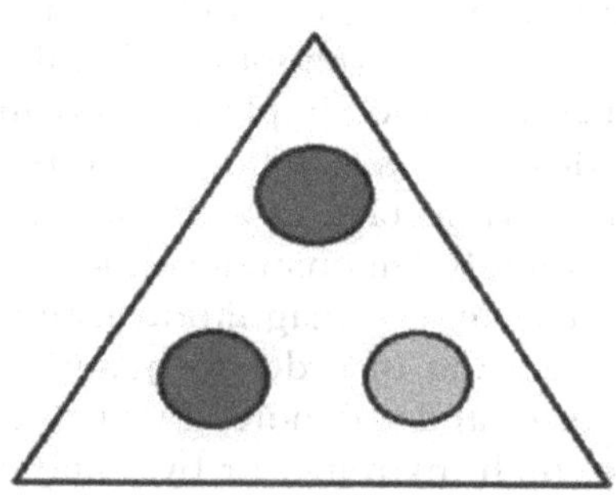

Figure 13.1 *Schematic representation of the triangular test method.*

	Triangle text typical form		
	Left	Center	Right
Triangle Set			
Different Sample			
Preference (1 or 2)			

Figure 13.2 *Example of a triangle test form.*

13.2.2 Color evaluation

Flavor analysis based on color evaluation, which is checked using Nessler's tube. The tube is filled with 50 mL of water and appropriate quantity of flavor component is added in water, both test and sample in each tube. Color is compared visually in both the tubes against white background and identified for any change. Result is noted and repeated for accuracy. **Figure 13.3** describes an example of changes in color with aging and use of Nessler's tube for color evaluation. Maraulja et al. (1973) demonstrated flavor deterioration in canned juices using Nesslers tube method.

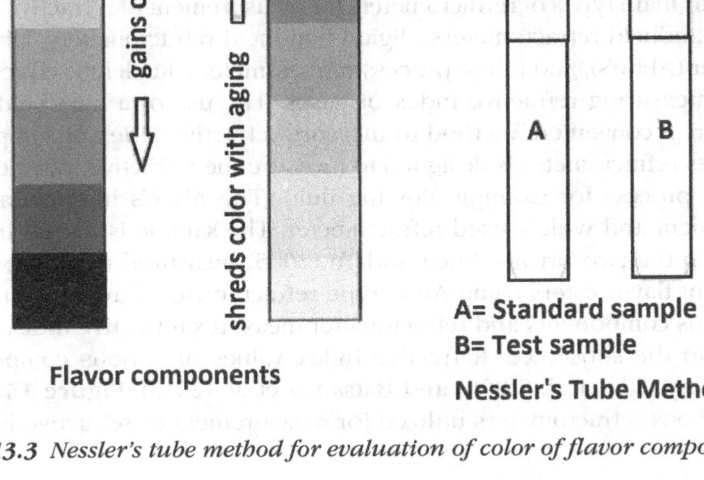

Figure 13.3 *Nessler's tube method for evaluation of color of flavor components.*

13.2.3 Sample rejection

The criteria for rejection of flavor sample depend on the difference of the color and flavor of the new flavor batch with the older retained sample batch. Evaluation of flavor considers flavor characteristics and strength as important parameters. If the flavor compound contains old juice materials then chances of difference in flavor and color is likely because of oxidation of citrus material leading to browning of compound. On rejection of the flavor sample with the older sample, the investigator will compare it with the first and second newest retained flavor sample. However, chances of mistakes by the compounder prevails the difference in flavor components, which may be accounted to use of older materials or wrong ratios of materials. Thus the compounder works on his own, leading to minimal chances of mistakes.

13.3 Analytical instrumentation methods

Instrumental analysis is carried out for evaluation of both physical and physiochemical characteristics of flavor components in flavor industry. Typical instruments are utilized for the purpose of evaluation. The aim of this analytical evaluation is to determine the composition of raw materials and to check whether the finished flavor product meets the specification.

13.3.1 Refractive index

Refractive index gives information regarding behavior of light passing through the sample. A refractometer is used to determine the index of refraction of light. It is one of the physicochemical properties of substances. It is ratio of velocity of light in the empty space to the velocity of light in the substance. When light travels through the substance, depending on the angle formed inside, the light bends according to refractive index of material. It is used to identify the material, confirm its purity, or measure concentration of substance (Koohyar 2012). There are four main types of refractometers for measurement of refractive indices: traditional handheld refractometers, digital handheld refractometers, laboratory refractometer (Abbe's), and inline process refractometers. Rayleigh refractometer is used for measuring refractive index of gases. The use of a handheld refractometer offers a convenient method to measure refractive index of samples. An inline process refractometer is designed to measure the refractive index continuously in the process for example flowing fluid. The Abbe's instrument is the most convenient and widely used refractometer. The sample is placed in a thin layer between the two prisms. Sheu and Tu (2005) measured refractive indices of 12 different flavor esters using Abbe type refractometer. Flavor material consists of various components and refractometer measures refractive index of each component in the substance. Refractive index values of various compound is documented by Lide et al. (2005) and Batsanov et al. (2016). **Figure 13.4** demonstrates various refractometers utilized for measurement of refractive index.

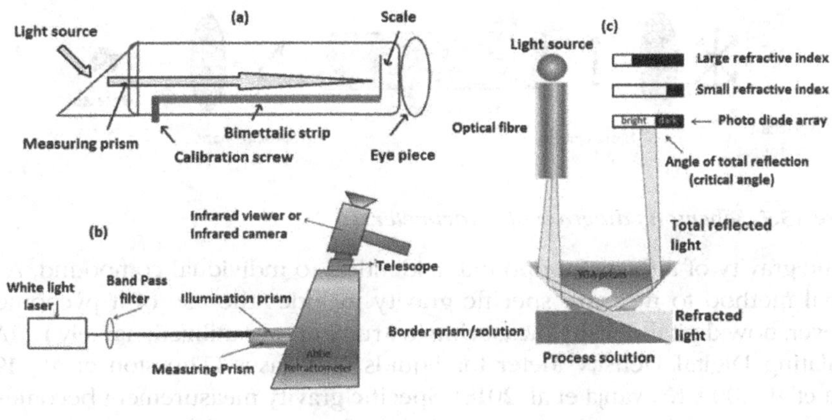

Figure 13.4 *Schematic diagram of (a) handheld refractometer, (b) Abbe's refractometer and (c) inline process refractometer.*

13.3.2 Optical rotation

Optical rotation (optical activity) is also referred as rotatory polarization. It is the efficiency of the substance to rotate the plane of polarization of linearly polarized light. It is because of a chiral center in the molecule; an asymmetric carbon. Optical activity is measured using polarimeter (**Figure 13.5**). Substances that possess such power to rotate the plane of polarization of light are known as optically active compounds. Substances that rotate the plane of polarization of light toward the right direction (clockwise) are called dextro-rotatory substances (+), while substances that rotate the plane of polarization of light toward left direction (anti-clockwise) are called levo-rotatory substances (–). A mixture containing both types of substances in equal proportion is found to be optically inactive known as racemic mixture. Optical rotation is the check of purity of the flavor substance as it is always the same for a pure compound in standard conditions of measurement (Mosandl and Guenther, 1989; Kubota et al. 1999, Saito et al. 2011, Aljaff et al. 2013, Eibler and Vetter, 2017). It gives more specific information not provided by general refractive index measurement. If any impurity is present in the test compound, then it will interfere in the optical activity thus presence of impurity can be claimed. It also gives idea of adulteration of flavor compound mixture as synthetic flavors, which are produced from petroleum sources and exist as a racemic mixture with an equal proportion of both dextro-rotatory and levo-rotatory substances, while natural flavor compounds have a chiral carbon and have a net rotation of light in either left or right direction (Armstrong et al., 1990).

13.3.3 Specific gravity

Specific gravity is the measurement of density of the substance relative to that of water density. Thus, it is also known as relative density. Specific gravity of flavor compound is the function of its components and their proportions. Thus

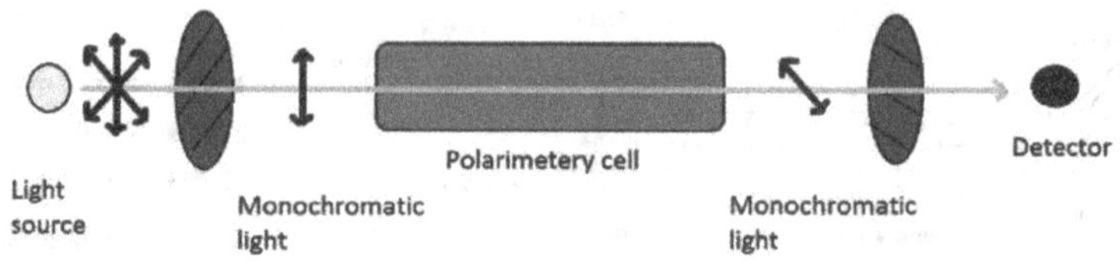

Figure 13.5 *Schematic diagram of polarimeter.*

specific gravity of a flavor compound is identical to individual compound. A traditional method to measure specific gravity includes the use of a pycnometer. However, now digital-density measuring instruments are utilized, namely DMA 45 Calculating Digital Density Meter for liquids and gases (Thurston et al., 1981, Vajjha et al. 2009, Kravanja et al. 2018). Specific gravity measurement becomes an important test method as it gives a unique information of flavor substance and can identify the presence of any adulteration. For example, specific gravity of grape fruit essential oil is between 0.848 and 0.856 g/mL at 25°C. If it is adulterated then it might be difficult to identify by odor and taste, but its specific gravity values will vary from the original value, thus adulteration can be identified.

13.3.4 Melting point and congealing point

Substances existing as solid and liquid at room temperature are evaluated for melting point and congealing point respectively. The melting point and congealing point are constant for respective substances. Generally crystalline substances shows sharp melting point whereas amorphous molecules have melting point in range. Dihydrocoumarin has a congealing point of 23°C–23.6°C. Impurities present in the substance will interfere with the structure of the molecule and thereby it will vary the melting or congealing point. Thus, measurement of melting or congealing point is an identification test method for presence of impurities in pure compound.

13.4 Methods for flavor emulsions

The flavor industry produces some flavor in forms of emulsions, thus cloud point becomes an important parameter indicating stability of prepared emulsion. Emulsions are thermodynamically unstable systems, flavor and water separates out causing ringing. Ringing can be checked visually by keeping the flavor emulsions for about two weeks in an upright position. Separation of two phases forming ringing explicates unstable emulsion formed (Taherian et al. 2006).

Another method to determine stability includes measurement of cloud point temperature (Tan and Holmes 1988, Souza et al. 2009, Linke and Drusch 2016). Cloud point is the maximum temperature at which the emulsion does not break and remains stable. It is one type of stress stability testing of emulsion. At that maximum temperature emulsion appears turbid. This turbidity depends upon particle/globule size, refractive index of dispersed phase and wavelength of light subjected.

Centrifugation of emulsion is a rapid method to determine its stability. However, centrifugation method does not always relate with the stability of flavor emulsion in real-time conditions (Tse and Reineccius 1997). Emulsion may get coalesce due to high shear rate of centrifugation leading to instability.

Turbidity of emulsion can be checked at wavelength of 400 nanometer or by taking ratio of absorption at wavelength 800 and 400 nanometer (Horie et al. 1978, Kauffman and Garti 1981, Kauffman and Garti 1984, Linke and Drusch 2016, Song et al. 2002, Rahn-Chique et al. 2012, Chang and McClements 2014). Turbidity is directly related with the particle/globule size of emulsion.

Particle/globule size of emulsion can be measured directly by microscope. Alternative methods to determine particle/globule size include an electron microscope and/or coulter counter method (Chilton and Laws 1980, Hillbrick et al. 1998, Hu et al. 2017). Particle/globule size of emulsion is one of the important parameter for a stable emulsion. **Figure 13.6** demonstrates schematic diagram of coulter counter and electron microscope technique.

13.4.1 Volatile oil content

Various flavor contains volatile oils which also determines the flavor strength. An example is various spices contain volatile oil, which is responsible for their particular flavor. Spray-dried volatile oils are commonly evaluated for content of volatile oil in order to determine the efficiency of encapsulation procedure and concentration of flavoring material in the compound. Volatile oil is generally measured using a distillation technique. The Clevenger trap is used for the measurement of volatile oil content in water insoluble flavor compounds (**Figure 13.7**).

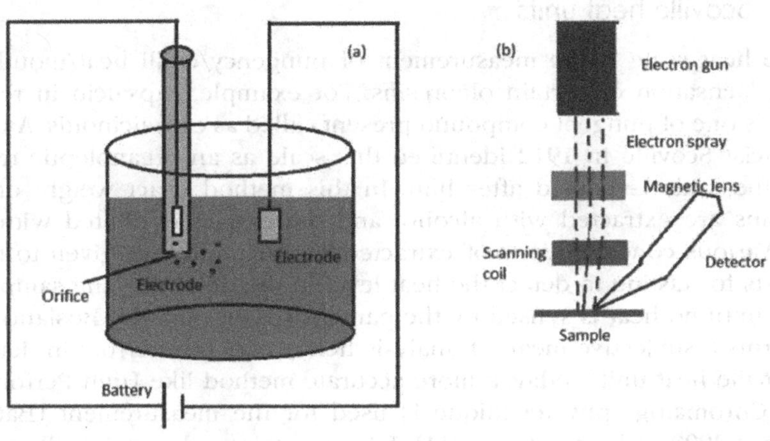

***Figure 13.6** Schematic diagram of (a) Coulter counter method, (b) electron microscope for particle size measuring.*

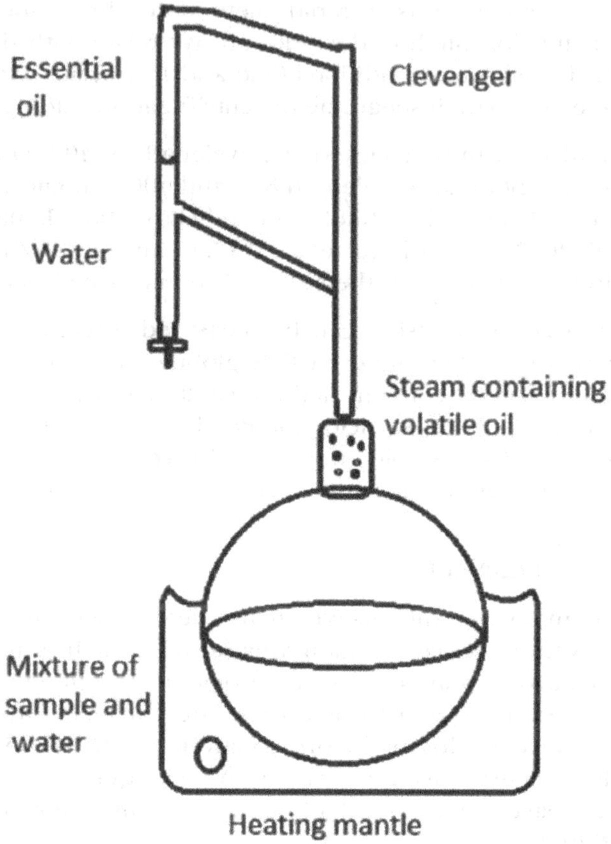

Figure 13.7 *Schematic diagram of Clevenger apparatus for volatile oil content.*

13.4.2 Scoville heat units

Scoville heat scale is the measurement of pungency/chilli heat/mouth bite/burning sensation of certain oleoresins. For example, capsaicin in red chili pepper is one of pungent compound present called as capsaicinoids. American pharmacist Scoville in 1912 identified this scale as an organoleptic test and hence the scale is named after him. In this method exact weight of dried oleoresins are extracted with alcohol and the extract is diluted with sugar water. Various concentrations of extracted capsaicinoids are given to trained panellists for tasting to detect the heat level in the dilutions. The samples are diluted until no heat is sensed by the panellists (Guzman and Bosland 2017). This forms a subjective mean of analysis hence may pose errors in determining Scoville heat unit. Today, a more accurate method like High Performance Liquid Chromatography technique is used for the measurement (Batchelor and Jones 2000, Othman et al. 2011). Jalapeno pepper has a Scoville score of 2500–5000). The heat level is based on dilution in multiples of 100 Scoville Heat Unit (SHU) (Peter 2012).

13.4.3 Gas chromatography

Chromatography is a physical technique for separation of components based on their affinity toward stationary and mobile phase. The difference in migration rate of components depends on differential adsorption, partition coefficients, ion-exchange, molecular sieving effect and so on. Forces like electrostatic between dipoles, vander Waals forces, inherent forces are also responsible for the separation of the components. Gas chromatography also known as gas liquid chromatography, which implies a liquid sample first needs to be vaporized and then injected onto the head of the chromatographic column. The sample is transported across the column with the help of inert, mobile gas phase. Commonly used inert gas mobile phase includes nitrogen, helium, argon and/or carbon dioxide. Selection of mobile gas phase depends on the choice of type of detector to be used. Flavor analysis by gas chromatography is associated with only volatile or semi-volatile components of flavor. Utilization of gas chromatography in flavor analysis is to identify individual flavor molecules at trace levels (Smith and Coffman 1960, Konig and Hochmuth 2004, Zellner et al. 2008, Kiene et al. 2015, Sfakianakis and Tzia 2017). Various types of gas chromatography methods have been developed for flavor analysis including gas chromatography combined with mass spectrometry, temperature programmed capillary gas chromatography and gas chromatography with olfactometry. **Figure 13.8** describes a schematic diagram of typical gas chromatography technique.

13.4.4 Mass spectrometry

Mass spectrometric method is an analytical technique that ionizes the component and sorts the ions according to their mass to charge ratios. A mass spectrum measures the mass in a sample. A mass spectrum is a plot of ion

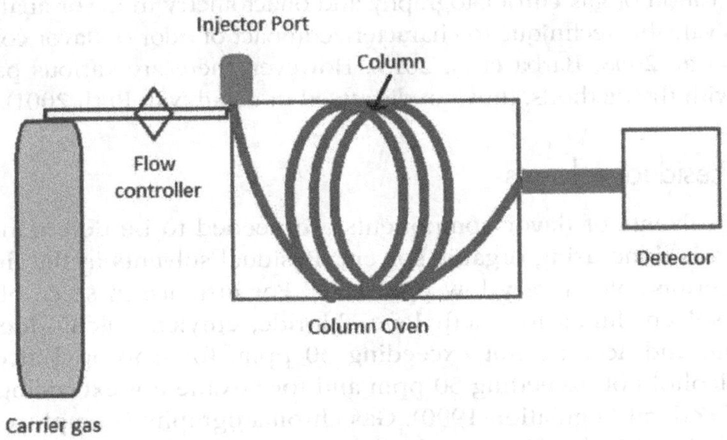

***Figure 13.8** Schematic diagram of gas chromatography method.*

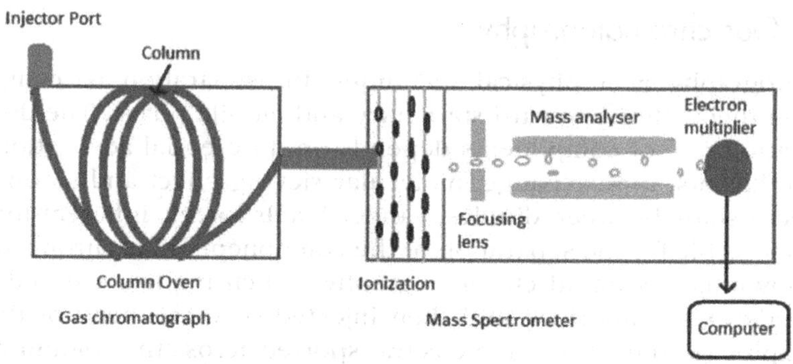

Figure 13.9 *Schematic diagram of gas chromatograph with mass spectrometer.*

signal as a function of the mass to charge ratio. Use of mass spectrometry in analysis of citrus flavors has been associated with the use of gas chromatography separation of the components. The separated components are then analyzed by mass spectrometry to identify the components on the basis of their masses (Issenberg et al. 1969, Shaw and Moshonas 1985, Zhang and Lee 1997, Begnaud and Chaintreau 2016, Jiao et al. 2017, Sakakura 2018). **Figure 13.9** demonstrates a typical schematic diagram of mass spectrometry analysis.

13.4.5 Olfactometry

Olfactometry in tie-up with gas chromatography is used to measure odor active compounds associated with taste characteristics in flavor analysis (Zellner et al. 2008, Barba et al. 2018). Gas chromatography with olfactometry (GC-O) was proposed by Fuller et al. (1964). The method is capable to measure the odor active compounds at very low concentrations. GC-O is limited to screen only for odor active volatile compounds. Methods for GC-O include dilution analysis, detection frequency methods, posterior intensity methods and time intensity methods. The application of gas chromatography and olfactometry in flavor analysis demonstrates valuable technique to characterize impact of odor of flavor compounds (Zellner et al. 2008, Barba et al. 2018). However, there are various parameters unclear with the methods, thus not discussed in detail (van Ruth 2001).

13.4.6 Residual solvents

Residual solvents of flavor components are needed to be determined when extraction is done using organic solvent. Residual solvents in the final product are permissible at very low ppm level. For instance in spice oleoresins, residual solvent limits for methylene chloride, ethylene dichloride, trichloroethylene and acetone not exceeding 30 ppm, for isopropyl alcohol and methyl alcohol not exceeding 50 ppm and for hexane not exceeding 25 ppm (Code of Federal Regulation 1990). Gas chromatography is employed for the detection of residual solvents in the final extracted flavor compound (Rayner et al. 1978, Uematsu et al. 1994, Uematsu et al. 1997).

13.4.7 Surface oil content

Surface oil is associated when flavor encapsulation particularly of essential oil is employed. This oil should be measured as it is likely to undergo oxidation and produce off flavor in final product. The method to determine surface oil is gravimetric. Organic solvent like hexane is added in the flavor compound and shaken for several minutes. Solvent is removed and residue is determined by simple weighing method (Anandaraman and Reineccius 1987, Kausadikar et al. 2015, Hermanto et al. 2016).

13.4.8 Moisture content

Moisture content is a concern for dried flavor components since flavor compounds contain volatile oils with some amount of moisture (Morgano et al. 2011). Here evaporative methods are not recommended for moisture content measurement. The Fischer method (1935) is utilized for the determination of moisture content. The method is based on the mechanism of chemical reaction between water and pyridine/iodine complex.

13.4.9 Electronic nose and electronic tongue

Electronic nose (e-nose) and electronic tongue (e-tongue) roughly mimic the human smell and taste senses (Wilson and Baietto 2009, Latha and Lakshmi 2012). Typical sensors to identify the odor and taste are employed in e-nose and e-tongue respectively (Cole et al. 2011). The stages of identification action are similar to that of human nose and tongue. Some type of signal is generated from the sensor, which determines patterns in data. Further the data is stored and quantified for the comparison of one sample from another based on headspace volatile compounds. Many developments have taken place in the devices that can fulfill industrial needs (Baldwin et al. 2011).

13.4.10 Microbiology test

Microbiology test finds not much place in flavor development. However, liquid flavor or flavor containing moisture need to be studied for microbial test. The test generally includes in-house methods like plate count using yeasts or moulds.

Overall the quality control laboratory carries out various tests on flavors during and after development of final product. Even quality control has to investigate any problems that may arise during process of preparation or post production. Problems likely to occur with the final preparation are adulteration of flavor, color changes in flavor, or flavor changes. A brief overview of various tests performed on flavor during and after preparation is tabulated in **Table 13.1**. However, the tests are not limited to the mentioned list; a quality control laboratory may have different in house procedurals methods to evaluate the flavors.

Table 13.1 Brief View of Various Tests Performed on Quality Control of Flavors

Test	Method/Type
Appearance	Visual
Ash value	Total, acid-insoluble, water soluble
Extractive material	Alcohol soluble, ether soluble, any other solvent soluble
Moisture content	Karl fischer, xylene co distillation method, oven drying
Volatile oil content	Distillation technique like Clavenger trap
Microscopic examination	Compound Microscope
Limit tests	As specified in Pharmacopoeia
Specific tests	Color index, iodine value, foreign matter, scoville pungency, lead number
Melting point, congealing point, boiling point	°C As specified in Pharmacopoeia
Specific gravity	At 25°C
Refractive index	Refractometer
Optical rotation	Polarimeter
Flash point	Open or closed cup (°C)
Solubility	In alcohol, in water
Adulteration/identification	Gas chromatography, infra red spectroscopy, UV-visible spectrometry
Specific tests for constituents	Acid value, heavy metals, aldehyde, ketone, chlorinated compounds, esters, phenol content
For citrus oils	Evaporation residue, peroxide value
Residual solvents	Gas chromatography
For emulsion	Particle/globule size, ringing test, centrifugation, turbidity, cloud point
Microbial test	Using plate count
Sensory evaluations	Triangle test, descriptive analysis, differential analysis

References

Aljaff, P., Rasheed, B.O., Omer, R.A. 2013. A comparison between natural and synthetic food flavoring extracts using infrared spectra and optical activity. *IOSR Journal of Applied Physics* 5 (3): 1–6.

Anandaraman, S., Reineccius, G.A. 1987. Analysis of encapsulated orange peel oil. *Perfume and Flavorist* 12 (2): 33–39.

Armstrong, A.W., Chang, C.D., Li, W.Y. 1990. Relevance of enantiomeric separations in food and beverage analyses. *Journal of Agricultural and Food Chemistry* 38 (8):1674–1677.

Baldwin, E.A., Bai, J., Plotto, A., Dea, S. 2011. Electronic noses and tongues: Applications for the food and pharmaceutical industries. *Sensors (Basel)* 11 (5): 4744–4766.

Barba, C., Beno, N., Guichard, E., Thomas-Danguin, T. 2018. Selecting odorant compounds to enhance sweet flavor perception by gas chromatography/olfactometry-associated taste (GC/O-AT). *Food Chemistry* 257: 172–181.

Barrett, D.M., Beaulieu, J.C., Shewfelt, R. 2010. Color, flavor, texture, and nutritional quality of fresh cut fruits and vegetables: Desirable levels, instrumental and sensory measurement, and the effects of processing. *Critical Reviews in Food Science and Nutrition* 50 (5): 369–389.

Batchelor, J.D., Jones, B.T. 2000. Determination of the Scoville heat value for hot sauces and chilies: An HPLC experiment. *Journal of Chemical Education* 77 (2): 266.

Batsanov, S.S., Ruchkin, E.D., Poroshina, I.A. 2016. *Refractive Indices of Solids*. Springer, Singapore, doi:10.1007/978-981-10-0797-2.

Begnaud, F., Chaintreau, A. 2016. Good quantification practices of flavors and fragrances by mass spectrometry. *Philosophical Transactions. Series A, Mathematical, Physical, and Engineering Sciences* 374 (2079): 20150365.

Burseg, K.M.M., Camacho, S.M., & Bult, J.H.F. 2012. Taste enhancement by pulsatile stimulation is receptor based but independent of receptor type. *Chemosensory Perception* 5 (2): 179–187.

Chang, Y., McClements, D.J. 2014. Optimization of orange oil nanoemulsion formation by isothermal low-energy methods: Influence of the oil phase, surfactant, and temperature. *Journal of Agricultural and Food Chemistry* 62 (10): 2306–2312.

Chilton, H.M., Laws, D.R.J. 1980. Stability of aqueous emulsions of the essential oil of hops. *Journal of the Institute of Brewing* 86: 126–130.

Clapperton, J.F. 1974. Profile analysis and flavor discrimination. *Journal of the Institute of Brewing* 80: 164–173.

Code of Federal Regulations, (1990). https://www.accessdata.fda.gov/scripts/cdrh/cfdocs/cfcfr/CFRSearch.cfm?CFRPart=173&showFR=1&subpartNode=21:3.0.1.1.4.3 (accessed April 4, 2018).

Cole, M., Covington, J.A., Gardner, J.W. 2011. Combined electronic nose and tongue for a flavor sensing system. *Sensors and Actuators B: Chemical* 156 (2): 832–839.

de la Luz Reyes Vega, M., Concepción Fahara Valdés, M., Valdés Silva, I.C., Saucedo Salazar, C.T., Peralta Rodriguez, R.D. 1995. Application of sensory evaluation triangle tests for quality control of liquid antacids. *Drug Development and Industrial Pharmacy* 21:10, 1203–1210.

Delwiche, J. 2009. Psychological considerations in sensory analysis. In *The Sensory Evaluation of Dairy Products*, ed. Clark, S., Costello, M., Drake, M.,Bodyfelt, F., pp. 7–15. Springer-Verlag, New York.

Eibler, D., Vetter, W. 2017. Enantioseparation and optical rotation of flavor-relevant 4–alkyl-branched fatty acids. *Journal of Chromatography A* 1505: 87–95.

Fischer, K. 1935. Neues Verfahren zur maβanalytischen Bestimmung des Wassergehaltes von Flüssigkeiten und festen Körpern. *Angewandte Chemie* 48: 394–396. doi:10.1002/ange.19350482605.

Forss, D.A. 1981. Sensory characterization. In *Flavor Research Recent Advances*, ed. Teranishi, R., Flath, R.A., Sugisawa, H., pp. 125–174. Marcel Dekker, New York.

Fuller, G.H., Steltenkamp, R., Tisser and, G.A. 1964. The gas chromatograph with human sensor: Perfumer model. *Annals of the New York Academy of Sciences* 116: 711–724.

Gillette, M. 1984. Applications of descriptive analysis. *Journal of Food Protection* 47: 403–409.

Guenther, S.E., Junkins, E.P., Corneli, H.M., Schunk, J.E. 2001. Taste test: Children rate flavoring agents used with activated charcoal. *Archives of Pediatrics and Adolescent Medicine* 155 (6): 683–686.

Guzman, I., Bosland, P.W. 2017. Sensory properties of chile pepper heat e and its importance to food quality and cultural preference. *Appetite* 117: 186–190.

Harris, E.K. 1960. Analysis of experiments measuring threshold taste. *Biometrics* 16: 245–260.

Hermanto, R.F., Khasanah, L.U., Atmaka, W., Manuhara, G.J., Utami, R. 2016. IOP Conference Series: Materials Science and Engineering 107, 012064. doi:10.1088/1757-899X/107/1/012064 (accessed May 24, 2018).

Hillbrick, G.C., McMahon, D.J., Deeth, H.C. 1998. Electrical impedance particle size method (Coulter Counter) detects the large fat globules in poorly homogenised UHT processed milk. *Australian Journal of Dairy Technology* 53 (1): 17–21.

Horie, K., Tanaka, S., and Akabori, T. 1978. Determination of emulsion stability by spectral absorption I: Relationship between surfactant type, concentration, and stability index. *Cosmetics and Toiletries* 93 (5): 53–62.

Hu, Y.T., Ting, Y., Hu, J.Y., Hsieh, S.C. 2017. Techniques and methods to study functional characteristics of emulsion systems. *Journal of Food and Drug Analysis* 25 (1): 16–26.

Issenberg P., Kobayashi, A., Mysliwy, T.J. 1969. Combined gas chromatography-mass spectrometry in flavor research. Methods and applications. *Journal of Agricultural and Food Chemistry* 17 (6): 1377–1386.

Jiao, A., Xu, X., Jin, Z. 2017. Study of the flavor volatiles in double-fried pork: Identification and analysis of their changes during frozen storage using headspace solid-phase microextraction and gas chromatography-mass spectrometry. *Journal of Food Technology and Preservation* 1 (1): 47–55.

Kauffman, V.R., Garti, N. 1981. Spectral absorption measurements for determination of ease of formation and stability of oil-in water emulsion. *Journal of Dispersion Science and Technology* 2 (4): 475–490.

Kauffman, V.R., Garti, N. 1984. Effect of cloudy agents on the stability and opacity of cloudy emulsions for soft drinks. *Journal of Food Science and Technology* 19: 255–261.

Kausadikar, S., Gadhave, A.D., Waghmare, J. 2015. Microencapsulation of lemon oil by spray drying and its application in flavour tea. *Advances in Applied Science Research* 6 (4): 69–78.

Keith, E.S., Powers, J.J. 1968. Determination of flavor threshold levels and sub-threshold levels, additive, and concentration effects. *Journal of Food Science* 33: 213–218.

Kiene, F.E., Pein, M., Thommes, M. 2015. Orientation to determine quality attributes of flavoring excipients containing volatile molecules. *Journal of Pharmaceutical and Biomedical Analysis* 110: 20–26.

Konig, W.A., Hochmuth, D.H. 2004. Enantioselective gas chromatography in flavor and fragrance analysis: Strategies for the identification of known and unknown plant volatiles. *Journal of Chromatographic Science* 42 (8): 423–39.

Koohyar, F. 2012. Refractive index and its applications. *Journal of Thermodynamics and Catalysis* 4: 117.

Kravanja, G., Knez, Z., Knez, M.H. 2018. Density, interfacial tension, and viscosity of polyethylene glycol 6000 and supercritical CO2. *The Journal of Supercritical Fluids* 39: 72–79. doi:10.1016/j.supflu.2018.05.012.

Kubota, K., Someya, Y., Kurobayashi, Y., Kobayashi, A. 1999. Flavor characteristics and stereochemistry of the volatile constituents of greater galangal (*Alpinia galanga* Willd.). In *Flavor Chemistry of Ethnic Foods*, ed. Shahidi, F., Ho, C.T., pp. 97–104, Springer, Boston, MA.

Lahne, J. 2016. Sensory science, the food industry, and the objectification of taste. *Anthropology of Food* 10: 2–14.

Latha, R.S., Lakshmi, P.K. 2012. Electronic tongue: An analytical gustatory tool. *Journal of Advanced Pharmaceutical Technology and Research* 3 (1): 3–8.

Lee, J., Chambers, D.H. 2007. A Lexicon for flavor descriptive analysis of green tea. *Journal of Sensory Studies* 22: 256–272.

Li, B., Hayes, J.E., Ziegler, G.R. 2014. Just-About-Right and ideal scaling provide similar insights into the influence of sensory attributes on liking. *Food Quality and Preference* 37: 71–78.

Lide, D.R. 2005. Physical constants of organic compounds. In *CRC Handbook of Chemistry and Physics*, 86th ed., CRC Press, Boca Raton, FL.

Liggett, R.E., Drake, M.A., Delwiche, J.F. 2008. Impact of flavor attributes on consumer liking of swiss cheese. *Journal of Dairy Science* 91: 466–476.

Linke, C., Drusch, S. 2016. Turbidity in oil-in-water-emulsions: Key factors and visual perception. *Food Research International* 89: 202–210.

Mamede, M.E.O., Kalschne, D.L., Santos, A.P.C., Benassi, M.T. 2015. Cajá-flavored drinks: A proposal for mixed flavor beverages and a study of the consumer profile. *Food Science and Technology* 35 (1): 143–149.

Maraulja, M.D., Blair, J.S., Olsen, R.W., Wenzel, F.W. 1973. Furfural as an indicator of flavor deterioration in canned citrus juices. *Florida State Horticultural Society* 270–275.

Meilgaard, M.C. 1993. Individual differences in sensory threshold for aroma chemicals added to beer. *Food Quality and Preference* 4 (3): 153–167.

Morgano, M.A., Milani, R.F., Martins, M.C.T., Rodriguez-Amaya, D.B. 2011. Determination of water content in Brazilian honeybee-collected pollen by Karl Fischer titration. *Food Control* 22 (10): 1604–1608.

Mosandl, A., Guenther, C. 1989. Stereoisomeric flavor compounds. 20. Structure and properties of .gamma.-lactone enantiomers. *Journal of Agricultural and Food Chemistry* 37: 413–418.

Moskowitz, H.R. 1977. Magnitude estimation: Notes on what, how, when, and why to use it. *Journal of Food Quality* 3: 195–227.

Moskowitz, H.R. 1983. *Product Testing and Sensory Evaluation of Foods*. Food and Nutrition Press, Westport, CT.

O'Mahony, M., Goldstein, L.R. 1986. Effectiveness of sensory difference tests: Sequential sensitivity analysis for liquid food stimuli. *Journal of Food Science* 51: 1550–1553.

Othman, Z.A.A., Ahmed, Y.B.H., Habila, M.A., Ghafar, A.A. 2011. Determination of capsaicin and dihydrocapsaicin in capsicum fruit samples using high performance liquid chromatography. *Molecules* 16: 8919–8929.

Peter, K.V. 2012. Handbook of herbs and spices. In *Food Science, Technology and Nutrition*, Vol. 1, Woodhead Publishing, Philadelphia, PA.

Radovich, T.J.K., Kleinhenz, M.D., Delwiche, J.F., Liggett, R.E. 2004. Triangle tests indicate that irrigation timing affects fresh cabbage sensory quality. *Food Quality and Preference* 15: 471–476.

Rahn-Chique, K., Puertas, A.M., Romero-Cano, M.S., Rojas, C., Urbina-Villalba, G. 2012. Nanoemulsion stability: Experimental evaluation of the flocculation rate from turbidity measurements. *Advances in Colloid and Interface Science* 178: 1–20.

Rayner, E.T., Wadsworth, J.I., Legendre, M.G., Dupuy, H.P. 1978. Analysis of flavor quality and residual solvent of soy protein products. *Journal of the American Oil Chemists' Society* 55 (5): 454–458.

Saavedra-Garcia, L., Bernabe-Ortiz, A., Gilman, R.H., Diez-Canseco, F., Cárdenas, M.K., Sacksteder, K.A., Miranda, J.J. 2015. Applying the triangle taste test to assess differences between low sodium salts and common salt: Evidence from peru. *PLoS ONE* 10 (7): e0134700. doi:10.1371/journal.pone.0134700.

Saito, K., Hosono, K., Kitazawa, N., Iwasaki, Y., Ito, R., Nakazawa, H. 2011. Development of quality assessment method for optically active food flavor chemicals. *Journal of AOAC International* 94: 923–930.

Sakakura, M. 2018. Application development of flavor analysis using real time mass spectrometry. *Journal of the Mass Spectrometry Society of Japan* 66 (1): 15–24.

Scoville, W.L. 1912. Note on capsicum. *Journal of the American Pharmacists Association* 1: 453–454.

Sfakianakis, P., Tzia, C. 2017. Flavour profiling by gas chromatography–mass spectrometry and sensory analysis of yoghurt derived from ultrasonicated and homogenised milk. *International Dairy Journal* 75: 120–128.

Shaw, P.E., Moshonas, M.G. 1985. Mass spectrometry for the identification of citrus flavor components. *Mass Spectrometry Reviews* 4 (4): 397–420.

Sheu, Y.W., Tu, C.H. 2005. Densities, viscosities, refractive indices, and surface tensions for 12 flavor Esters from T = 288.15 K to T = 358.15 K. *Journal of Chemical & Engineering Data* 50: 1706–1710.

Smith, D.E., Coffman, J.R. 1960. Analysis of food flavors by gas-liquid chromatography. Separation and identification of the neutral components from bread pre-ferment liquid. *Analytical Chemistry* 32 (13): 1733–1737.

Snyder, D.J., Prescott, J., Bartoshuk, L.M. 2006. Modern psychophysics and the assessment of human oral sensation. In *Taste and Smell, An Update* (Advances in Otorhinolaryngology), ed., Hummel, T., Welge-Lüssen, A., 63: 221–241, Karger, Basel, Switzerland.

Song, M.G., Cho, S.H., Kim, J.Y., Kim J.D. 2002. Novel evaluation method for the water-in-oil (w/o) emulsion stability by turbidity ratio measurements. *Korean Journal of Chemical Engineering* 19 (3): 425–430.

Souza, V.B., Spinelli, L.S., Gonzalez, G., Mansur, C.R.E. 2009. Determination of the phase inversion temperature of orange oil/water emulsions by rheology and microcalorimetry. *Journal of Analytical Letters* 42 (17): 2864–2878.

Taherian, A.R., Fustier, P., Ramaswamy, H.S. 2006. Effect of added oil and modified starch on rheological properties, droplet size distribution, opacity and stability of beverage cloud emulsions. *Journal of Food Engineering* 77: 687–696.

Tan, C.T., Holmes, J.W. 1988. Stability of beverage flavor emulsions. *Perfumes and Flavorist* 13 (1): 23–41.

Thurston, P.A., Taylor, R., Ahvenainen, J. 1981. Effects of linoleic acid supplements on the synthesis by yeast of lipids and acetate esters. *Journal of the Institute of Brewing* 87: 92–95.

Tse, K.Y., Reineccius, G.A. 1997. Methods to predict the physical stability of flavor: cloud emulsion. In *Flavor Technology Physical Chemistry, Modification, and Process*, ed. Ho, C.T., Tan, C.T., Tong, C.H., pp. 172–182. American Chemical Society, Washington, DC.

Uematsu, Y., Hatooka, Y., Hirata, K., Hirokado, M., Kazama, M. 1994. Determination of residual organic-solvents in flavor by standard addition headspace gas-chromatography. *Journal of the Food Hygienic Society of Japan* 35 (6): 645–651.

Uematsu, Y., Sadamasu, Y., Hirata, K., Hirokado, M., Yasuda, K., Suzuki, S. 1997. Analysis of residual solvents in flavor preparations by head-space gas chromatography. *Food Hygiene and Safety Science* 38 (2): 97–104.

Vajjha, R.S., Das, D.K., Mahagaonkar, B.M. 2009. Density measurement of different nanofluids and their comparison with theory. *Petroleum and Science Technology* 27 (6): 612–624.

van Ruth, S.M. 2001. Methods for gas chromatography-olfactometry: A review. *Biomolecular Engineering* 17: 121–128.

Wilson, A.D., Baietto, M. 2009. Applications and advances in electronic-nose technologies. *Sensors* 9: 5099–5148.

Wise, P.M., Breslin, P.A.S. 2013. Individual differences in sour and salt sensitivity: Detection and quality recognition thresholds for citric acid and sodium chloride. *Chemical Senses* 38 (4): 333–342.

Zellner, B.A., Dugo, P., Dugo, G., Mondello, L. 2008. Gas chromatography: Olfactometry in food flavour analysis. *Journal of Chromatography A* 1186 (1–2): 123–143.

Zhang, H.Z., Lee, T.C. 1997. Gas chromatography-mass spectrometry analysis of volatile flavor compounds in Mackerel for assessment of fish quality. In *Flavor and Lipid Chemistry of Seafoods*, ed. Shahidi, F., Cadwallader, K.R., pp. 55–63, American Chemical Society, Washington, DC.

Zhang, L.L., Zhao, L., Wang, H.Y., Zhi, R.C., Shi, B.L., Xie, N. 2016. Determination of recognition threshold and just noticeable difference in the sensory perception of pungency of *Zanthoxylum bangeanum*. *International Journal of Food Properties* 19:1044–1052.

14

Legislation for Natural and Synthetic Flavors Used in Functional Foods and Nutraceuticals

Priyanka Bhatt, Priya Narvekar, Gulimirerouzi Fnu,
Dewi Melani Hariyadi, Umi Athiyah, and Yashwant V. Pathak

Contents

14.1 Introduction

14.1.1 Functional foods

There is no universally accepted definition for functional foods; all foods in general are functional to some extent since they provide taste, aroma and nutritive value. However, functional foods are considered those foods that apart from providing the usual benefits also provide physiologic benefits, reduce risk of having chronic diseases and are useful to optimize health. The functional foods concept was first coined in Japan in the 1980s by the Ministry of Health and Welfare for improving the health of the nation's aging population by initiating regulatory system for approving certain foods that have health benefits (Arai 1996). These foods are now recognized as Foods for Specified Health Use (FOSHU) and contain a special seal. To date, Japan is the only country that has formulated a specific regulatory approval process for functional foods. Currently, 100 products are licensed as FOSHU foods in Japan (Hasler 2002).

There is no regulatory identity of functional foods in United States. Irrespective of this, many organizations have proposed definitions for this new and emerging area of the food and nutrition sciences. According to the Food and Nutrition Board of National Academy of Sciences, functional foods are "any modified food or food ingredient that may provide a health benefit beyond the traditional nutrients it contains" (Medicine 1994).

The International Life Sciences Institute defines them as "foods that, by virtue of the presence of physiologically active components, provide a health benefit beyond basic nutrition" (1999). According to American Dietetic Association (Hasler 2002), functional foods are the foods which is "whole, fortified, enriched, or enhanced," but more importantly, it states that such foods must be consumed as "… part of a varied diet on a regular basis, at effective levels" for consumers to enjoy their potential health benefits (Hasler and Brown 2009).

14.1.2 Nutraceuticals

In 1989, Stephen DeFelice, MD, founder and chairman of the Foundation for Innovation in Medicine (FIM), Cranford, NJ, coined the term "nutraceutical" which was from "nutrition" and "pharmaceutical." He defined nutraceutical as "a food (or part of a food) that provides medical or health benefits, including the prevention and/or treatment of a disease" (Brower 1998). Functional foods are called nutraceutical, when they aid in the prevention and/or treatment of disease(s) and/or disorder(s) other than anemia. Development of the functional food concept and nutraceuticals are based on dietary components, as they play beneficial roles beyond basic nutrition (Laparra and Sanz 2010). Nutraceutical, term commonly used in marketing, but like functional foods it does not have any regulatory definition. The following points make the nutraceuticals differ from dietary supplements—(1) nutraceuticals, other than supplement the diet must also aid in the prevention and/or treatment of disease and/or disorder, and (2) nutraceuticals are used as conventional foods or as sole items of a meal or diet (Kalra 2003).

14.1.3 Challenge and opportunities

The realm of nutraceuticals and functional foods is an emerging research field and many gaps exist in the knowledge base; for example, it is widely accepted that single components play an active part in health-promoting properties, but rather a few or several active ingredients are responsible for this. Due to the lack of handling measurement, many of the bioactive phytochemicals under investigation have long been ignored. In general, manufacturers wish to make specific claims of health benefits on their product. However, such claims must be based on solid scientific evidence and laboratory testing, which unfortunately is sometimes lacking in today's climate (Hasler 2002). As this new category of health products lies between foods and drugs, government regulatory bodies also face challenges, even though they share the desire to improve personal and public health through diet modification, to reap the consequent social and economic benefits. The research field of nutraceuticals and functional foods is at times confounding, or it seems like, combination of biotechnology and genetic modification (Ghosh 2010). The two areas are distinctly different, although there is some potential for overlap. To enhance the phytochemical content of food and nonfood plants, the genetic modification techniques may be applicable. The complex series of biochemical reactions which take place in plants to synthesize specific phytochemicals is often hard to understand. However, if the plant's sophisticated biochemical machinery can be modified or controlled by any way, that can lead to the synthesis of valuable compounds that may ultimately enhance human health (Weingärtner et al. 2009). Getting a new product into the food market is always faced by numerous challenges and critics. People who understand the significant correlation among diet, consumption of quality food and good health are currently the target customers for the companies, as customers always emphasize on name-brand products to assure superior quality, even if the price may be a little higher (El Sohaimy 2012).

14.2 What is flavor?

To choose a food, all the senses come into play. These senses can be called as "gatekeepers," because they guide the selection of foods. There are five widely known tastes: sweet, salty, bitter, sour and umami. They are considered to be the evolutionary bases for taste discrimination. Salt is a necessary nutrient and responsible for salty taste, sugar-sources of energy, which gives sweet taste and umami is an indicator of protein. Sometimes spoilage in a food can be indicated by the sour taste whereas, bitter taste indicates the presence of toxic compounds.

Foods usually consist of certain volatile chemicals components. Some of these compounds act as the stimulation of receptors associated with the olfactory nerve and are perceived as aroma. Basic tastes are usually detected by the taste bud when the food is being ingested, simultaneously olfactory nerve detects the air containing volatile compounds, which swept to back of the mouth and also into the nasal cavity. Sense of smell makes a lion share contribution for discriminating the diversity of flavor sensation. Since taste and smell occur simultaneously people often confound taste with smell. This difference can be observed while eating food with the nose pinched closed, this will block the nasal passage.

Chemethesis are referred to as the tactile response of some compounds that can trigger the pain receptor and light touch receptor, which is also known as mechanoreceptors. Heat or pungency of chili pepper, cooling of mint from menthol are some of the well-known examples.

14.3 Natural and artificial flavors

14.3.1 Natural flavor

The term "natural flavor" refers to the essential oil, oleoresin, essence or extractive, protein hydrolysate, distillate, or any product of roasting, heating or enzymolysis that contains the flavoring constituents derived from a spice, fruit or fruit juice, vegetable or vegetable juice, edible yeast, herb, bark, bud, root, leaf or similar plant material, meat, seafood, poultry, eggs, dairy products, or fermentation products. The primary role for natural flavor in food is to contribute flavoring part rather than nutritional. **Figure 14.1** shows various types of flavoring preparations and their method of preparations. **Table 14.1** describes examples of artificial and natural flavors.

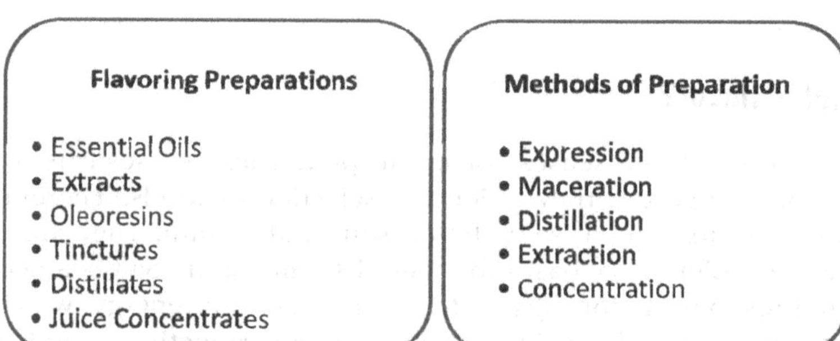

Figure 14.1 *Flavoring preparations and methods of preparation.*

Table 14.1 Examples of Natural and Artificial Flavors

Type	Example	Properties
Natural	Peppermint, natural honey, castoreum extract, spices and herb	Less stable
Artificial	Vanilla, chocolate, butter, cheese	Highly stable
Natural and artificial	Strawberry, orange, lemon, lime, grape, apple	Effective at low concentration

14.3.2 Artificial flavor

The term "artificial flavor" refers to any substance, the function of which is to impart flavor that is similar to natural flavor. This artificial flavor is not derived from a spice, fruit or fruit juice, vegetable or vegetable juice, edible yeast, herb, bark, bud, root, leaf or similar plant material, meat, fish, poultry, eggs, dairy products, or fermentation products (Burdock 2010).

14.4 Flavor legislation

Flavor legislation refers to the review of the system regarding flavor used, and which is currently in force worldwide; permitted flavoring agents such as positive, negative and mixed lists; labeling; toxicology, use level, flavoring materials, testing its safety for use, patent protection and so on (Heath 1981).

14.4.1 Importance of flavors in functional food and nutraceuticals

Flavors are an indispensable piece in the field of nutraceuticals. According to Reineccius (2000) despite the heath functionality for functional foods, consumers are unlikely to buy functional foods over the long term as they give importance of food taste (flavoring system for functional foods). In 2011, consumer trending survey by International Food Information Council on functional foods found that the Americans' decisions to consume functional foods were mainly governed by price and taste.

To understand functional foods, it should be studied as separate products within the various food categories rather than one homogeneous group. To enhance a functional food's functionality, such as increasing bioactive compounds, one may lose its taste, which in turns may cause aversive customer reaction (Siro et al. 2008). A study conducted by Tuorila and Cardello in 2002 found that consumers' degree of liking and likelihood of consumption decreased with increasing severity of the flavor defect, even with knowledge of the health claims (Tuorila and Cardello 2002). However, no strong evidence was found; the theory-a slight off-flavor was perceived as a positive marker of health benefits (McDonald 2017).

Even in medical foods, flavor still, it is very important. Besides making food taste good including stimulation of digestion and the wellbeing of physiology, the flavor has several benefits (Reineccius 2000).

Flavors are mainly added to functional foods:

1. For inducing flavor to a bland product
2. For modifying or complementing an existing flavor base
3. To disguise or cover undesirable flavor attributes or off-notes, either by blocking, masking, or otherwise tricking the taste buds into further not recognizing the off-flavor (McDonald 2017)

14.4.2 How are flavors created, and how do they work?

Our taste buds respond immediately to the quality of flavors. We never need to think somehow about whether our "taste buds" are requesting that we acknowledge or dismiss a nourishment. One conceivable special case to this is for flavors that we have not experienced before where, in the event that we get over our underlying carefulness in regard to (Pliner and Pelchat 1991), we have a tendency to effectively look for some perspective—that is, for comparable, recognizable flavors—preceding settling on issues of adequacy. Nonetheless, the obviously quick nature of flavor hedonics shrouds an extensive variety of scholarly reactions that depend on the coordination of the scent segment of flavors with other jolts—tastes—that are as of now hedonically valanced. Imperatively, since scents shape the trademark quality that distinguishes flavors, there is to prove that the fundamental measurement of olfaction is hedonic in nature (Khan et al. 2007). This tells that although the odor is in contrast from each other in a synthetic structure, it is their level of pleasantness that recognizes their distinctive smell.

Flavors are characterized here as continually incorporating the two tastes and smells, although perceiving that somatosensory characteristics (surfaces, for instance), and visual and sound-related characteristics can likewise add to season discernments and inclinations in various settings. Flavors require not be, but rather quite often are, likewise joined by consumption and metabolic use of vitality, supplements, or pharmaco dynamic specialists exhibit in food and drinks. These components add to season hedonics by means of affiliation.

For developing liking toward a flavor, a back-to-back exposure of generally neutral or maybe even detested, flavor or scent is fundamental and now and again adequate (Methven et al. 2012). The way that presentation to the smell mixes of flavors in utero prompts expanded enjoying for those individual flavors, which (Mennella et al. 2001) gives off an impression of being a key case of this procedure. Since liking can create pairing with another positive stimulus (Zajonc 1968), there remains an essential inquiry of the procedures fundamental such mere exposure (ME) impacts. The first of these, developed largely from studies that have used rapidly presented, brief visual stimuli, is that the repeated stimulus becomes increasingly more easily processed by the brain. This processing fluency hypothesis suggests that since processing a stimulus is effortful, any reduction in such effort will be rewarding, resulting in positive evaluation of the stimulus (Reber et al. 1998). It has also been suggested that novel stimuli initially produce unpleasant levels of arousal, and that repeated exposure is rewarding because of the reduction in arousal that comes with increased familiarity (Berlyne 1970). This may explain why rapidly repeated stimuli are associated with positive mood states in research participants, with these mood states consequently becoming associated with the stimulus (Monahan, Murphy et al. 2000).

Procedures for advancement of flavor references are important or are not clear yet and subject to the end use. But what is more important is that what we know about the repetition of a stimulus that is novel to reduce the drawbacks. Regardless of whether through the diminishment of the dread of disease or of a disagreeable taste, the kinds of new flavors wind up, which is acknowledged through repeated tasting (Pliner et al. 1993).

Because of brief visual jolts, dynamic consideration seems to have no effect on the improvement of preferring for the repeated exposed stimulus. Notwithstanding, preparing investigations of longer visual stimulus have demonstrated that went to jolts are assessed more emphatically than those that are disregarded (Raymond et al. 2003). Development of a liking for odors has also been shown to be facilitated by explicit attention to a repeated odor in contrast to a similarly exposed, but ignored, odor. This may be another example of liking being derived from the pairing of the stimulus with a positive emotional state that arises from the strong reciprocal interactions between attention and emotion. Yet another example of this may be the fact that choice itself—that is, choosing between two equally liked alternatives—acts to increase liking for the chosen option. One rationale for this effect is that choices are an indication of exerting control over our environment and it is this perception of having such control that gives rise to positive affect (Leotti and Delgado 2011).

Minor impacts of the exposure can likewise be viewed as a type of accidental exchange of hedonics from a variety of relevant properties that go with any repeated introduction of stimulus. Thus, a new food flavors could be accompanied by a pleasant environment, enjoyable company, a general state of relaxation, and so on. It has been suggested, for example, that comfort foods, which appear to be a widespread phenomenon across cultural boundaries, are liked because they are foods that have most often been eaten in the company of those with whom we have important relationships, including parents, siblings, and partners. One possibility is that the feelings of psychological comfort that were experienced at the time of eating such foods become "encoded" with the flavor of those foods, thus providing a positive emotional experience while eating.

14.4.3 Chemistry of the flavors

Chemicals can be either volatile or nonvolatile (or somewhere in between). Chemicals that cause taste sensations tend to be nonvolatile, and water-soluble.

For example, sucrose is dissolved in the saliva or beverage and is detected by certain taste buds, to be perceived as sweet. A certain stereochemical arrangement of functional groups on the molecule will fit (bind) into the receptor in the taste bud to cause the sweet perception. While sucrose is the standard for sweetness, other molecules are detected as sweet, and many give a much stronger sweetness response, such as aspartame and saccharine. Similarly, bitter receptors require molecules with a certain arrangement to elicit bitter response. Unfortunately, similarities between the sweetness receptors and the bitterness receptors result in molecules, especially many high intensity sweeteners (HIS), being both sweet and bitter. Bitterness, sweetness, and umami are all due to these molecules interacting with G-protein receptors. Saltiness and sour perception are associated with ion channels. Sour is associated with the hydrogen protons of acids. Salt taste is elicited by the movement of sodium ions, although some other monovalent cations can also generate salty response, such as lithium and potassium.

Chemesthesis can be triggered by nonvolatile and volatile compounds. Examples of nonvolatile pungent (hot) chemesthetic agents include capsaicinoids from chili peppers; gingerols, shogaols, and paradols from ginger; and piperine from black and white pepper. Sanshools are tingling compounds from Szechuan peppers and spilanthol from jambu. Incidentally, Szechuan peppers, chili peppers, and black and white peppers are not related.

Volatile chemesthetic compounds include the pungent compounds i.e., cinnamaldehyde (from cinnamon and cassia), isothiocyanate (mustard and horseradish), and eugenol (numbing compound from cloves).

Most of what is perceived as flavor arises from volatile compounds that impact the sense of smell. This encompasses a huge variety of various types of compounds: phenolics, carbonyls, acids, esters, heterocyclic, and "aromatic" (the organic chemistry meaning, as typical of a benzene ring) compounds, sulfur- and nitrogen-containing chemicals, and more. The only characteristics they have in common are that they are volatile to some extent and can elicit a response from olfactory receptors.

14.5 Legal regulations for functional foods

14.5.1 United States

The regulation of functional foods in the United States differs somewhat. The US Food and Drug Administration (FDA) is responsible for regulation and supervision of foods and drugs, and it was the first of more than 200 laws that constituted one of the world's most comprehensive and effective networks of public health and consumer protection. Today the FDA regulates \$1 trillion worth of products a year. It ensures the safety of all food except for meat, poultry, and some egg products; ensures the safety and effectiveness of all drugs, biological products (including blood, vaccines, and tissues for transplantation), medical devices, and animal drugs and feed; and makes sure that cosmetics, medical and consumer products that emit radiation do no harm. There is not a category of functional foods as in Europe, but accordance exists in that the health claims may be based on an authoritative statement of a scientific body (FDA Modernization Act of 1997).

FDA both assesses health claims and makes final decisions about their approval/refusal, differently from Europe, where two entities are involved, EFSA (European Food Safety Authority) and the EU (European Union) Commission. The United States only allows nutrient deficiency health claims, while Europe allows reduction of disease claims. The main difference is that the United States may approve health claims based on authoritative statements of a scientific body or institution and may even allow claims that are new and for which solid scientific evidence does not exist yet (Moors 2012). The criteria are not as strict as in Europe and health claims are easier to use for marketing purposes.

With respect to regulations of food and flexibility for regulations regarding food claims (Section 301), FDA determinations provide for information necessary to:

- enable consumers to develop and maintain healthy dietary practices;
- enable consumers to be informed promptly and effectively of important new knowledge regarding nutritional and health benefits of food;
- ensure that scientifically sound nutritional and health information is provided to consumers as soon as possible;
- enable the FDA Secretary to act promptly to ban or modify a claim.

The FDA provides guidance and regulatory information, with links to Federal Register documents, in their Web page. Guidance documents represent FDA's current thinking on a topic. They do not create or confer any rights for or on any person and do not operate to bind the FDA or the public. Manufacturers can use an alternative approach if the approach satisfies the requirements of the applicable statutes and regulations. The FDA also issues regulations to implement its statutory authority. The regulations can create binding obligations and have the force of law. Links to Federal Register documents (advance notices of proposed rulemaking, proposed rules, interim final rules, and final rules) are posted in their Web page.

14.5.2 Recent FDA-banned flavorants

The US Food and Drug Administration has restricted the use of six artificial flavorants in sweet, treats, frozen yogurt, and every other foods and drinks that has appeared to cause malignancy in animal testing. The synthetic substances like

- benzophenone
- ethyl acrylate

- eugenyl methyl ether
- myrcene
- pulegone
- pyridine

These are commonly called as artificial flavors, which are being mentioned on the labels of the consumer products. Hence, the customers have no chance to get them out of knowing which items contain those above chemical substances.

The FDA has additionally prohibited the utilization of styrene as an artificial seasoning. The US National Toxicology Program grouped styrene as a "sensibly foreseen" human cancer-causing agent in 2011.

The candidates gave the FDA proof that the flavorings are cancer-causing in research center creatures.

The FDA guarantees that the synthetic compounds "don't represent a hazard to general health of the consumers under the states of their planned utilize," yet the organization can't legitimately approve the utilization of these flavorants added substances that have been appeared to cause malignancy in human beings.

The bans go into effect on October 9 when FDA publishes the "final rule" in the *Federal Register*, but FDA does not intend to enforce the rule until October 9, 2020, to give manufacturers time to reformulate their products (Erickson 2018).

14.5.3 Europe

The claims are a form of communication about functional foods and suggest that the foods possess characteristics. In Europe, health claims of food products need to be authorized under Regulation (EC) N° 1924/2006 before being used for labeling and marketing. The assessment to the European Commission on the approval/refusal of the claim is carried out by the European Food Safety Authority (EFSA) by means of its Panel on Dietetic Products, Nutrition, and Allergies, which verifies the scientific substantiation of the health claim. The European Commission (DG Health and Consumers) makes the final decision whether to accept or reject a claim.

The Regulation defines a claim as "any message or representation that is not mandatory under Community or national Legislation, including pictorial, graphic or symbolic representation, in any form, which states, suggests or implies that a food has a particular characteristic" (EC 2006, Article 2.2.1 Regulation (EC) N° 1924/2006). The aim is to ensure that any claim made about a food is clear, accurate, truthful, not misleading to the consumer, and based on scientific evidence.

The Regulation classifies three different claims: nutrition claim, health claim, and reduction of disease risk claim.

"Nutrition claim" refers to any claim that states, implies, or suggests that a food has a beneficial nutritional property due to:

- the energy it provides, does not provide, or provides at reduced/increased rate and/or,
- the nutrients or other substances it contains, does not contain, or contains at reduced/increased proportions.

"Health claim" refers to any claim that states, implies, or suggests that a relationship exists between a food or one of its constituents and health.

"Reduction of disease risk claim" refers to any claim that states, implies, or suggests that the consumption of a food or one of its constituents reduces a risk factor in the development of a human disease.

Health claims are divided into Article 13 and Article 14 claims. Article 13 includes health claims that describe (1) the role of a nutrient or other substance in the growth, development, and functions of the body, (2) psychological and behavioral functions, (3) slimming or weight control or reduction in the sense of hunger or increase of satiety or reduction in available energy from diet. Article 14 includes health claims specifically referring to the reduction of risk of disease, as well as to children's development and health.

The general conditions for the use of nutrition and health claims are specified in Article 4 of Regulation (EC) N° 1924/2006. It states that functional foods must meet the criteria of an appropriate "nutrient profile" to bear claims. It means that the food must comply with quantities of specific nutrients and other substances contained so that a claim can be made. The nutrient profile depends on the potential of the food to adversely affect the overall dietary balance of nutrients. In this way it is ensured that the consumer makes healthy choices of food products with good nutritional quality. The establishment of the nutrient profile for a category of foods considers the nutritional composition of the foods (main nutrients as carbohydrates, fats, proteins, vitamins, etc.), the dietary recommendations, as well as the dietary habits of the population. Some nutrients for which intakes might exceed the recommended levels include energy density, total fat, saturated fatty acids (SFA), *trans*-fatty acids (TFAs), sugars, and sodium. Nutrients for which intakes might be inadequate in relation to recommended levels include dietary fiber, unsaturated fatty acids, vitamins, and minerals. These nutrient profiles have been set by the Commission. Besides, the nutrient or other substance for which the claim is made must be provided by a quantity of food reasonably expected to be consumed and must be present in the body in quantities sufficient to exert the claimed effect.

From 20 January 2011, Regulation (EC) No 1334/2008 lays down **general requirements** for safe use of flavorings and provides definitions for several types of flavorings. **The Union list of flavoring substances**, approved for use in and on foods, was adopted on 1 October 2012 and was introduced in Annex I of this Regulation.

The Union list of flavorings has been amended as follows:

- Commission Regulation (EU) No 545/2013 of 14 June 2013
- Commission Regulation (EU) No 985/2013 of 14 October 2013
- Commission Regulation (EU) No 246/2014 of 13 March 2014
- Commission Regulation (EU) No 1098/2014 of 17 October 2014
- Commission Regulation (EU) 2015/648 of 24 April 2015
- Commission Regulation (EU) 2015/1102 of 8 July 2015
- Commission Regulation (EU) 2015/1760 of 1 October 2015
- Commission Regulation (EU) 2016/54 of 19 January 2016
- Commission Regulation (EU) 2016/55 of 19 January 2016
- Commission Regulation (EU) 2016/178 of 10 February 2016
- Commission Regulation (EU) 2016/637 of 22 April 2016
- Commission Regulation (EU) 2016/692 of 4 May 2016
- Commission Regulation (EU) 2016/1244 of 28 July 2016
- Commission Regulation (EU) 2017/378 of 3 March 2017
- Commission Regulation (EU) 2017/1250 of 11 July 2017
- Commission Regulation (EU) 2018/1246 of 18 September 2018

The Regulation prohibits the addition of certain natural undesirable substances as such to food and lays down maximum levels for certain substances, which arc naturally present in flavorings and in food ingredients with flavorings properties, but which may raise concern for human health (**Table 14.2**).

Table 14.2 Repealed Legislations on Flavoring Substances

1. **Directive 88/388/EEC**	States approximation of the laws of the Member States relating to flavorings for use in foodstuffs and to source materials for their production.
2. **Decision 88/389/EEC**	The establishment, by the Commission, of an inventory of the source materials and substances used in the preparation of flavorings.
3. **Directive 91/71/EEC**	16 January 1991 completing Directive 88/388/EEC on the approximation of the laws of the Member States relating to flavorings for use in foodstuffs and to source materials for their production.
4. **Decision 1999/217/EC**	23 February 1999 adopting a register of flavoring substances used in or on foodstuffs drawn up in application of Regulation (EC) No 2232/96 of the European Parliament and of the Council of 28 October 1996 (Decision 1999/217/EC). The register was amended several times (**Decision 2000/489/EC, Decision (EC) No 2002/113, Decision (EC) No 2004/357, Decision (EC) No 2005/389, Decision (EC) No 2006/252, Decision (EC) No 2008/478, Decision (EC) No 2009/163**).
5. **Regulation (EC) No 622/2002**	11 April 2002 establishing deadlines for the submission of information for the evaluation of chemically defined flavoring substances used in and on foodstuffs.

Note: EU Legislation, https://ec.europa.eu/food/safety/food_improvement_agents/flavourings/eu_rules_en.

14.5.4 Japan

Functional foods in Japan is regulated by Food for Specified Health Uses (FOSHU). On their Web page, the definition of FOSHU refers to foods containing an ingredient with function(s) for health and officially approved to claim its physiological effects on the human body. FOSHU is also intended to be consumed for the maintenance/promotion of health or special health uses by people who wish to control health conditions, including blood pressure or blood cholesterol. In order to sell a food as FOSHU, the assessment for the safety of the food and effectiveness of the functions for health is required, and the claim must be approved by the Ministry of Health, Labor, and Welfare.

There are some crucial requirements for FOSHU approval, also specified in their Web page, and these include:

- effectiveness on the human body is clearly proven;
- absence of any safety issues (animal toxicity tests, confirmation of effects in the cases of excess intake, etc.);
- use of nutritionally appropriate ingredients (e.g., no excessive use of salt, etc.);
- guarantee of compatibility with product specifications by the time of consumption;
- established quality control methods, such as specifications of products and ingredients, processes, and methods of analysis.

Prohibited Substances for Use as Food Additives: The food additive is controlled by the designation system (positive list) and there is no list of prohibited substances as food additives (negative list). However, among 55 JAS (Japan Agriculture Standards) Mark Standards under the Law concerning Standardization and Proper Labeling of Agricultural and Forest Products (JAS Law), use of some food additives are restricted in the food so described. JAS Mark is voluntary certification system (**Table 14.3**).

Table 14.3 List of Newly Designated Food Additives from 2010 to 2012

Year	Examples
2010	2-ethylpyrazine, sodium stearoyllactylate, calcium sorbate, propionaldehyde, etc.
2011	5-ethyl-2-methylpyridine, pyrazine, 1-penten-3-ol, 3-methyl-2-butenal, isoquinoline, etc.
2012	*trans*-2-methyl-2-butanol, trimethylamine, saccharin calcium, 2-ethyl-6-methylpyrazine, 3-amino-3-carboxypropyl, dimethyl sulfonium chloride, etc.

Note: http://www.ilsijapan.org/ILSIJapan/COM/EastAsia/EastAsia/PDF/EastAsia(4)2012E_Japan.pdf.

14.6 Conclusion

In conclusion, functional foods and nutraceuticals have proved to provide potential health benefits; however, major challenges associated with their scale up and development is low productivity and lesser yields after purification as well as lack of validated procedures and legislative guidelines. Thus there is a need of clear legislations on use of the natural as well as food additives such as flavors in food, pharmaceutical and nutraceutical products that can be helpful in future.

References

Arai, S. (1996). "Studies on functional foods in Japan—state of the art." *Biosci Biotechnol Biochem* **60**(1): 9–15.

Berlyne, D. E. (1970). "Novelty, complexity, and hedonic value." *Percept Psychophys* **8**(5): 279–286.

Brower, V. (1998). "Nutraceuticals: Poised for a healthy slice of the healthcare market?" *Nat Biotechnol* **16**(8): 728–731.

Burdock, G. (2010). *Fenaroli's Handbook of Flavor Ingredients*. Boca Raton, FL, CRC Press.

Erickson, B. E. Agency responds to evidence that chemicals can cause cancer in animals FDA bans 7 synthetic food flavorings, October 9, 2018, https://cen.acs.org/safety/consumer-safety/FDA-bans-7-synthetic-food/96/web/2018/10.

El Sohaimy, S. A. (2012). "Functional foods and nutraceuticals-modern approach to food science." *World Appl Sci J* **20**(5): 691–708.

Ghosh, D. (2010). "Personalised food: how personal is it?" *Genes Nutr* **5**(1): 51–53.

Hasler, C. M. (2002). "Functional foods: Benefits, concerns and challenges-a position paper from the american council on science and health." *J Nutr* **132**(12): 3772–3781.

Hasler, C. M. and A. C. Brown. (2009). "Position of the American Dietetic Association: functional foods." *J Am Diet Assoc* **109**(4): 735–746.

Heath, H. B. *Source Book of Flavors (AVI Sourcebook and Handbook Series)*. The Netherlands, Springer.

Kalra, E. K. (2003). "Nutraceutical—definition and introduction." *AAPS PharmSci* **5**(3): E25.

Khan, R. M., C. H. Luk, A. Flinker, A. Aggarwal, H. Lapid, R. Haddad and N. Sobel. (2007). "Predicting odor pleasantness from odorant structure: Pleasantness as a reflection of the physical world." *J Neurosci* **27**(37): 10015–10023.

Laparra, J. M. and Y. Sanz. (2010). "Interactions of gut microbiota with functional food components and nutraceuticals." *Pharmacol Res* **61**(3): 219–225.

Leotti, L. A. and M. R. Delgado. (2011). "The Inherent Reward of Choice." *Psychol Sci* **22**(10): 1310–1318.

McDonald, S. T. (2017). Chapter 13—Concepts of flavor creation in novel nutraceuticals and functional food formulations. *Developing New Functional Food and Nutraceutical Products*. D. Bagchi and S. Nair. San Diego, CL, Academic Press: 231–247.

Medicine, I. O. (1994). *Opportunities in the Nutrition and Food Sciences: Research Challenges and the Next Generation of Investigators*. Washington, DC, The National Academies Press.

Mennella, J. A., C. P. Jagnow and G. K. Beauchamp. (2001). "Prenatal and postnatal flavor learning by human infants." *Pediatrics* **107**(6): E88.

Methven, L., E. Langreney and J. Prescott. (2012). "Changes in liking for a no added salt soup as a function of exposure." *Food Qual Prefer* **26**(2): 135–140.

Monahan, J. L., S. T. Murphy and R. B. Zajonc (2000). "Subliminal mere exposure: Specific, general, and diffuse effects." *Psychol Sci* **11**(6): 462–466.

Moors, E. H. M. (2012). "Functional foods: regulation and innovations in the EU." *Innovation: Eur J Soc Sci Res* **25**(4): 424–440.
Pliner, P., M. Pelchat and M. Grabski. (1993). "Reduction of neophobia in humans by exposure to novel foods." *Appetite* **20**(2): 111–123.
Pliner, P. and M. L. Pelchat (1991). "Neophobia in humans and the special status of foods of animal origin." *Appetite* **16**(3): 205–218.
Prescott, J., Kim, H. and Kim, K. O. (2008). Cognitive mediation of hedonic changes to odors following exposure. *Chem Percept* **1**: 2. doi:10.1007/s12078-007-9004-y
Rawson, N. E., and X. Li. (2007). The cellular basis of flavour perception: Taste and aroma. In *Flavor Perception*, pp. 57–85. doi:10.1002/9780470995716.ch3
Raymond, J. E., M. J. Fenske and N. T. Tavassoli. (2003). "Selective attention determines emotional responses to novel visual stimuli." *Psychol Sci* **14**(6): 537–542.
Reber, R., P. Winkielman and N. Schwarz. (1998). "Effects of perceptual fluency on affective judgments." *Psychol Sci* **9**(1): 45–48.
Reineccius, G. A. (2000). Flavoring Systems for Functional Foods. In *Essentials of Functional Foods*. M. K. Schmidl and T. P. Labuza (Eds.), pp. 87–95, Gaithersburg, MD: Aspen Publishers.
Siro, I., E. Kapolna, B. Kapolna and A. Lugasi. (2008). "Functional food. Product development, marketing and consumer acceptance—A review." *Appetite* **51**(3): 456–467.
The Sensory Systems and Food Palatability. Nutrition and Metabolism. (1999). "Safety assessment and potential health benefits of food components based on selected scientific criteria. ILSI North America Technical Committee on Food Components for Health Promotion." *Crit Rev Food Sci Nutr* **39**(3): 203–316.
Tuorila, H. and A. V. Cardello. (2002). "Consumer responses to an off-flavor in juice in the presence of specific health claims." *Food Qual Prefer* **13**(7): 561–569.
Weingärtner, O., M. Böhm and U. Laufs. (2009). "Controversial role of plant sterol esters in the management of hypercholesterolaemia." *Eur Heart J* **30**(4): 404–409.
Zajonc, R. B. (1968). "Attitudinal effects of mere exposure." *J Pers Soc Psychol* **9**(2, Pt.2): 1–27.

Index

Note: Page numbers in italic and bold refer to figures and tables, respectively.

C

D

E

F

G

H

I

J

K

O

P

Q

R

S

T

U

V

W

X

Y

Z